# The Biology and Management of
# Red Alder

# The Biology and Management of
# Red Alder

EDITED BY DAVID E. HIBBS

DEAN S. DEBELL

& ROBERT F. TARRANT

Oregon State University Press • Corvallis • Oregon

*The Washington Hardwood Commission provided substantial encouragement and financial support to the creation of this book. We, the editors, want to express our thanks and appreciation for this important contribution.*
*David E. Hibbs*
*Dean S. DeBell*
*Robert F. Tarrant*

*Disclaimer*
The use of trade or firm names in this publication is for reader information and does not imply endorsement of any product or service by the publisher or author.

The paper in this book meets the guidelines for permanence and durability of the Committee on Production Guidelines for Book Longevity of the Council on Library Resources and the minimum requirements of the American National Standard for Permanence of Paper for Printed Library Materials Z39.48-1984.

**Library of Congress Cataloging-in-Publication Data**
The Biology and management of red alder / edited by David E. Hibbs, Dean S. DeBell & Robert F. Tarrant.
    p.  cm.
  Includes bibliographical references and index.
      ISBN 0-87071-382-5 (cloth : alk. paper)
  1. Red alder—Pacific States. 2. Red alder—utilization—Pacific States. I. Hibbs, David E. II. DeBell,
Dean S. III. Tarrant, Robert F., 1918-  .
  SD397.R18B56  1994
  634.9'73976—dc20                         93-42131
                                          CIP

# Contents

SECTION 2: MANAGEMENT OF ALDER

# Preface

We offer this book as a synthesis of current knowledge about the biology and management of red alder. The first section covers topics in the basic biology of alder, on a scale of microorganism-to-landscape extent. The second section offers principles and practices of alder management founded upon biological knowledge.

Each chapter is a synthesis of previously published information plus interim results of ongoing research. To insure an up-to-date review, chapters have been written by inter-institutional teams whenever possible. With this information, forest managers can weave red alder into the improved forest management systems demanded by an increasingly well-informed public.

Metric units are generally used throughout the book, but where publications or studies that use English units are cited the original units are maintained.

*David E. Hibbs*
*Dean S. DeBell*
*Robert F. Tarrant*

# Red Alder and the Pacific Northwest

ROBERT F. TARRANT, DAVID E. HIBBS, & DEAN S. DEBELL

The alder, whose fat shadow nourisheth—
Each plant set neere him long flourisheth.

William Browne, 1613
*Brittania's Pastorals*, Book I, Song 2

---

Alder (*Alnus* spp.) has long been observed to possess unique biological properties. Virgil (70—19 B.C.), author of the *Aeneid,* reported his observations on the ecology of alder, making particular reference to its occurrence in moist places (Kellogg 1882). In *Forest Trees of Britain,* Johns (1849) wrote, "It has been observed that their [alders'] shade is much less injurious to vegetation than that of other trees," and quoted Browne's jingle, above.

The first "modern" scientist to report on nodulated alder roots was Meyen (1829). Following this clue, Hiltner (1896) performed a classic study in which seeded black alder (*Alnus glutinosa* Gaertn.) grew vigorously in sterile, nitrogen-free soil inoculated with alder root materials while similar plants in uninoculated soil remained small and nitrogen-deficient. Other scientists then became interested and developed sufficient information to include alder in a list of "nitrogen-gathering" plants published in the *Yearbook of the United States Department of Agriculture 1910* (Kellerman 1911).

This early research was done with species of alder other than red alder (*Alnus rubra* Bong.), the major deciduous (hardwood) forest tree of the Pacific Coast of the United States and Canada. During the latter half of the twentieth century, however, great advances have been made in understanding the biology, management, and utilization of red alder (Trappe et al. 1968; Briggs et al. 1978), and there now are hundreds of published research findings (Heebner and Bergener 1983). Much of this information describes the significant role of red al-

der in ecosystem function, including documentation that the actinomycete endophyte (*Frankia* spp.) is responsible for nitrogen fixation in nodules on alder roots (Berry and Torrey 1979).

Only a few years ago, red alder was regarded by most Pacific Northwest forest managers as a weed to be eradicated with herbicides. Today, red alder provides the raw material for a thriving industry, producing fine furniture, cabinetry, specialized veneers and plywoods, shipping pallets, and paper products. Present markets for these products are limited primarily by the supply of alder logs.

Western Oregon has more than nine billion board feet (Scribner rule) of red alder sawtimber, almost two-thirds of which is in non-federal ownership as trees of 12 to 16 inches diameter; slightly more than one-third is in trees 18 inches or larger (Gedney 1990). In the state of Washington, the alder resource is even greater (Beachy and McMahon 1987), but there is very little young alder in either state because of "weed-control" programs of the past several decades. The scarcity of young alder is now a cause of concern to an industry that continues to grow.

A hardwood policy report being prepared by T. Raettig, G. Ahrens, and K. Connoughton (pers. com.) provides a good overview of the current role of the alder industry in the regional and global economy. In 1991, the harvest of hardwoods in Oregon and Washington was 635 million board feet. In Oregon, 76 percent of the hardwood harvest was red alder; the percentage in Washington is not known but probably is similar or higher. Eighty-eight percent of this harvest came from private lands. The 1991 hardwood harvest level has risen 28 percent since 1987; during the same period, the softwood harvest decreased 31 percent.

Domestic markets for alder are diverse. In 1990, for example, 60 percent of the alder lumber produced in Washington was used in the manufacture of cabinets and furniture, both being products whose finished value is much higher than the value of raw materials they contain (high value-added products). In 1988, hardwood chips were produced from 8.6 million board feet of roundwood and 196,000 tons of mill residue. The market for both interior ply and face veneer has grown rapidly as softwood sources have declined, but no figures are available on the size of this industry which competes with both the chip and sawlog markets.

From 1982 to 1992, alder exports to Asia and Europe have grown from almost nil to over 65 million board feet. In 1991, alder accounted for 10 percent by volume of all U.S. hardwood lumber exports to Europe and Asia. These exports accounted for 15 percent of the total hardwood lumber volume produced in the Pacific Northwest. Including chips, the 1990 value of hardwood exports for Oregon and Washington was $91 million.

The hardwood industry in Oregon and Washington employed 5600 people in 1990 and paid wages of over $150 million, an increase over previous years. Nationwide, alder provided production-line jobs to 23,500 people in secondary manufacturing industries. The Pacific Northwest alder industry is dynamic and important. It has become global and continues to grow, providing an element of stability in the general forest industry.

But what about the bottom line: Does alder have the economic value of softwoods? Recent research concludes that although volume recovery from alder logs is less than that from softwoods, alder lumber value increases dramatically as log diameter increases. Alder log values compare favorably with those for Douglas-fir on a net scale basis (Plank et al. 1990). There have been a number of formal and informal economic comparisons of growing alder and Douglas-fir (e.g., Tarrant et al. 1983). Some analyses show Douglas-fir to be a superior investment; others reach the opposite conclusion. Almost all, however, find little difference between the two options, in spite of the different assumptions of value and growth that each makes.

The market performance of red alder is also impressive (Western Hardwood Association 1992). Since 1980, alder prices have increased steadily, in contrast to fluctuating softwood prices. Alder logs are priced in the range of $300 to $350 per million board feet delivered to the mill. Select-and-Better grade alder lumber (4/4 kiln-dried) commands a price of $1000 per million board feet in a market that has expanded to Pacific Rim and European countries. As a bonus, the domestic economy benefits because the hardwood industry is labor intensive, requiring ten times more jobs per unit of raw material than the softwood industry.

For many additional reasons, interest in managing and utilizing red alder is at an all-time high. One very basic reason is that 40 years of research by scientists in universities, public agencies, and private industry has paid off. In a rare instance in the history of forestry, biological information is available, *in advance of management,* as the basis for growing a biologically and economically valuable tree species in short-rotation systems.

Another reason is the change in how the world's forests are to be managed. These biological systems have been greatly altered by intensive harvesting under yesterday's technology. Like the people of central Europe (Plochman 1989), the U.S. public, especially in the Pacific Northwest, seriously questions current forest management practice and calls for new, better approaches to sustaining ecosystems.

The high incidence of laminated root rot in the Douglas-fir region offers another reason for interest. Douglas-fir growth is reduced by the fungus, and trees are eventually killed. Conifer tree species show varied levels of resistance to the disease, but all hardwoods are immune. On appropriate sites, red alder is the species of choice in reforesting lands infected with laminated root rot.

The motivating interest of small private forest land owners is often the shorter time to grow and harvest a crop of alder than is generally found with conifers. Most forest landowners are over 50 years old. Thus, a short 30-year alder rotation is often much more appealing than a 60- to 70-year Douglas-fir rotation.

Finally, red alder enjoys significant political and scientific support in the Pacific Coast states. In 1987, the Oregon State Legislature created the Hardwood Forest Products Resources Committee to assess the future of the industry and work toward its best interests. In 1991, the Oregon Legislature directed that this committee present to the 1993 Legislative Assembly an operational plan for converting the committee to commodity commission status. In 1989, the Washington State Legislature authorized development of the Washington Hardwoods Commission. In full function now, this commission is supported by a small levy on hardwood processors to help support and expand their industry. In California, interest also is developing in forming a hardwood commission. The California Timber Industry Revitalization Committee and the Hardwood Foundation have joined forces to support hardwoods as an important resource in modern forest management.

The Western Hardwood Association, since 1955 a champion of hardwood resources, plays a strong role in all these developments. Along with providing a forum for hardwood interests, the Association disseminates technical information through an array of publications and sponsors technical meetings to advance knowledge in the field.

Red alder research is being strengthened. At Oregon State University, the Hardwood Research Cooperative coordinates and conducts research programs according to user needs. The USDA Forest Service, Pacific Northwest Forest Research Station provides excellent information on alder supply, management, and utilization. In western Washington, Weyerhaeuser Corporation has installed extensive field studies of regeneration and growth of alder. All this activity helps ensure that alder and associated hardwood forest tree species can continue to contribute their economic and biological values to sound forest management systems.

## Literature Cited

Beachy, D. L., and R. O. McMahon. 1987. Unpublished report. Forest Research Laboratory, Oregon State Univ., Corvallis, and Western Hardwood Association, Portland, OR.

Berry, A. M., and J. G. Torrey. 1979. Isolation and characterization *in vivo* and *in vitro* of an actinomycetous endophyte from *Alnus rubra* Bong. *In* Symbiotic nitrogen fixation in the management of temperate forests. *Edited by* J. C. Gordon, C. T. Wheeler, and D. A. Perry. Oregon State Univ. Press, Corvallis. 69—83.

Briggs, D. G., D. S. DeBell, and W. A. Atkinson, *comps.* 1978. Utilization and management of alder. USDA For. Serv., Gen. Tech. Rep. PNW-70.

Gedney, D. R. 1990. Red alder harvesting opportunities in western Oregon. USDA For. Serv., Res. Bull. PNW-173.

Heebner, C. F., and M. J. Bergener. 1983. Red alder: a bibliography with abstracts. USDA For. Serv., Gen. Tech. Rep. PNW-161.

Hiltner, L. 1896. Über die Bedeutung der Wurzelknöllchen von *Alnus glutinosa* für die Stickstoffernährung dieser Pflanze. Die Landwirtschaftlichen Versuchs-Stationen 46:153—161.

Johns, C. A. 1849. Forest trees of Britain, vol. 2. Samuel Bentley and Co., London. Seen in later ed. 1911. British Trees. *Edited by* E. T. Cook. George Routledge and Sons, London.

Kellerman, K. F. 1911. Nitrogen-gathering plants. *In* Yearbook of the United States Department of Agriculture 1910. U.S. Govt. Printing Office, Washington, D.C. 213—218.

Kellogg, A. 1882. Forest trees of California. California State Mining Bureau, Sacramento.

Meyen, J. 1829. Über das herauswachsen parasitischer gewachse aus den wurzeln anderer pflanzen. Flora (Jena) 12:49—63.

Plank, M. E., T. A. Snellgrove, and S. Willits. 1990. Product values dispel "weed species" myth of red alder. For. Prod. J. 40(2):23—28.

Plochman, R. 1989. The forests of Central Europe: a changing view. *In* the 1989 Starker Lectures. Oregon's forestry outlook: an uncertain future. College of Forestry, Oregon State Univ., Corvallis. 1—9.

Tarrant, R. F., B. T. Bormann, D. S. DeBell, and W. A. Atkinson. 1983. Managing red alder in the Douglas-fir region: some possibilities. J. For. 81(12):787—792.

Trappe, J. M., J. F. Franklin, R. F. Tarrant, and G. M. Hansen, eds. 1968. Biology of alder. USDA For. Serv., PNW For. Range Exp. Sta., Portland, OR.

Western Hardwood Association (WHA). April 1992. Unpublished Fact Sheet. WHA, Portland, OR.

# Section 1
# Biology of Alder

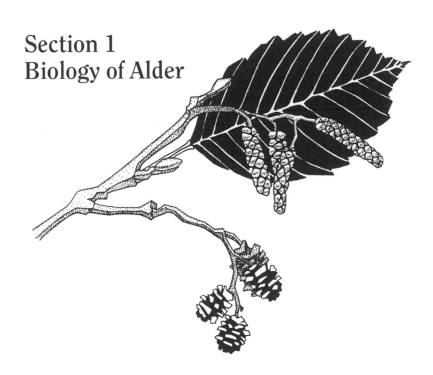

# 1

# Biology of Red Alder (*Alnus rubra* Bong.)
CONSTANCE A. HARRINGTON, JOHN C. ZASADA,
& ERIC A. ALLEN

Red alder (*Alnus rubra*), also called Oregon alder, western alder, and Pacific coast alder, is the most common hardwood in the Pacific Northwest. It is a relatively short-lived, intolerant pioneer with rapid juvenile growth, the capability to fix atmospheric nitrogen, and tolerance of wet soil conditions. The species is favored by disturbance and often increases after harvesting and burning. Because the commercial value of alder has traditionally been lower than that of its associated conifers, many forest managers have tried to eliminate the species from conifer stands. On the other hand, red alder is the major commercial hardwood tree species in the region; its wood is used for furniture, cabinets, pallets, and to make paper (Harrington 1984b). Its value has increased substantially in recent years and interest in the management of the species has increased accordingly, which in turn has led to an increased need for detailed information on the biology of the species. This chapter summarizes published information as well as unpublished data and observations on the biology and ecology of red alder with emphasis on topics not covered by other authors in this volume and on information that has become available since previous summaries were published (Trappe et al. 1968; Briggs et al. 1978; Heebner and Bergener 1983). Much of our knowledge on many aspects of the biology of red alder is based on limited information; thus, where it is appropriate, we point out subject areas where common assumptions or beliefs about red alder are not supported by specific data.

## Taxonomy

Red alder (genus *Alnus*) is a member of the family Betulaceae. Other common western North American genera in this family are the birches (*Betula*) and hazelnuts (*Corylus*). The most conspicuous feature which these three genera have in common is the presence of male catkins (compact aggregates of staminate flowers) (Brayshaw 1976). The seed-bearing catkins in alder and birch are similar when immature, but the birch catkin disintegrates as seeds are dispersed, while the alder catkin remains intact and attached to the plant during seed dispersal and for a time after dispersal is completed. More detailed information on the taxonomy and evolution of the species is presented in Ager and Stettler (Chapter 6). When information is not available on flowering or seed production and dissemination of red alder, we present information from studies of other species in Betulaceae which have similar reproductive structures.

## Genetic Variation

No races of red alder have been described, though they may exist, especially in the disjunct populations or in the extremes of the range. One researcher has divided the species into three populations (northern, central, and southern) based on vegetative and reproductive features from herbarium specimens (Furlow 1974).

Geographic variation in growth rates, sensitivity to frost, and other characteristics has been reported (DeBell and Wilson 1978; Lester and DeBell 1989; Ager and Stettler, Chapter 6). In one study, provenances from areas with cold winters (i.e., Alaska, Idaho, high elevations in Washington and Oregon) had the poorest growth but the greatest resistance to frost damage. Specific gravity did not differ significantly among provenances, nor was it correlated with growth rate (Harrington and DeBell 1980). In another study which compared

*3*

families from coastal sources, it was possible to identify families with high growth rates and low sensitivity to spring frosts (Peeler and DeBell 1987; DeBell et al. 1990). A 24-family progeny trial in western Washington also demonstrated family variation in height-growth response to water-table depth (Hook et al. 1987).

Phenotypic variation between trees is also high. Differences in form and in characteristics of branch, bark, and wood were assessed for eight stands in western Washington; only bark thickness, a branch diameter index, branch angle, and a crown-width index differed significantly among stands (DeBell and Wilson 1978). Variation among trees in seed production has also been reported (see discussion below).

A cut-leaf variety (*Alnus rubra* var. *pinnatisecta*) is found in a few isolated areas in British Columbia, Washington, and Oregon. The cut-leaf characteristic is caused by a single recessive gene (Wilson and Stettler 1981), and thus this variety can be used as a marker in genetic breeding studies (Stettler 1978).

As forest managers plant red alder on increasing acreage, we anticipate the need for additional information on genetic variation in the species. Preliminary recommendations are available on seed zones for red alder (Hibbs and Ager 1989; Ager et al., Chapter 11) and the major tree improvement options and long-term breeding prospects for the species are discussed later in this proceedings (Ager and Stettler, Chapter 6). Information, however, is lacking on the variation within the species in its tolerance of low nutrient or low soil moisture conditions and on the possible interactions among silvicultural practices, genotype, and wood quality characteristics.

## Habitat

### Native Range

Red alder occurs most commonly as a lowland species along the northern Pacific coast. Its range extends from southern California (lat. 34° N) to southeastern Alaska (60° N). Red alder is generally found within 200 km of the ocean and at elevations below 750 m. Tree development is best on elevations below 450 m in northern Oregon, Washington, and British Columbia. In Alaska, red alder generally occurs close to sea level. Farther south, scattered trees are found as high as 1100 m, but most stands are at elevations below 750 m. Red alder seldom grows east of the Cascade Range in Oregon and Washington or the Sierra Nevada in California, although several isolated populations exist in northern Idaho (Johnson 1968a, 1968b).

### Climate

Red alder grows in climates varying from humid to superhumid. Annual precipitation ranges from 400 to 5600 mm, most of it as rain in winter. Summers generally are warm and dry in the southern part of the range and cooler and wetter in the northern portion. Temperature extremes range from -30°C in Alaska and Idaho to 46°C in California. Low winter temperatures and lack of precipitation during the growing season appear to be the main limits to the range of red alder. For good development of trees, either annual precipitation should exceed 630 mm or tree roots should have access to ground water.

### Soils and Topography

Red alder is found on a wide range of soils, from well-drained gravels or sands to poorly drained clays or organic soils. In Washington and Oregon it grows primarily on soils of the orders Inceptisols and Entisols but is also found on some Andisols, Alfisols, Ultisols, Spodosols, and Histosols (Harrington and Courtin, Chapter 10). In British Columbia, alder occurs on Brunisols, Gleysols, Organics, Podzols, and Regosols (Harrington and Courtin, Chapter 10). Best stands are found on deep alluvial soils in river and stream flood plains; however, some excellent stands are also found on upland sites on residual or colluvial soils derived from volcanic materials.

Soil moisture during the growing season appears to influence where the species grows. Alder can tolerate poor drainage conditions and some flooding during the growing season; consequently, it prevails on soils where drainage is restricted—along stream bottoms or in swamps or marshes. It is not commonly found on droughty soils, however; and in areas of low precipitation, it seldom grows on steep south- or southwest-facing slopes. In Idaho and California, stands are usually limited to borders of streams or lakes.

## Associated Forest Cover

Red alder grows in both pure and mixed stands. Pure stands are typically confined to stream bottoms and lower slopes. Red alder is, however, much more widely distributed as a component of mixed stands. It is a major component of the Red Alder cover type (Society of American Foresters Type 221) and occurs as a minor component in most of the other North Pacific cover types (Eyre 1980).

Common tree associates are Douglas-fir (*Pseudotsuga menziesii*), western hemlock (*Tsuga heterophylla*), western redcedar (*Thuja plicata*), grand fir (*Abies grandis*), Sitka spruce (*Picea sitchensis*), black cottonwood (*Populus trichocarpa*), bigleaf maple (*Acer macrophyllum*), and willow (*Salix* spp.). Occasional tree associates include cascara buckthorn (*Rhamnus purshiana*), Pacific dogwood (*Cornus nuttallii*), and Oregon ash (*Fraxinus latifolia*). Western paper birch (*Betula papyrifera* var. *commutata*) is an occasional associate in the northern portion of the range of alder, and redwood (*Sequoia sempervirens*) in the southern portion.

Common shrub associates include vine maple (*Acer circinatum*), red and blue elder (*Sambucus callicarpa, S. cerulea*), Indian plum (*Osmaronia cerasiformis*), salmonberry (*Rubus spectabilis*), western thimbleberry (*R. parviflorus*), devilsclub (*Oplopanax horridum*), Oregongrape (*Berberis nervosa*), and salal (*Gaultheria shallon*).

Herbaceous associates include stinging nettle (*Urtica dioica*), skunkcabbage (*Lysichitum americanum*), blackberries (*Rubus laciniatus, R. leucodermis*), California dewberry (*R. ursinus*), swordfern (*Polystichum munitum*), lady fern (*Athyrium filix-femina*), Pacific water parsley (*Oenanthe sarmentosa*), youth-on-age (*Tolmiea menziesii*), Oregon oxalis (*Oxalis oregana*), and western springbeauty (*Montia sibirica*).

## Other Associates

General associations or interactions between red alder and wildlife species are discussed elsewhere in this volume (McComb, Chapter 9). Interactions between red alder and animals, insects, and fungi that can result in tree damage are discussed later in this chapter.

## Forest Succession

Red alder is a pioneer species favored by high light levels and exposed mineral soil. Its ability to fix atmospheric nitrogen permits establishment on geologically young or disturbed sites with low levels of soil nitrogen. It can form pure stands on alluvium, avalanche paths, and other disturbed sites. Alder has been favored by harvesting and burning; pollen records indicate that alder stands are more extensive in the twentieth century than they were for several centuries before that time (Heusser 1964; Davis 1973). Red alder pollen, however, was also abundant between 9000 and 4800 B.C.; this has been interpreted as indicating a somewhat warmer climate accompanied by an increase in fire frequency (Cwynar 1987).

Observations of mature forests in the Pacific Northwest suggest that alder stands are ultimately replaced by longer-lived, more tolerant conifers having sustained growth rates at older ages than does alder. This is undoubtedly true in most cases (see discussion below for exceptions), but the time required for this to occur in unmanaged forests is not well documented. Rapid growth and high stem densities of alder in younger stands make it difficult for conifers (especially shade-intolerant species) to regenerate and grow if they do not become established at the same time or shortly before alder invades a disturbed area. Douglas-fir can be easily eliminated in dense young alder stands while more tolerant species (western hemlock, redcedar, and Sitka spruce) can survive and over time grow into the alder canopy and ultimately dominate the site.

Many alder stands in western Oregon have few associated conifers, leading some researchers to conclude that those alder stands will be replaced by shrubs and that without disturbance a shrub-dominated community may persist for an extended period of time (Newton et al. 1968; Carlton 1988; Tappeiner et al. 1991; O'Dea 1992). Clonal shrubs, particularly salmonberry, but also thimbleberry and vine maple, often form a dense shrub canopy which makes it difficult for conifers to invade and become established from seed. These shrub species can expand rapidly by vegetative reproduction as space becomes available due to death of the alder overstory (Tappeiner et al. 1991; O'Dea 1992; Zasada et al. 1992).

Experience suggests that alder replacement by conifers will have a high degree of spatial and temporal variation if the process is left to proceed naturally. For example, when the senior author established 40 plots in western Washington and western Oregon, with domination by red alder one of the stand selection criteria, 24 of the sites had shade-tolerant conifers present (western hemlock in 18 stands, western redcedar in 8 stands, and grand fir and Sitka spruce in 6 stands each). Bigleaf maple (a tolerant hardwood) was present in eight of the stands, half of which lacked tolerant conifers. Thus, 28 of the 40 stands contained tolerant, long-lived tree species which would be expected to grow through the alder canopy and eventually dominate. The presence and abundance of shade-tolerant tree species are obviously important in influencing which successional trajectory will be followed; however, the specific ecological factors that determine the successional sequence in alder stands are not known.

## Life History

### Reproduction and Early Growth

**Flowering and fruiting.** Red alder reaches sexual maturity at age 3 to 4 years for individual trees and at age 6 to 8 for most dominant trees in a stand (Stettler 1978). The species is generally monoecious, with separate male and female catkins developing on the previous year's twigs (Hitchcock et al. 1964). Staminate catkins, which occur in pendulous groups and are usually in a terminal position on a short shoot, elongate in late winter, changing from green to reddish brown and from 2 to 3 cm long to about 7 or 8 cm. Pollen grains are small (20 to 23 µM in diameter, Owens and Simpson, N.d.), lightweight, and produced in abundance. Several pistillate catkins are borne per floral bud and are commonly located on a bud proximal to the staminate catkins. They are 5 to 8 mm long and reddish green when receptive. Both male and female catkins usually occur in groups of three to six; variation in the ratio of male to female catkins in terms of numbers and weights has been described for three elevational transects (Brown 1986). Flowering occurs in late winter or early spring; peak shedding of pollen generally precedes peak receptivity by a few days but synchrony in pollen shed and receptivity have been observed in some trees (Brown 1985). Pistillate catkins are upright at the time of flowering but become pendulous as they mature. Most alder seed is probably the result of outcrossing, but some self-pollination occurs (Stettler 1978).

Limited information is available on the effects of cultural practices on flowering of red alder. In an experimental plantation on a well-drained site near Olympia, Washington, both male and female flower production were decreased by irrigation during the growing season (data on file, Olympia Forestry Sciences Laboratory); this is consistent with observations for other species that moisture stress increases flowering (Owens 1991). Within an irrigation regime, flower production was generally concentrated on the larger trees; thus, the

Table 1. Effects of spacing and irrigation regime on the percentage of red alder trees with male catkins. The trees were five years old from seed and had been planted in replicated treatment plots near Olympia, Washington.

| Spacing (m) | Irrigation regime | |
| --- | --- | --- |
| | Low | High |
| | (% with male catkins) | |
| 0.5 x 0.5 | 8.0 | 2.9 |
| 1.0 x 1.0 | 12.4 | 8.7 |
| 2.0 x 2.0 | 41.8 | 12.1 |

percentage of trees flowering was greatest at the widest spacing (Table 1), and within each spacing the percentage of trees flowering increased as tree size increased. Genotypic differences in flowering were also observed. Families that did not grow well did not flower much, but differences in the percentage of trees flowering were also observed among families with relatively similar growth rates. Although inferences can be drawn from these types of observations, the physiological factors controlling flower production in red alder have not been determined.

**Seed production and dispersal.** Seeds are small, winged nutlets borne in pairs on the bracts of woody, conelike strobili (Schopmeyer 1974). The seeds are without endosperm and contain only small cotyledons (Brown 1986). The strobili are 11 to 32 mm long and 8 to 15 mm wide.

Red alder is believed to be a prolific and consistent producer of seed. There is, however, a paucity of data to support this generally held belief. We were unable to find any long-term records of red alder seed production that provided a quantitative assessment of annual variation in seed quantity and quality, timing of dispersal, and distance of dispersal. Seed production in red alder varies substantially among trees; Brown (1985) reported production rates for 45 mature trees of similar size from 0 to 5.4 million seeds per tree. She reported within-stand variation to be much greater than variation among stands. Based on a two-year study of seedfall on two sites in British Columbia, McGee (1988) reported substantial variation between sites and years. Maximum total production was 1550 seeds/$m^2$; in the more productive year, production was four to seven times greater than in the less productive year. In the more productive year, one site produced 1.8 times more seed than the other, while in the less productive year seedfall was similar at both sites. Seed-crop quality (percent viable seeds) was similar between sites with 40 to 50 percent of the seeds viable in the good year and less than 10 percent in the poorer year. Worthington (1957) concluded that moderate seed crops are produced almost annually and bumper crops occur every three to five years; however, no specific studies were cited. Complete failure of a seed crop is rare, but after a severe freeze in 1955, almost no seed was produced in 1956 (Worthington 1957). Additional information on variation in red alder seed production is presented in Ager et al. (Chapter 11).

The most similar species for which long-term seed production records were available was black alder (*Alnus glutinosa*, a European species occurring in the same section of the genus) in Finland (Koski and Tallquist 1978). Maximum seed production over a 13-year period did not exceed 350 seeds/$m^2$, a level substantially lower than the highest total production reported for red alder (Lewis 1985; McGee 1988), but seeds of black alder are about two times larger than those of red alder. Maximum seed production in various birch species with seeds of equal or smaller size than those of red alder are 10 to 15 times greater than those reported for black alder (Brinkman 1974; Koski and Tallquist 1978; Zasada et al. 1991). McGee (1988) reported that red alder and paper birch (*Betula papyrifera* var. *commutata*) seed production followed similar patterns of annual variation and had similar levels of seedfall over a two-year period on one site where the species occurred together. The large differences in seed-crop quality reported for alder (Brown 1985; Lewis 1985; McGee 1988; Ager et al., Chapter 11) are also characteristic of seed production in birch (Safford et al. 1990; Zasada et al. 1991).

The annual pattern of red alder seed dispersal and how it varies over its range is not well documented. In general it is believed to begin in the middle of September in the center of its range and slightly earlier or later to the north and south respectively. McGee (1988), working in British Columbia, found that small amounts of seed were dispersed in September, but that the major dispersal events occurred from November to February. Lewis (1985) reported that major seedfall of red alder in Washington occurred during winter and spring, but that some seedfall was observed throughout the year. Major dispersal events occurred in consecutive months as well as in months separated by several months of low dispersal, a pattern similar to that in other members of the *Betulaceae* (McGee 1988; Zasada et al. 1991).

The nature of the catkin suggests that the timing of seed dispersal is regulated by factors similar to those regulating the release of seeds from the cones of conifers; that is, once catkins are mature, dispersal is determined by the occurrence of weather which dries them, thus opening the scales and allowing the seeds to be released. In general, wet weather keeps catkins closed and wet weather following dry weather closes catkins, thus terminating a dispersal event. Nonetheless, heavy seedfall can occur during wet weather under certain catkin conditions (Lewis 1985) and dispersal will not occur if ice freezes the seed in the catkin (Lewis 1985; Brown 1986). The quantity of seed dispersed during a dispersal event is dependent on the length of the period during which suitable weather occurs, the condition of the catkins that retain seeds, and the amount of catkin movement caused by wind and other agents that shake seeds loose. The dispersal patterns reported for red alder (Lewis 1985; McGee 1988) are consistent with the hypothesized mechanisms, but more frequent observations of seed dispersal in relation to weather and catkin condition need to be made. In coastal Alaska, drying trends brought by high-pressure weather systems are important to seed dispersal of Sitka spruce and western hemlock (Harris 1969), two common associates of alder. In addition, generally similar patterns of seed dispersal have been observed for alder and hemlock growing in the same stand in western British Columbia (McGee 1988).

Red alder seeds are very light, numbering 800 to 3000/g, and wind dissemination is effective over long distances. Lewis (1985) documented dispersal of red alder seeds for a two-year period and found amount of seed, seed weight, percentage of filled seed, and viability all to be inversely correlated with distance from the seed source. Amount of seedfall 100 m from the edge of an alder stand was 2 to 3 percent of the seedfall density inside the stand. Additional information on distance of seed dispersal in the Betulaceae is available for paper birch (*Betula papyrifera*) from New England and Alaska (e.g., Bjorkbom 1971; Zasada 1986). Those studies report that birch seed as a percent of within-stand production ranges from 5 to 20 percent at 55 m from the seed source and drops to about 1 percent at 100 m.

Although seeds are dispersed primarily by wind, some dispersal may occur by water (Brown 1986) and by birds or other animals. Birds are commonly seen around catkins, and alder seeds have been shown to be an important source of food for some species (White and West 1977). The role birds play in dispersal is both passive and active. Passive dispersal occurs simply by movement of the catkins as birds work in the crown of alders; active dispersal occurs as birds extract seeds from the catkins while feeding.

**Germination, seedling survival and development.** Seed germination, seed predation, and first-year seedling survival were the subject of a detailed two-year study in the central Oregon Coast Range by Haeussler (1988; see also Haeussler and Tappeiner 1993). The following briefly summarizes the main findings from this study, which was conducted on north and south aspects in forested and clearcut environments located near the coast (westside, relatively wet microenvironment) and on the eastside of the Coast Range (relatively dry). Seeds were sown on protected and unprotected microsites, which were disturbed (exposed mineral soil) or undisturbed.

Seed germination in clearcut environments began in late February and early March and was completed by mid-April. Differences among north and south aspects were small compared to the forested environments where the onset of germination was delayed, relative to that in clearcuts, for a month and continued into June. On average, the number of germinants emerging was higher on disturbed than on undisturbed seedbeds. There was no clear difference in germinant appearance between forested and clearcut environments for either seedbed type. A positive relationship between spring soil moisture conditions and germinant appearance was stronger in the clearcut environment than in the modified light and temperatures prevailing under forested conditions (Haeussler 1988).

**Seedling establishment.** The number of seeds required to produce a seedling one growing-season old differed dramatically between the westside and eastside Coast Range environments. Under the drier conditions on the eastside of the Coast Range, no seedlings survived through the growing season in either of the years of study. On the north aspect of the westside coast site, 1 seedling was produced per 32 seeds sown; on an adjacent south aspect, 1 seedling was produced per 181 seeds (Haeussler 1988). In another study on a southwest-facing coastal site, sowings in each of two years on newly created mineral soil seedbeds at a rate of 1000 to 1500 seeds/m$^2$ failed to produce any surviving seedlings at the end of one growing season (J. Zasada, unpub. data). In a third study, sowing of alder seed on dry Coast Range sites similar to those studied by Haeussler (1988) resulted in germinants but no surviving seedlings after one growing season (J. Tappeiner and J. Zasada, unpub. data). These three studies suggest that alder establishment is not assured even when large quantities of seeds are sown on what are believed to be desirable seedbeds. This seems contrary to operational experience where alder occurs everywhere it is not wanted, but as M. Newton of Oregon State University (pers. com. to J. Zasada) has observed, it is difficult to predict where natural alder regeneration will be successful and often alder fails to appear where it would seem most likely to occur.

A number of environmental factors result in high mortality of seeds and seedlings between the time seeds arrive on the seedbeds and the end of the first growing season (Haeussler 1988), and these certainly contribute to the temporal and spatial variation in alder regeneration. In unprotected microsites, seedling emergence was 75 percent on disturbed seedbeds and 38 percent on undisturbed seedbeds on protected microsites. Loss of seeds to soil biota was greater under forest conditions than in clearcuts. On undisturbed and mineral soil microsites, 60 and 20 percent of the seed population were destroyed by soil organisms, respectively. Causes of seedling mortality included drought and heat injury, pathogens, animals, erosion, frost and smothering by organic debris. Drought and heat-related mortality were the major causes of mortality in clearcuts, whereas damping-off fungi and other pathogens were most important under forest conditions (Haeussler 1988).

Alder seeds are most commonly described as having little or no dormancy. This is based on studies which have shown that germination of stored seeds under optimum germination temperatures is not improved by stratification (Radwan and DeBell 1981; Berry and Torrey 1985); however, one provenance from British Columbia was reported as having a physiological dormancy that was released by stratification (Elliott and Taylor 1981a). Germi-

nation under suboptimum temperatures, such as may prevail at the time of germination under field conditions, is enhanced by stratification (Tanaka et al. 1991). Studies with other species of the family *Betulaceae* and with common associates (e.g., bigleaf maple; J. Zasada, pers. obs.) have also shown that germination rates at lower temperatures are higher for stratified seed.

Bormann (1983) and Haeussler (1988) demonstrated that alder seeds do not germinate in the dark and that the phytochrome system is very sensitive (i.e., germination is inhibited by exposure to far-red light). In a field study, Haeussler (1988) showed that seed germination under optimum moisture conditions created in sealed germination chambers was greater than 90 percent in a clearcut and Douglas-fir stand but less than 50 percent under an alder canopy. Furthermore, under all overstory conditions, germination was reduced to varying degrees by the presence of understory vegetation. This study strongly suggests that alder germination is controlled by light quality and that a type of light-enforced dormancy may prevent seeds from germinating when other conditions appear optimal; thus a persistent alder seedbank may be present under some conditions. Because of the high seed mortality rate caused by soil organisms, however, Haeussler (1988) concluded that a buried seedbank was not important in alder. Although we concur with this conclusion, it is important to note that seeds of some species of *Betula* remain viable in the soil for much longer than would be expected based on seedcoat structure and general seed germination characteristics (Granstrom 1982; Perala and Alm 1989).

Assuming that site conditions are suitable, red alder can be regenerated by any silvicultural system that provides full sunlight and exposed mineral soil. The species is an aggressive pioneer on avalanche paths, road cuts, log landings, skid trails, or other areas where mineral soil has been freshly exposed to seedfall. For example, shortly after a heavy thinning (removal of 50 percent of the basal area) in a 62-year-old Douglas-fir stand, an alder understory became established and grew rapidly (Berg and Doerksen 1975). Clearcutting and large-group selection are feasible regeneration systems. During harvesting or in a subsequent site preparation treatment, the site must be disturbed sufficiently to expose mineral soil. Fire probably can substitute for mechanical disturbance on most sites. To exclude red alder from the next rotation stand, some forest managers try to reduce the supply of alder seed by cutting possible alder seed trees in the vicinity before or at the time of final harvest; to avoid creating favorable seedbed conditions, they also disturb the site as little as possible during logging and, if feasible, do not burn the logging slash (Lousier and Bancroft 1990).

Artificial regeneration can be accomplished with either bare-root or containerized seedlings, and guidelines for producing planting stock are available (Berry and Torrey 1985; Radwan et al. 1992; Ahrens, Chapter 12). Survival and growth of planted seedlings are usually excellent (Radwan et al. 1992), but can vary significantly with slope, slope position, and aspect within a given clearcut. For example, Zasada (unpub. data) followed the fates of seedlings planted on different sites within a clearcut and observed nearly 100 percent survival on steep north aspects over a three-year period while immediately adjacent south-facing and stream bottom sites (with higher soil moisture stress and a higher probability of early season frosts, respectively) suffered as much as 60 percent mortality.

Height growth of red alder seedlings is generally rapid. On favorable sites, seedlings can grow 1 m or more the first year, and on all but the poorest sites, seedlings surpass breast height (1.3 m) the second year (Smith 1968; Harrington and Curtis 1986). Even on some frost-prone sites, seedlings affected by frost shortly after outplanting attained breast height in two years (J. Zasada, pers. obs.). Maximum annual height growth of more than 3 m a year can be achieved by 2- to 5-year-old seedlings (Harrington and Curtis 1986).

Seasonal growth of red alder is under strong climatic control and consequently quite variable. The timing of radial growth is similar for red alder and its common associate Douglas-fir; in the Puget Sound area of Washington State, growth begins about mid-April and continues until mid-September (Reukema 1965). Height growth begins slightly later in the season than radial growth. Red alder has indeterminate height growth; thus, height growth continues through the growing season until soil moisture, temperature, or light conditions become unfavorable (see DeBell and Giordano, Chapter 8). The specific environmental conditions that control root and shoot growth have not been determined.

**Vegetative reproduction.** Red alder sprouts vigorously from the stump when young. It can be repeatedly coppiced on short cycles but rootstock mortality increases with each harvest (Harrington and DeBell 1984). Age, time of year, and cutting height influence the likelihood of obtaining stump sprouts and the vigor of the sprouts (Harrington 1984a). Stumps sprout best when trees are cut in the winter and when stump height exceeds 10 cm. Older trees rarely sprout and coppice regeneration cannot be expected after pole-size or sawlog-size material is harvested (Harrington 1984a). Because of reduced vigor of sprouting, manual cutting of alder as a means of competition control in conifer plantations can be an effective vegetation management practice (DeBell and Turpin 1989); however, results from cuts at different times during the summer can be variable (Pendl and D'Anjou 1990).

Greenwood cuttings from young trees can be readily rooted. More than 50 percent of cuttings from 1- to 3-year-old plants took root within six weeks after treatment with 4000 to 8000 ppm indole-3-butyric acid and 10 percent benomyl (Monaco et al. 1980). The cuttings were set in a well-aerated planting mix and placed in a warm environment (22° to 25°C) in the daytime and 16° to 22°C at night with high relative humidity and a 16-hour photoperiod.

Cuttings of succulent new spring growth from shoots of 3- to 6-year-old trees and epicormic sprouts from 27- to 34-year-old trees have also been rooted successfully (Radwan et al. 1989). Best results were obtained with a 10-second dip in 2000 or 4000 ppm indole-3-butyric acid. The extent of rooting and root vigor on the cuttings varied greatly among ortets and treatments.

Red alder can be propagated by mound layering (Wilson and Jewett, unpub. data). For this technique the seedlings are first coppiced. When the sprouts are a few months old, their stumps and bases are covered with soil. The sprouts soon form roots and can be severed from the stump and planted at the end of the first growing season.

## Sapling and Pole Stages to Maturity

**Growth and yield.** Alder growth form is strongly excurrent during the period of rapid height growth. Crown form becomes moderately to strongly deliquescent as the trees mature. Growth of vegetative shoots is primarily monopodial (e.g., branching with the apical bud a persistent leader and new branches arising laterally below the apex, Swartz 1971); however, shoots producing flowers exhibit sympodial growth (e.g., the terminal bud withers and the main axis of branching is made up of a series of lateral branches, Swartz 1971). Young, rapidly growing trees often exhibit sylleptic branching as current-year buds produce branches. The physiological factors that determine the amount of apical control on branch growth and angle have not been studied for alder.

Growth of primary shoots of alder can be phototropic (i.e., differential elongation of cells resulting in growth toward light; Zimmerman and Brown 1971). Continued growth of branches into openings, however, occurs when high light levels result in enough photosynthate production to support continued elongation and secondary branch growth (i.e., thickening). As branch growth continues in a favorable light environment, the center of gravity moves outward, thus increasing the bending force on the branch. This causes the formation of reaction or tension wood and results in the downward movement or curvature of the branch (Zimmerman and Brown 1971). Presumably as branch development (and thus crown weight) is favored on the side toward an opening, the same process occurs in the main stem, resulting in trees that lean. Alder trees can exhibit substantial amounts of lean when grown in irregularly spaced stands or when located along roads, streams, stand boundaries, or other areas with unequal light distribution on all sides of the tree. Other changes in stem form may occur as the result of heavy snow or if gravity causes all or part of the tree to shift abruptly (e.g., as a result of soil slumping or high winds when soils are saturated). If juvenile red alder is grown at wide and fairly even spacing,

however, lean and sweep will be minimized (Bormann 1985; DeBell and Giordano, Chapter 8).

Red alder has rapid juvenile growth. Among its associates, only black cottonwood grows as much or more during the juvenile phase. On good sites, alder trees may be 9 m at age 5, 16 m at age 10, and 24 m at age 20. One tree was 9.8 m tall and 16.3 cm in dbh 5 years from seed (Smith 1968).

Growth slows after the juvenile stage, the decrease beginning much sooner on poor sites. Site index as determined at base age 20 years ranges from 10 to 25 m (Harrington and Curtis 1986); at base age 50, it ranges from 18 to 37 m (Worthington et al. 1960). Associated conifers have much slower juvenile growth, but they sustain height growth years longer than alder. On an average upland site, both Douglas-fir and red alder can attain the same height at about age 45 (Williamson 1968). Beyond that age, Douglas-fir surpasses red alder in height. Because the two species have different site tolerances, their relative performances will be site-specific as well as age-specific (Harrington and Courtin, Chapter 10).

Red alder is a relatively short-lived species, maturing at about 60 to 70 years; maximum age is usually about 100 years (Worthington et al. 1962). On favorable sites, trees can be 30 to 40 m tall and 55 to 75 cm in diameter. A record-size tree measured 198 cm in dbh, but trees over 90 cm in diameter are rare. Maximum cubic volume is attained at age 50 to 70 (500 m$^3$/ha [DeBell et al. 1978; Worthington et al. 1960; Chambers 1983]). In pure stands on good sites, it has been estimated that red alder can achieve annual cubic volume growth rates of 21 m$^3$/ha in pulpwood rotations of 10 to 12 years, and 14 m$^3$/ha in sawlog rotations of 30 to 32 years (DeBell et al. 1978). Most of the existing alder volume is in naturally regenerated mixed-species stands where growth and yield are variable.

**Rooting habit.** Red alder forms extensive, fibrous root systems. Root system distribution is primarily controlled by soil drainage, soil structure, and compaction (C. Harrington, unpub. data on file, Olympia Forestry Sciences Laboratory). In poorly drained soils, most rooting is surface-oriented, and rooting is often prolific in the boundary between the lower organic layer and the uppermost mineral horizon. In wet soils, the uppermost mineral horizon usually is rooted heavily, as is the lower organic horizon if it is thick enough. On well-drained sites, root distribution is strongly influenced by water availability; increased rooting is common at horizon boundaries when changes in soil texture slow downward water movement through the profile. Because rooting also follows the path of least resistance, it is greatest in old root channels or, especially if the soil is compacted and soil structure well developed, between units of soil structure (C. Harrington, pers. obs.). Root system extent is a function of soil characteristics and tree size. Smith (1964) showed tree diameter and average root length to be significantly correlated; larger trees also tended to have deeper roots than smaller trees. Root growth of seedlings is rapid; 2-year-old nursery-grown seedlings have been planted using a shovel because of their wide-spreading, large, woody roots.

Red alder, especially when young, forms adventitious roots when flooded. In a greenhouse study, alder seedlings previously growing under well-drained conditions produced adventitious roots when the soil was saturated (Harrington 1987). These roots emerged at or near the root collar and grew on top of the saturated soil surface; when the soil was drained, the root exteriors suberized and many of the longer roots turned downward into the soil where they continued to grow. Minore (1968) also reported formation of adventitious roots when seedlings in pots were flooded, noting that seedlings that did not form adventitious roots did not survive. Although it has not been documented, formation of adventitious roots may be an important adaptive trait on floodplain sites.

The sensitivity of red alder root growth to environmental conditions is not well known, but recent studies provide some information (also see Shainsky et al., Chapter 5). Under soil moisture stress, red alder saplings shifted carbon allocation from leaf and stem biomass to root biomass (Chan 1990). In a companion study, root biomass decreased with increasing density of alder stems (Shainsky et al. 1992). Ratios of root to shoot were significantly affected by density; however, most of the variation in root biomass was directly attributable to variation in shoot biomass. When grown

in pots in a growth chamber, these ratios were decreased by fertilization and were lower in sandy soil than in loam or sandy loam (Elliott and Taylor 1981b).

Red alder roots are commonly ectomycorrhizal, although only a few species of fungi form ectomycorrhizal associations with alder. Fungal symbionts include alder-specific fungi and fungi capable of mycorrhizal associations with other hosts (Molina 1979; Molina et al., Chapter 2).

Red alder also has root nodules that fix atmospheric nitrogen. The nodules are a symbiotic association between the tree and an actinomycete (*Frankia* spp.). In natural stands nodulation occurs soon after seed germination; root systems of seedlings a few months old commonly have dozens of visible nodules, ranging from the size of a pinhead up to 25 mm in diameter. Mature trees have nodules on both the large woody roots and the smaller new roots. Nodules found on large trees can be 80 or 90 mm in diameter. Rates of nitrogen fixation and the effects of these nitrogen additions on soil chemistry are discussed in other chapters (Binkley et al., Chapter 4; Bormann et al., Chapter 3).

**Reaction to competition.** Red alder requires more light than any of its tree associates except black cottonwood and is classed as intolerant of shade (Minore 1979). Light quality has been shown to be important in germination (Bormann 1983; Haeussler 1988); though its role in seedling development has not been documented. Young seedlings can withstand partial shade for a few years but will grow very little; if not released, they will die. The only trees that survive are those that maintain dominant or codominant crown positions. Self-thinning or mortality caused by competition is rapid in red alder stands; densities in natural stands may be as high as 124,000 seedlings/ha at age 5 (DeBell 1972) and fully stocked stands at age 20 averaged 1665 seedlings/ha (Worthington et al. 1960).

Red alder also self-prunes extremely well when grown in dense stands. Shaded lower branches rapidly die and fall off, resulting in clear and slightly tapered boles. Live crown ratios in crowded, pure stands are very low, and narrow, domelike crowns are characteristic. As would be expected for a shade-intolerant species, branch retention and crown shape are strongly related to light levels in the canopy. Trees grown at low densities develop large lower branches that live longer and take longer to fall off after death than do branches that develop under higher stand densities.

Early control of spacing is necessary to keep live crown ratios high enough to maintain good growth beyond the juvenile phase. Sawlog yields can be maximized on short rotations by combining early spacing control with pulpwood thinnings (DeBell et al. 1978). Thinnings in previously unthinned stands are most effective in stimulating growth of residual trees if done before height growth slows— about age 15 to 20 (Warrack 1949; Olson et al. 1967; Smith 1978). Thinning in older stands can salvage mortality and help maintain the vigor of residual trees, but usually does not accelerate diameter growth (Lloyd 1955; Warrack 1964).

Epicormic branching has been reported after thinning, especially when thinning has been late or drastic (Warrack 1964; Smith 1978). If epicormic sprouting occurs after thinning, it is most common on the south or west side of stressed trees (C. Harrington, pers. obs.); however, trees drastically opened up (e.g., via construction activities) may have epicormic branches on any or all sides. Epicormic branches appearing after early thinning are probably ephemeral, but this has not been documented. Epicormic branches were reported after pruning 21-year-old trees (Berntsen 1961), but not after pruning of younger trees (C. Harrington, data on file, Olympia Forestry Sciences Laboratory).

Red alder can be managed in pure stands or as part of a mixture with other intolerant species, such as Douglas-fir and black cottonwood (or *Populus* hybrids), or with more shade-tolerant species, such as western redcedar or western hemlock. Knowledge of site-specific growth rates and relative shade tolerances of each component in a mixture is critical to achieving the potential benefits from mixed stands. Alder must be kept in the upper canopy to survive in mixed stands. Even if alder is shaded out in a mixed stand, however, it may make substantial contributions to soil nitrogen prior to that time (Berg and Doerksen 1975).

Reaction of alder to competition is influenced by many factors including the size, species composition, and density of the competing vegetation (other alder stems, non-alder stems in the upper canopy, and plants in the understory) as well as soil and site factors. For example, growth of closely spaced, dominant alder was decreased with increasing density of subordinate Douglas-fir (Shainsky and Radosevich 1991; Shainsky et al., Chapter 5). The high densities of Douglas-fir decreased soil moisture availability for alder. This caused alder to shift carbon from leaf area production to root growth, resulting in a more favorable light environment for the understory species, which was less moisture-limited (Shainsky and Radosevich 1991). Thus, the interactions among plants can be complex and may influence both current growth rates of alder and long-term stand development and succession.

## Damaging Agents

There are relatively few instances where damaging agents kill enough red alder trees to result in large openings in a natural stand. Forest managers may be concerned, however, at lower levels of mortality and when growth rates are depressed or tree form or wood quality is affected. In addition, problems will likely increase as management is intensified, particularly in nurseries and plantations.

**Fungi.** Red alder is fairly free from most disease problems, especially when young and uninjured (Worthington et al. 1962; Hepting 1971). Many species of fungi have been reported growing in association with alder (Lowe 1969; Shaw 1973; Farr et al. 1989), but few have been shown to cause ecologically or economically important levels of damage in natural stands. Several canker-causing stem diseases—*Didymosphaeria oregonensis*, *Hymenochaete agglutinans*, and *Nectria* spp.— cause some damage, especially in young stands, but overall their impact is slight. Nonetheless, the potential exists for more damaging levels of disease should conditions occur that favor their rapid development and spread. For example, naturally occurring weak pathogens have been used as biocontrol agents for juvenile red alder (i.e., to kill unwanted alder) (Dorworth, in press), thus indicating their ability to cause mortality under certain conditions.

*Table 2. Incidence of decay with age in red alder stands on Vancouver Island, British Columbia.*

| Age class (years) | Trees with decay (%) | Decay events per tree |
|---|---|---|
| 15—24 | 18.8 | 1.7 |
| 25—34 | 14.1 | 1.3 |
| 35—44 | 35.0 | 1.4 |
| 45—54 | 50.4 | 1.5 |
| 55—64 | 65.0 | 1.4 |
| 65—74 | 63.6 | 1.4 |
| 75+ | 93.8 | 1.7 |

*Table 3. Occurrence of decay associated with pathological indicators in red alder trees on Vancouver Island, British Columbia.*

| Pathological indicator | Occurrence of decay (%) |
|---|---|
| Scars | 36 |
| Forks[1] | 13 |
| Crooks[1] | 9 |
| Branches | 12 |
| Dead tops[1] | 6 |
| Other | 2 |
| Unknown[2] | 22 |
| Total | 100 |

[1]Trees that sustain top damage may be identified years later by the presence of broken or dead tops, multiple or forked tops, or stem crooks; since the amount of decay could be associated with tree appearance, these indicators of past top damage were recorded separately.

[2]During dissection it was not always possible to trace decay to a recognized pathological indicator. In such cases, a designation of "unknown" was assigned. These decay events are likely associated with branches or scars.

The primary disease of concern in nursery production is *Septoria alnifolia*, a disease that causes leaf spots and stem cankers. Although some infected seedlings grow over stem cankers caused by *S. alnifolia*, on other trees the cankers result in top die-back, stem breakage, reduced growth, or mortality. Thus, infected nursery stock should be graded out as cull (i.e., discarded and thus not available for planting). This disease can be controlled with monthly applications of Benlate® (a fungicide) and by locating alder nursery beds in areas not adjacent to alder stands (W. Littke, pers. com. 1992).

Fungi diseases of alder catkins (*Taphrina occidentalis* and *T. alni*) cause enlargements of the bracts of female catkins (Mix 1949) and thus prevent or hinder normal fertilization and seed development. Although currently these fungi are not important economically, they could become so if alder seed production is desired.

Compared with other hardwood species, living red alder trees have very little decay. In a study currently underway in British Columbia, Allen (1992) examined 383 alder trees on Vancouver Island, ranging from 20 to 120 years old. Dissected trees were assessed for decay volume and the relation of decay columns to externally visible pathological indicators. A total of 243 decay "events" were observed in the sample trees. Decay losses of merchantable volume were less than 4 percent in all trees sampled. Although the incidence of decay increased with age (Table 2), decay volume was poorly correlated with age, and susceptibility to decay in older trees does not appear to be as severe as suggested in previous reports (Johnson et al. 1926; Worthington 1957). For example, trees harvested at 60 to 80 years would have predicted decay losses under 3.5 percent.

Much of the decay present in living alder results from injury to standing trees due to broken tops and branches, or scars from falling trees (Table 3). Once trees are injured, decay organisms gain entry through the damaged tissue. Alder, however, is very efficient in its ability to compartmentalize decay, and most decay events do not spread much beyond the injured tissue. For example, the dead tissue of stubs formed from self-pruned branches was colonized by fungi and sometimes developed into a decay column in the main stem. Most branch stubs, however, were overgrown by healthy wood with no further decay development. In general, individual decay columns were not large, with a median volume of 0.0024 m$^3$.

In the past, some foresters have suggested that red alder would have to be managed on short rotations due to increasing disease problems with age (so-called pathological rotations) or that thinning and pruning were risky due to the increased probability of inducing stem damage; based on our current information on alder's ability to compartmentalize decay, these suggestions appear unwarranted. During intermediate cuts, however, care should be taken to avoid injuring residual trees.

A number of decay fungi have been isolated from living alder trees in British Columbia, including *Heterobasidion annosum*, *Sistrotrema brinkmannii*, *Pholiota adiposa*, *Trametes* sp., and *Meruliopsis corium*. A previous report suggested that white heart rot, caused by *Phellinus* (syn. *Fomes*) *igniarius*, is the most destructive disease of living alder trees (Worthington 1957). This statement seems to have originated from Johnson et al. (1926), who made a similar claim that was unsupported by data or reference citation. *Phellinus igniarius* has been found only rarely on living alder in British Columbia, although the pathogen may be more common in other parts of the alder range. As indicated above, these decay fungi do not appear to result in serious losses in living trees.

Special mention should be made regarding the potential hazards of the root rot pathogen *Heterobasidion annosum*, which was observed growing on alder in mixed-species stands in which both alder and conifers were present (E. A. Allen, pers. obs.). Since the fungus infects both hardwoods and softwoods, it is possible that alder could become infected when planted on sites previously occupied by infected conifers, or that infected alder could serve as an inoculum source for subsequently planted conifers. Thus, preharvest surveys for root-rot fungi should be considered for sites where alder is a possible species to manage as well as on sites where conversion from alder to other species is being considered.

Wood stain and decay proceed rapidly in cut alder trees. Losses due to stain resulting from fungal infection that occurs during the time *between* harvesting and milling are much greater than losses from decay in living trees. For this reason, logs should be processed as soon as possible after harvest, particularly in warm summer months. The development of stain and decay is retarded in winter months and in logs stored in fresh water (Worthington 1957).

***Phellinus weirii.*** All hardwoods are immune to *Phellinus weirii* (a widespread conifer root rot) and red alder occurs naturally and has been planted on sites where *P. weirii* infection levels are high. The absence of susceptible species will eventually "starve-out" the fungus as suitable substrates decay and are not replaced. It has been hypothesized that red alder alters the soil environment to the detriment of *P. weirii* by enhancing the growth of microbes (e.g., *Trichoderma*) antagonistic to the pathogen (Nelson et al. 1978) and/or by inhibiting the growth of the fungus (Li et al. 1969, 1972; Hansen 1979).

Many foresters, following the suggestion that alder may serve as a biological control agent for *Phellinus weirii* (Trappe 1972; Nelson et al. 1978), have recommended the planting of alder as a root-rot control measure. This may have contributed to much of the recent increase in planting of red alder on public lands, but the interactions between red alder and *P. weirii* are not fully understood and therefore caution should be exercised in making management decisions. For example, Hansen (1979) observed vigorous and extensive development of *P. weirii* ectotrophic mycelium on infected roots of large Douglas-fir stumps in a 20-year-old alder stand that established following logging of the Douglas-fir. He concluded that red alder apparently "does not shorten the time required to reduce the inoculum."

Long-term trials are underway to quantify the long-term effects of alder stands on *Phellinus weirii*. At this time all we can conclude is that alder should be considered as one of several immune species to plant on sites with high levels of *P. weirii*. Foresters planting red alder on sites with high levels of *P. weirii* inoculum that are considered poor or unsuitable sites for alder should expect alder growth to be poor and problems with damaging agents to increase.

**Insects.** Numerous insects have been reported feeding on or associated with red alder (Furniss and Carolin 1977; Gara and Jaeck 1978; Dolan 1984). Insect pests are not usually a major concern, but serious outbreaks of some defoliators can cause growth reductions. The forest tent caterpillar (*Malacosoma disstria*), western tent caterpillar (*M. californicum*), alder woolly sawfly (*Eriocampa ovata*), striped alder sawfly (*Hemichroa crocea*), the alder flea beetle (*Altica ambiens*), and a leaf beetle (*Pyrrhalta punctipennis*) have caused substantial damage, but reports of mortality are rare (Worthington et al. 1962; Furniss and Carolin 1977; Briggs et al. 1978). Mortality, however, was observed when a forest tent caterpillar outbreak overlapped a drought period (Russell 1991) and probably was substantially greater than would have occurred if only one stress was present.

A flatheaded wood borer (*Agrilus burkei*) can kill twigs and branches (Furniss and Carolin 1977; Briggs et al. 1978). The alder leaf miner, *Lithocolletis alnicolella*, can cause necrotic spots up to 30 mm in diameter on leaves but does not apparently affect growth (W. Littke, pers. com. 1992). An epidemic of grasshoppers was reported to only slow growth slightly (Russell 1986). The fall webworm (*Hyphantria cunea*) skeletonizes or consumes leaf blades, but its damage is usually minor (Furniss and Carolin 1977). The alder bark beetle (*Alniphagus aspericollis*) breeds primarily in slash and in young stressed trees; however, healthy trees can be attacked when bark beetle populations are high (Briggs et al. 1978).

The alder aphid (*Pterocaulis alni*) feeds on tender shoots (Furniss and Carolin 1977) and on foliage with high nitrogen content (Dolan 1984). Aphids are common associates in many young alder stands and generally are not considered to cause much damage, although a severe aphid epidemic was reported in a young alder plantation (Dolan 1984). Under those epidemic conditions, plots sprayed with insecticide had diameter growth increases up to 38 percent over unsprayed plots. Poor vigor was suspected of predisposing the trees to supporting an outbreak (Dolan 1984).

Ambrosia beetles (*Gnathotrichus retusus*, *Trypodendron lineatum*, *Xyleborus saxeseni*) attack logs and slash left on the ground, causing rapid degrade in quality. Insect holes can also serve as entry sites for fungi. Merchantable material should be removed rapidly, and large accumulations of slash should be avoided.

**Animals.** In general, animals cause only minor damage in alder stands; however, under some circumstances animal damage can be significant. Alder is not a highly preferred browse species for black-tailed deer (*Odocoileus hemionus columbianus*) or Roosevelt elk (*Cervus elaphus roosevelti*) during most of the year. Young trees are occasionally browsed by deer and elk, especially during the late summer and fall (Brown 1961), and browsing begins earlier in the growing season when weather conditions are dry or when other food sources are not available (C. Harrington, pers. obs.).

Abscising or freshly abscised leaves were documented as being a major component of deer and elk diets in old-growth forests on the Olympic Peninsula (Leslie et al. 1984) and penned black-tailed deer have been observed eating freshly abscised alder leaves in the fall when other food sources were available (D. L. Campbell, pers. com. 1992). Seasonal changes in deer and elk browsing may be related to changes in foliar chemical composition; alder foliage in the fall is higher in crude fat content and lower in total phenols than during the summer (Radwan et al. 1978). Elk repeatedly browsed red alder planted on a debris flow associated with the 1981 eruption of Mt. St. Helens (Russell 1986) when alternative food sources were limited. Most browsed trees resprouted vigorously and very little mortality was associated with the heavy browse damage; however, the repeated browsing resulted in trees with shrublike forms.

Deer and elk can cause stem deformation, reduce growth, and provide entry sites for decay organisms when they rub their antlers against tree trunks; in localized pockets this type of damage can be common. In a young spacing trial near Centralia, Washington, the incidence of stem rubbing was greatest in the narrowest spacings, presumably because the closer spacings had higher rates of branch mortality which resulted in easier access to the main stems. Cole (see Newton and Cole, Chapter 7), however, has observed greater damage in wide spacings. The relationship between spacing and damage may change with age or other factors. Deer and elk occasionally strip and eat alder bark, especially during winter and spring.

Mountain beaver (*Aplodontia rufa*) clip small alder stems and branches; only the bark is eaten from stems 5 to 20 mm in diameter while smaller pieces are consumed whole (D. L. Campbell, pers. com. 1992). Although mountain beaver only clip small-diameter pieces, they climb trees and can continue to clip branches and terminals as trees increase in size. In an artificial feeding trial in which several plant species were made available at the same time, mountain beaver consistently selected alder stems all months of the year (data on file, USDA APHIS Animal Damage Research Unit, Olympia, Washington). Thus, alder appears to be a regular item in mountain beaver diets and problems in stand establishment should be anticipated on sites with established mountain beaver populations (D. L. Campbell, pers. com. 1992). Mountain beaver use of alder *foliage* for food is minor except when other food sources are not available or in late September when use is fairly heavy (Voth 1968).

Observations of other animals damaging red alder are limited. Beaver (*Castor canadensis*) will cut any species of tree near their ponds to support their construction activities. As a food source, beaver prefer red alder over Douglas-fir, but other plants are selected before alder if they are equally available (D. L. Campbell, pers. com. 1992). In years of high populations, meadow mice (*Microtus* sp.) girdle young stems; this type of damage has been most commonly observed in grassy or very wet areas. Deer mice (*Peromyscus maniculatus*) eat alder seed from the surface of snowpacks when other food is difficult to obtain (Moore 1949); however, alder seed is not usually a preferred food source. Individual trees can be heavily damaged by red-breasted sapsuckers (*Sphyrapicus ruber*); if the damage encircles all or most of the stem, the top may break off during periods of wind or snow.

**Extremes in physical factors.** Extremes in temperature, wind, or fire can damage red alder. Mortality and top damage have been documented in natural stands after ice storms or unseasonable frosts (Duffield 1956; Worthington et al. 1962).

Widespread cold damage was observed in bare-root nurseries after a prolonged cold period in December 1990; this caused terminal die-back in many trees and some mortality, especially trees in exposed areas. Recently planted trees are also susceptible to cold damage; late spring frosts and early fall frosts have caused top die-back and mortality (DeBell and Wilson 1978; Peeler and DeBell 1987). When grown at a common location, alder sources from northerly locations or higher elevations set bud and became cold tolerant earlier in the fall than sources from more southerly locations or lower elevations (Cannell et al. 1987; Ager and Stettler, Chapter 6); geographic variation in spring budbreak and frost hardiness were more complex. Ager (1987) reported that variation in budbreak could be predicted from growing season thermal sums (see Ager and Stettler, Chapter 6). The winter dormancy requirement for red alder has not been studied, however, and the causal factors controlling timing of spring budbreak are not known. Presumably, once chilling requirements (if any exist) are met or day length is permissive, budbreak is temperature dependent. This assumption is consistent with the observation of Peeler and DeBell (1987) that cold damage occurred when late frosts followed a period of warmer-than-normal temperature.

Other temperature-related problems observed on alder are sunscald and frost cracks. As is generally true for other species, this type of damage is most common on the south and west side of exposed trees.

Fire is rarely a damaging agent because of the scarcity of flammable debris in alder stands; in fact, the species has been planted as a firebreak to protect adjacent conifers (Worthington et al. 1962). Alder bark is thin but sufficiently fire-resistant to prevent damage during light surface fires (Worthington 1957).

Windthrow is not common in alder because of the intermingling of roots and branches, the absence of leaves during winter storms when soils can be waterlogged, and the relatively deep-rooting habit of the species on well-drained soils. Uprooted trees are most commonly observed along cutting boundaries or where established root systems have been undercut by flooding or erosion. High winds, heavy snow, and ice storms will break alder tops and branches, but these problems are generally less for alder than for associated species which are foliated during the winter. Exposed windy sites, however, such as those near the ocean or mountain passes, have top breakage and reductions in height growth consistently enough to reduce site index (Harrington 1986).

Red alder has evolved to survive in climates with low summer rainfall. The greater stomatal control of red alder as compared to black cottonwood (Harrington 1987) is probably a key feature that allows the species to grow on upland sites. When the two species were planted on a droughty site that was irrigated during establishment, subsequent red alder growth without irrigation was much less than that of black cottonwood (data on file, Olympia Forestry Sciences Laboratory). The poorer growth by red alder could be due to shallower root system development, but this hypothesis has not been tested. In general, red alder is probably not as drought tolerant as most of its coniferous associates (Shainsky et al., Chapter 5).

During the summer of 1987, rainfall in the Puget Sound area of Washington was less than one-third of normal; for red alder this resulted in widespread leaf yellowing and premature abscission, terminal die-back, and—on droughty sites or new plantings —mortality (Russell 1991). Prior to 1987, the Puget Sound area experienced several decades without back-to-back dry summers and many years of above-normal rainfall. Combining these weather patterns with high levels of harvesting activity that created seedbed conditions favorable to alder establishment may have increased the percentage of alder stands growing on drought-sensitive sites (K. Russell, pers. com. 1992). Every summer (1 June to 30 September) from 1987 through 1992, the Puget Sound region had below-normal precipitation. Thus trees stressed by the extreme drought in 1987 may have been further stressed in subsequent years; presumably these back-to-back dry summers are one of the causes of the wide-spread instances of alder top die-back and mortality in the Puget Sound region in the late 1980s and early 1990s (K. Russell, pers. com. 1992).

The sensitivity of red alder to stress factors other than those discussed above is not documented. Al-der is found on sites close to the ocean and presumably is fairly tolerant of salt spray. Alder has also been observed adjacent to pulp mills and other industrial plants and thus exhibits tolerance for at least some components of air pollution.

## Future Research Needs

Most of the other chapters in this volume include recommendations for future research needs on specific topics; thus, we will only mention broad information needs not included in those chapters. Even with that restriction, there are many areas where additional research is needed. Almost 15 years ago Minore (1979) commented on the surprising lack of information on autecological characteristics of red alder. Although available information has increased since then, much of our knowledge of the biology of red alder is still based on casual or short-term observations and not on detailed life histories or controlled experiments. Additional research on the silvics of red alder is warranted to provide a firm knowledge base from which to make management recommendations. Specific topics of interest include the physiological or ecological factors that control: (1) forest succession, (2) alder seed production, dispersal and germination, (3) spatial distribution, timing of, and interrelationships between root and shoot growth, (4) tree responses to changes in light, nutrients, moisture, or temperature regimes, and (5) the occurrence and significance of biotic damaging agents. In 1979, Minore concluded that red alder "should be neglected no longer." We think it is time to follow-up on that recommendation.

## Acknowledgments

Portions of this manuscript were taken from Harrington (1990). Some of the work summarized in this chapter was supported with funding provided by U.S. Department of Energy, Wood Crops Program, under interagency agreement DE-AI05-810R20914. We thank D. Campbell (Project Leader, USDA Animal, Plant and Health Inspection Service, Animal Damage Research Unit, Olympia, Washington), W. Littke (Research Pathologist, Weyerhaeuser Company, Centralia, Washington), and K. Russell (Forest Pathologist, Washington State Department of Natural Resources, Olympia, Washington) for sharing unpublished observations.

# Literature Cited

Ager, A. A. 1987. Genetic variation in red alder (*Alnus rubra* Bong.) in relation to climate and geography in the Pacific Northwest. Ph.D. thesis, Univ. of Washington, Seattle.

Allen, E. A. 1992. Unpublished data on file. Pacific Forestry Centre, Victoria, B.C. (Manuscript in prep.)

Berg, A., and A. Doerksen. 1975. Natural fertilization of a heavily thinned Douglas-fir stand by understory red alder. Forest Research Laboratory, Oregon State Univ., Corvallis, Res. Note 56.

Berntsen, C. M. 1961. Pruning and epicormic branching in red alder. J. For. 59(9):675-676.

Berry, A. M., and J. G. Torrey. 1985. Seed germination, seedling inoculation and establishment of *Alnus* spp. in containers in greenhouse trials. Plant and Soil 87(1):161-173.

Bormann, B. T. 1983. Ecological implications of phytochrome mediated seed germination in red alder. For. Sci. 29(4):734-738.

———. 1985. Early wide spacing in red alder (*Alnus rubra* Bong.): effects on stem form and stem growth. USDA For. Serv., Res. Note PNW-423.

Bjorkbom, J. 1971. Production and germination of paper birch seed and its dispersal into a forest opening. USDA For. Serv., Northeastern For. Exp. Sta., Res. Pap. NE-209.

Brayshaw, T. C. 1976. Catkin-bearing plants (*Amentiferae*) of British Columbia. Occasional papers, British Columbia Provincial Museum. No. 18.

Briggs, D. G., D. S. DeBell, and W. A. Atkinson, *comps.* 1978. Utilization and management of alder. USDA For. Serv., Gen. Tech. Rep. PNW-70.

Brinkman, K. A. 1974. *Betula* L.—Birch. *In* Seeds of woody plants in the United States. *Technically coordinated by* C. S. Schopmeyer. USDA For. Serv., Agric. Handb. 450. 252-257.

Brown, E. R. 1961. The black-tailed deer of western Washington. Washington State Dept. of Game, Biol. Bull. 13. Olympia, WA.

Brown, S. M. 1985. A study of reproductive biology of *Alnus rubra* along three elevational transects in Washington and Oregon. Report on file, USDA For. Serv., PNW Res. Sta., Olympia, WA.

———. 1986. Sexual allocation patterns in red alder (*Alnus rubra* Bong.) along three elevational transects. M.S. thesis, Univ. of Washington, Seattle.

Cannell, M. G. R., M. B. Murray, and L. J. Sheppard. 1987. Frost hardiness of red alder (*Alnus rubra*) provenances in Britain. Forestry 60:57-67.

Carlton, G. C. 1988. The structure and dynamics of red alder communities in the central Coast Range of western Oregon. M.S. thesis, Oregon State Univ., Corvallis.

Chambers, C. J. 1983. Empirical yield tables for predominantly alder stands in western Washington. 4th print. Washington State Dept. of Natural Resources, DNR Rep. 31. Olympia, WA.

Cwynar, L. C. 1987. Fire and the forest history of the North Cascade Range. Ecology 68(4):791-802.

Chan, S. S. 1990. Effects of light and soil moisture availability on Douglas-fir and red alder sapling development, carbon allocations, and physiology. Dissertation, Oregon State Univ., Corvallis.

Davis, M. B. 1973. Pollen evidence of changing land use around the shores of Lake Washington. Northwest Sci. 47(3):133-148.

DeBell, D. S. 1972. Potential productivity of dense, young thickets of red alder. Crown Zellerbach, Forest Res. Note 2. Camas, WA.

DeBell, D. S., M. A. Radwan, C. A. Harrington, G. W. Clendenen, J. C. Zasada, W. R. Harms, and M. R. McKevlin. 1990. Increasing the productivity of biomass plantations of cottonwood and alder in the Pacific Northwest. Annual technical report submitted to the U. S. Dept. of Energy, Woody Crops Program.

DeBell, D. S., R. F. Strand, and D. L. Reukema. 1978. Short-rotation production of red alder: some options for future forest management. *In* Utilization and management of alder. *Compiled by* D. G. Briggs, D. S. DeBell, and W. A. Atkinson. USDA For. Serv., Gen. Tech. Rep. PNW-70. 231-244.

DeBell, D. S., and T. C. Turpin. 1989. Control of red alder by cutting. USDA For. Serv., Res. Pap. PNW-414.

DeBell, D. S., and B. C. Wilson. 1978. Natural variation in red alder. *In* Utilization and management of alder. *Compiled by* D. G. Briggs, D. S. DeBell, and W. A. Atkinson. USDA For. Serv., Gen. Tech. Rep. PNW-70. 193-208.

Dolan, L. S. 1984. The cultural treatment of selected species for woody biomass production in the Pacific Northwest. Final report prepared by Seattle City Light Dept. for U.S. Dept. of Energy. Grant DE-FG-79-78BP35773.

Dorworth, C. E. Augmentation of biological control in Canada's forests. *In* Proc. Forest biological control in the Great Plains, 13-16 July 1992, Bismarck, ND. *Edited by* M. E. Dix. (In press.)

Duffield, J. W. 1956. Damage to western Washington forests from November 1955 cold wave. USDA For. Serv., Res. Note 129.

Elliot, D. M., and I. E. P. Taylor. 1981a. Germination of red alder (*Alnus rubra*) seed from several locations in its natural range. Can. J. For. Res. 11:517-521.

———. 1981b. The importance of fertility and physical characteristics of soil in early development of red alder seedlings grown under controlled environmental conditions. Can. J. For. Res. 11:522-529.

Eyre, F. H., *ed.* 1980. Forest cover types of the United States and Canada. Society of American Foresters, Washington, D.C.

Farr, D. F., G. F. Bills, G. P. Chamuris, and A. Y. Rossman. 1989. Fungi on plants and plant products in the United States. American Phytopathological Society Press. 92-97.

Furlow, J. 1974. A systematic study of the American species of *Alnus* (Betulaceae). Ph.D. thesis, Michigan State Univ., East Lansing.

Furniss, R., and V. M. Carolin. 1977. Western forest insects. USDA For. Serv., Misc. Publ. 1339. Washington, D.C.

Gara, R. I., and L. L. Jaeck. 1978. Insect pests of red alder: potential problems. *In* Utilization and management of alder. *Compiled by* D. G. Briggs, D. S. DeBell, and W. A. Atkinson. USDA For. Serv., Gen. Tech. Rep. PNW-70. 265-269.

Granstrom, A. 1982. Seed viability of fourteen species during five years of storage in forest soil. J. Ecol. 75:321-331.

Haeussler, S. 1988. Germination and first-year survival of red alder seedlings in the central Coast Range of Oregon. M.S. thesis, Oregon State Univ., Corvallis.

Haeussler, S., and J. C. Tappeiner II. 1993. Effect of the light environment on seed germination of red alder (*Alnus rubra*). Can. J. For. Res. 23:1487-1491.

Hansen, G. M. 1979. Survival of *Phellinus weirii* in Douglas-fir stumps after logging. Can. J. For. Res. 9:484-488.

Harrington, C. A. 1984a. Factors influencing sprouting of red alder. Can. J. For. Res. 14(3):357-361.

———. 1984b. Red alder: an American wood. USDA For. Serv., Publ. FS-215.

———. 1986. A method of site quality evaluation for red alder. USDA For. Serv., Gen. Tech. Rep. PNW-192.

———. 1987. Responses of red alder and black cottonwood seedlings to flooding. Physiol. Plant. 69:35-48.

———. 1990. *Alnus rubra* Bong.—red alder. *In* Silvics of North America, vol. 2, Hardwoods. *Technically coordinated by* R. M. Burns and B. H. Honkala. USDA For. Serv., Agric. Handb. 654. Washington, D.C. 116-123.

Harrington, C. A., and R. O. Curtis. 1986. Height growth and site index curves for red alder. USDA For. Serv., Res. Pap. PNW-358.

Harrington, C. A., and D. S. DeBell. 1980. Variation in specific gravity of red alder (*Alnus rubra* Bong.). Can. J. For. Res. 10(3):293-299.

———. 1984. Effects of irrigation, pulp mill sludge, and repeated coppicing on growth and yield of black cottonwood and red alder. Can. J. For. Res. 14(6):844-849.

Harris, A. S. 1969. Ripening and dispersal of a bumper western hemlock-Sitka spruce seed crop in southeast Alaska. USDA For. Serv., Res. Note PNW-105.

Heebner, C. F., and M. J. Bergener. 1983. Red alder: a bibliography with abstracts. USDA For. Serv., Gen. Tech. Rep. PNW-161.

Hepting, G. H. 1971. Diseases of forest and shade trees of the United States. USDA For. Serv., Agric. Handb. 386. Washington, D.C.

Heusser, C. J. 1964. Palynology of four bog sections from the western Olympic Peninsula, Washington. Ecology 45:23-40.

Hibbs, D. E., and A. A. Ager. 1989. Red alder: guidelines for seed collection, handling, and storage. Forest Research Laboratory, Oregon State Univ., Corvallis, Spec. Publ. 18.

Hitchcock, C. L., A. Cronquist, M. Ownbey, and J. W. Thompson. 1964. Vascular plants of the Pacific Northwest. Part 2: *Salicaceae* to *Saxifragaceae*. Univ. of Washington Press, Seattle.

Hook, D. D., M. D. Murray, D. S. DeBell, and B. C. Wilson. 1987. Variation in growth of red alder families in relation to shallow water-table levels. For. Sci. 33(1):224-229.

Johnson, F. D. 1968a. Disjunct populations of red alder in Idaho. *In* Biology of alder. *Edited by* J. M. Trappe, J. F. Franklin, R. F. Tarrant, and G. M. Hansen. USDA For. Serv., PNW For. Range Exp. Sta., Portland, OR. 1-8.

———. 1968b. Taxonomy and distribution of northwestern alders. *In* Biology of alder. *Edited by* J. M. Trappe, J. F. Franklin, R. F. Tarrant, and G. M. Hansen. USDA For. Serv., PNW For. Range Exp. Sta., Portland, OR. 9-22.

Johnson, H. M., E. J. Hanzlik, and W. H. Gibbons. 1926. Red alder of the Pacific Northwest: its utilization, with notes on growth and management. USDA Dept. Bull. 1437.

Koski, V., and R. Tallquist. 1978. Results of long-time measurements of the quantity of flowering and seed crop of forest trees. Folia Forestalia 364.

Leslie, D. M., Jr., E. E. Starkey, and M. Vaura. 1984. Elk and deer diets in old-growth forests in western Washington. J. Wildl. Manage. 48(3):762-775.

Lester, D. T., and D. S. DeBell. 1989. Geographic variation in red alder. USDA For. Serv., Res. Pap. PNW-409.

Lewis, S. J. 1985. Seedfall, germination, and early survival of red alder. M.S. thesis, Univ. of Washington, Seattle.

Li, C. Y., K. C. Lu, E. E. Nelson, W. B. Bollen, and J. M. Trappe. 1969. Effect of phenolic and other compounds on growth of *Poria weirii in vitro*. Microbios 3:305-311.

Li, C. Y., K. C. Lu, J. M. Trappe, and W. B. Bollen. 1972. *Poria weirii*-inhibiting and other phenolic compounds in roots of red alder and Douglas-fir. Microbios 5:65-68.

Lloyd, W. J. 1955. Alder thinning—progress report. USDA Soil Conservation Serv., West Area Woodland Conservation Tech. Note 3. Portland, OR.

Lousier, J. D., and B. Bancroft. 1990. Guidelines for alder seed tree control. Canadian Forestry Service and British Columbia Ministry of Forests, FRDA Memo No. 132.

Lowe, D. P. 1969. Checklist and host index of bacteria, fungi, and mistletoes of British Columbia. Environment Canada, Canadian Forestry Service, Forest Research Laboratory, Publ. BC-X-32. 241-244.

McGee, A. B. 1988. Vegetation response to right-of-way clearing procedures in coastal British Columbia. Ph.D. thesis, Univ. of British Columbia, Vancouver.

Minore, D. 1968. Effects of artificial flooding on seedling survival and growth of six northwestern species. USDA For. Serv., Res. Note PNW-92.

———. 1979. Comparative autecological characteristics of northwestern tree species—a literature review. USDA For. Serv., Gen. Tech. Rep. PNW-87.

Mix, A. J. A. 1949. A monograph of the genus *Taphrina*. Univ. Kansas Sci. Bull. 33, part 1: 3-167.

Molina, R. 1979. Pure culture synthesis and host specificity of red alder mycorrhizae. Can. J. Bot. 57(11):1223-1228.

Monaco, P. A., T. M. Ching, and K. K. Ching. 1980. Rooting of *Alnus rubra* cuttings. Tree Planters' Notes 31(3):22-24.

Moore, A. W. 1949. Forest tree-seed-eaters and methods used to measure their populations in the Pacific Northwest Douglas-fir region. Washington (State) Univ. Forest Club Quarterly 23(1): 7-11, 25.

Nelson, E. E., E. M. Hansen, C. Y. Li, and J. M. Trappe. 1978. The role of red alder in reducing losses from laminated root rot. *In* Utilization and management of alder. *Compiled by* D. G. Briggs, D. S. DeBell, and W. A. Atkinson. USDA For. Serv., Gen. Tech. Rep. PNW-70. 273-282.

Newton, M., B. A. El Hassan, and J. Zavitkovski. 1968. Role of red alder in western Oregon forest succession. *In* Biology of alder. *Edited by* J. M. Trappe, J. F. Franklin, R. F. Tarrant, and G. M. Hansen. USDA For. Serv., PNW For. Range Exp. Sta., Portland, OR. 73-84.

O'Dea, M. E. 1992. The clonal development of vine maple during Douglas-fir development in the Coast Range of Oregon. M.S. thesis, Oregon State Univ., Corvallis.

Olson, R., D. Hintz, and E. Kittila, E. 1967. Thinning young stands of alder. USDA Soil Conservation Serv., West Area Woodland Conservation Tech. Note 122. Portland, OR.

Owens, J. N. 1991. Flowering and seed set. *In* Physiology of trees. *Edited by* A. S. Raghavendra. Wiley, New York. 247-271.

Owens, J. N., and S. J. Simpson.n.d. Manual of conifer pollen from British Columbia conifers. Biology Dept., Univ. of Victoria, Victoria, B.C.

Peeler, K. C., and D. S. DeBell. 1987. Variation in damage from growing-season frosts among open-pollinated families of red alder. USDA For. Serv., Res. Note PNW-464.

Pendl, F., and B. D'Anjou. 1990. Effect of manual treatment timing on red alder regrowth and conifer release. British Columbia Ministry of Forests, FRDA Rep. 112.

Perala, D. A., and A. Alm. 1989. Regenerating paper birch in the Lake States with the shelterwood method. Northern J. Applied For. 6:151-153.

Radwan, M. A., and D. S. DeBell. 1981. Germination of red alder seed. USDA For. Serv., Res. Note PNW-370.

Radwan, M. A., W. D. Ellis, and G. L. Crouch. 1978. Chemical composition and deer browsing of red alder foliage. USDA For. Serv., Res. Pap. PNW-246.

Radwan, M. A., T. A. Max, and D. W. Johnson. 1989. Softwood cuttings for propagation of red alder. New For. 3:21-30.

Radwan, M. A., Y. Tanaka, A. Dobkowski, and W. Fangen. 1992. Production and assessment of red alder planting stock. USDA For. Serv., Res. Pap. PNW-450.

Reukema, D. L. 1965. Seasonal progress of radial growth of Douglas-fir, western redcedar, and red alder. USDA For. Serv., Res. Pap. PNW-26.

Russell, K. 1986. Revegetation trials in a Mount St. Helens eruption debris flow. *In* Mt. St. Helens: five years later. Proceedings of a Symposium, 16-18 May 1985, Eastern Washington Univ., Cheney, WA. 231-248.

———. 1991. Drought injury to trees related to summer and winter weather extremes. Washington State Dept. of Natural Resources, Forest Health Alert #3. Olympia, WA.

Safford, L. O., J. C. Bjorkbom, and J. C. Zasada. 1990. *Betula papyrifera* Marsh.—paper birch. *In* Silvics of North America, vol. 2, Hardwoods. *Technically coordinated by* R. M. Burns and B. H. Honkala. USDA For. Serv., Agric. Handb. 654. Washington, D.C. 158-171.

Schopmeyer, C. S. 1974. *Alnus* B. Ehrh.—alder. *In* Seeds of woody plants in the United States. *Technically coordinated by* C. S. Schopmeyer. USDA For. Serv., Agric. Handb. 450. 206-211.

Shainsky, L. J., and S. R. Radosevich. 1991. Analysis of yield-density relationships in experimental stands of Douglas-fir and red alder seedlings. For. Sci. 37:574-592.

Shainsky, L. J., M. Newton, and S. R. Radosevich. 1992. Effects of intra- and inter-specific competition on root and shoot biomass of young Douglas-fir and red alder. Can. J. For. Res. 22:101-110.

Shaw, C. G. 1973. Host fungus index for the Pacific Northwest, vol. 1, Host. Washington State Univ., Washington Agric. Exp. Sta. Bull. 766. 14-15.

Smith, J. H. G. 1964. Root spread can be estimated from crown width of Douglas fir, lodgepole pine, and other British Columbia tree species. For. Chron. 40:456-473.

———. 1968. Growth and yield of red alder in British Columbia. *In* Biology of alder. *Edited by* J. M. Trappe, J. F. Franklin, R. F. Tarrant, and G. M. Hansen. USDA For. Serv., PNW For. Range Exp. Sta., Portland, OR. 273-286.

———. 1978. Growth and yield of red alder: effects of spacing and thinning. *In* Utilization and management of alder. *Compiled by* D. G. Briggs, D. S. DeBell, and W. A. Atkinson. USDA For. Serv., Gen. Tech. Rep. PNW-70. 245-263.

Stettler, R. F. 1978. Biological aspects of red alder pertinent to potential breeding programs. *In* Utilization and management of alder. *Compiled by* D. G. Briggs, D. S. DeBell, and W. A. Atkinson. USDA For. Serv., Gen. Tech. Rep. PNW-70. 209-222.

Swartz, D. 1971. Collegiate dictionary of botany. Ronald Press Co., New York.

Tanaka, Y., P. J. Brotherton, A. Dobkowski, and P. C. Cameron. 1991. Germination of stratified and non-stratified seeds of red alder at two germination temperatures. New For. 5:67-75.

Tappeiner, J., J. Zasada, P. Ryan, and M. Newton. 1991. Salmonberry clonal and population structure: the basis for a persistent cover. Ecology 72(2):609-618.

Trappe, J. M. 1972. Regulation of soil organisms by red alder—a potential biological system for control of *Poria weirii*. *In* Managing young forests in the Douglas-fir region. *Edited by* A. B. Berg. Oregon State Univ., Corvallis. 35-46.

Trappe, J. M., J. F. Franklin, R. F. Tarrant, and G. M. Hansen, *eds*. 1968. Biology of alder. USDA For. Serv., PNW For. Range Exp. Sta., Portland, OR.

Voth, E. H. 1968. Food habits of the Pacific mountain beaver, *Aplodontia rufa pacifica* Merriam. Ph.D. thesis, Oregon State Univ., Corvallis.

Warrack, G. 1949. Treatment of red alder in the coastal region of British Columbia. British Columbia For. Serv., Res. Note 14. Victoria.

Warrack, G. C. 1964. Thinning effects in red alder. British Columbia For. Serv., Res. Div., Victoria.

White, C. M., and G. C. West. 1977. The annual lipid cycle and feeding behavior of Alaskan redpolls. Oecologia 27:227-238.

Williamson, R. L. 1968. Productivity of red alder in western Oregon and Washington. *In* Biology of alder. *Edited by* J. M. Trappe, J. F. Franklin, R. F. Tarrant, and G. M. Hansen. USDA For. Serv., PNW For. Range Exp. Sta., Portland, OR. 287-292.

Wilson, B. C., and N. W. Jewett. Propagation of red alder by mound layering. Unpublished report. Washington State Dept. of Natural Resources, Olympia.

Wilson, B. C., and R. F. Stettler. 1981. Cut-leaf red alder. Unpublished report. Univ. of Washington, Seattle.

Worthington, N. P. 1957. Silvical characteristics of red alder. USDA For. Serv., PNW For. Range Exp. Sta., Silvical Series 1.

Worthington, N. P., F. A. Johnson, G. R. Staebler, and W. J. Lloyd. 1960. Normal yield tables for red alder. USDA For. Serv., Res. Pap. PNW-36.

Worthington, N. P., R. H. Ruth, and E. E. Matson. 1962. Red alder: its management and utilization. USDA For. Serv., Misc. Publ. 881.

Zasada, J. 1986. Natural regeneration of trees and tall shrubs on forest sites in interior Alaska. *In* Forest ecosystems in the Alaska taiga: a synthesis of structure and function. *Edited by* K. Van Cleve, F. S. Chapin, III, P. W. Flanagan, L. A. Viereck, and C. T. Dyrness. Springer-Verlag. 44-73.

Zasada, J., T. Sharik, and M. Nygren. 1991. The reproductive process in boreal forest trees. *In* A systems analysis of the global boreal forest. *Edited by* H. H. Shugart, R. Leemans, and G. B. Bonan. Cambridge Univ. Press, Cambridge, 85-125.

Zasada, J., J. Tappeiner, and M. O'Dea. 1992. Clonal structure of salmonberry and vine maple in the Oregon Coast Range. *In* Ecology and management of riparian shrub communities. *Compiled by* W. Clary, E. McArthur, D. Bedunah, and C. Wambolt. Proceedings of a symposium held in Sun Valley, ID, 29-31 May 1991. USDA For. Serv., Gen. Tech. Rep. INT-289. 56-61.

Zimmerman, M. H., and C. L. Brown. 1971. Trees: structure and function. Springer-Verlag, New York.

## 2

# Root Symbioses of Red Alder: Technological Opportunities for Enhanced Regeneration and Soil Improvement

RANDY MOLINA, DAVID MYROLD, & C. Y. LI

Red alder is unique among forest trees of the Pacific Northwest because it forms a tripartite symbiosis among roots, nitrogen-fixing actinomycetes in root nodules (actinorhizae), and mycorrhizal fungi. The amount of nitrogen fixed by the actinorhizal associations, the positive impact of this fixed nitrogen on growth of red alder, and the contribution made to soil fertility are well appreciated by foresters as reviewed in Binkley et al. (Chapter 4).

Past reviews of alder symbioses (Tarrant and Trappe 1971; Hall et al. 1979; Trappe 1979) called for intensive research on the biology of these root symbionts so that their benefits could be optimized through biotechnology. Consequently, over the last decade, characterization of the biology and ecology of the symbioses, especially the actinorhizal symbioses, has progressed considerably. The isolation of symbiotic actinomycetes in the late 1970s provided the biological framework for intensive study of their physiology and ecology; nearly 1000 papers have been published and several international symposia conducted over the last decade. Schwintzer and Tjepkema (1990) produced a comprehensive review on their systematics, host relations, and strain variations.

Mycorrhizal symbioses of alder have received less attention than actinorhizae, but identities of fungal symbionts, degree of host specificity, effects of mycorrhizae on seedling growth, and abundance of fungal symbionts in forest soils have been studied (Molina 1979, 1981; Rose 1980; Koo 1989; Miller et al. 1991, 1992). Most importantly, interactions between the two symbioses have been examined, resulting in a holistic appreciation of the tripartite symbiosis. Three general concepts are evident: each symbiosis provides unique benefits to host nutri-

tion, the two symbioses interact synergistically to improve host nutrition; and the root endophytes are widespread in west-side forests of the Pacific Northwest.

In this chapter we describe the root symbionts, discuss their ecologies and interactions, and address current efforts to artificially inoculate red alder seedlings with selected isolates of root symbionts. Our overall objective is to emphasize how strongly red alder depends on the root symbionts and to develop strategies for using these symbioses in future management applications.

## Actinorhizal (*Frankia*) Associations of Red Alder

*Frankia* species are sporulating actinomycetes capable of fixing nitrogen. They form symbiotic root nodules on several dicotyledonous plant genera, including alders. This symbiosis between *Frankia* and plant roots was termed "actinorhiza" by analogy to mycorrhizal associations (Tjepkema and Torrey 1979). Several review articles (Silvester 1977; Baker and Seling 1984; Baker 1988, 1989), proceedings of international conferences (Table 1), and a book (Schwintzer and Tjepkema 1990) on *Frankia* and actinorhizal plants are available for additional background.

### Historical Perspective

Nonleguminous root nodules have been the object of observation and study for more than 100 years. Quispel (1990) recounted the history of actinorhizal studies before the successful isolation of *Frankia* in 1978. Although these earlier studies were mainly observational, they yielded important landmarks: Brunchorst (1886-88) concluded that root nodules are associated with a microorganism

*Figure 1. Alder root nodules.*
*Photo courtesy of K. Huss-Danell,*
*University of Umeå.*

and suggested that these organisms be assigned to a new genus, *Frankia* (after B. Frank, his major professor); Hiltner (1895) demonstrated that nodulated *Alnus glutinosa* can grow in nitrogen-free soil and thus presumably fixed nitrogen; and Schaede (1933) confirmed the actinomycetous nature of the microsymbiont. In the 1950s, work on actinorhizae intensified and became increasingly quantitative. Nitrogen fixation was measured with $^{15}N_2$ for the first time in *Alnus* (Virtanen et al. 1954) and other actinorhizal genera (Bond et al. 1954). The first successful isolation of the microsymbiont was accomplished about this time by Pommer (1959). Unfortunately, the culture was lost, and it was not until 1978 that Callaham et al. (1978) once again isolated *Frankia* in pure culture. Since 1978, research on *Frankia* has increased rapidly, as evidenced by more than 900 citations in the bibliography published by the Program in Forest Microbiology (1988).

### Root Nodules and Nodulation

*Alnus* species almost universally harbor root nodules. Even tiny alder seedlings quickly become nodulated in nature. Exceptions to rapid nodulation occur on sites not previously inhabited by actinorhizal plants (e.g., mine wastes and peat bogs) (Huss-Danell and Frey 1986).

**Root nodule initiation.** Nodulation of alder roots begins by entry of *Frankia* through a root hair. Only deformed root hairs, those which have curled, presumably in response to *Frankia* or other rhizosphere bacteria, are colonized. *Frankia* grows within the root hair and is encapsulated by host tissue. Although *Frankia* may penetrate plant cell walls, it does not penetrate the plant cell membrane. Proliferation of *Frankia* within the root coincides with increased production of cortical cells, which give rise to a macroscopic swelling on the root known as a prenodule. Nodule development continues with the formation of nodule lobe primordia. Nodule lobes are modified lateral

*Table 1. List of international conferences on Frankia and actinorhizal plants, and resulting proceedings.*

| Year | Meeting site | Proceedings reference |
|------|-------------|----------------------|
| 1987 | Harvard Forest | *Botanical Gazette* 140(S) |
| 1979 | Oregon State University, Corvallis | *Symbiotic Nitrogen Fixation in the Management of Temperate Forests* (J. C. Gordon, C. T. Wheeler, and D. A. Perry, eds.), 1979 |
| 1982 | University of Wisconsin, Madison | *Canadian Journal of Botany* 61(11), 1983 |
| 1983 | Landbouwhogeschool Wageningen, The Netherlands | *Plant and Soil* 78(1/2), 1984 |
| 1984 | Laval University, Ste-Foy, Quebec, Canada | *Plant and Soil* 87(1), 1985 |
| 1986 | University of Umeå, Umeå, Sweden | *Physiologia Plantarum* 70(2), 1987 |
| 1988 | University of Connecticut, Stoors | *Plant and Soil* 118(1/2), 1989 |
| 1991 | University of Lyon, Lyon, France | *Acta Oecologia* |
| 1993 | University of Waikato, Hamilton, New Zealand | *Soil Biology and Biochemistry* |

roots. Details of morphogenesis of actinorhizal nodules are reviewed by Newcomb and Wood (1987) and Berry and Sunnell (1990).

**Root nodule structure.** Actinorhizal root nodules are perennial structures with yearly cycles of growth and senescence. The youngest, most active tissue is located on the periphery of nodule lobes. Nodules come in a variety of shapes and sizes; some have discrete, branched lobes, others branch profusely and develop into compact clusters. Alder nodules are typically compact and become quite large, with diameters of several centimeters (Fig. 1). Nodule morphology is probably under host plant control but may reflect environmental conditions (e.g., presence of rocks, aeration of soil).

The anatomy of alder nodules resembles that of lateral roots: a central stele is surrounded by the endodermis, layers of cortical cells, and a periderm (Fig. 2). The periderm of alder nodules contains numerous lenticels, presumably to facilitate gas transport. Cells colonized by *Frankia* abound throughout the cortical region of the nodule. Unlike actinorhizal nodules of other plant genera, alder nodules contain relatively low concentrations of hemoglobin (Silvester et al. 1990).

Within alder nodules, *Frankia* can display all three morphological structures observed in pure culture: hyphae, vesicles, and sporangia (Fig. 3). Hyphae differentiate to form spherical, septate vesicles (Newcomb and Wood 1987). Nitrogen fixation occurs within the vesicles. The thick-walled vesicles serve as a diffusion barrier to oxygen and protect the nitrogenase enzyme from inactivation (Silvester et al. 1990). The formation of vesicles and

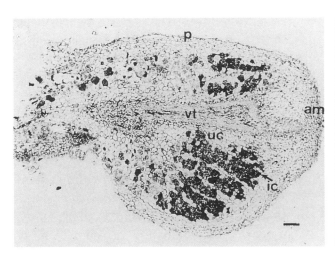

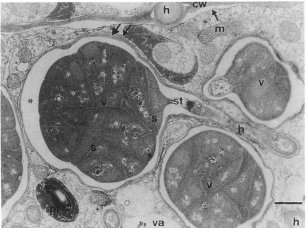

*Figures 2 and 3. Anatomy and ultrastructure of alder nodules.*

*Figure 2 (left). Longitudinal photomicrograph of an alder nodule lobe surrounded by periderm (p), with vascular tissue (vt) in the center. Dark cortex cells are infected (ic) with* Frankia. *Many cortex cells are uninfected (uc). The apical meristem (am) gives rise to new nodule cells. Bar = 100 μm. Photo courtesy of K. Huss-Danell.*

*Figure 3 (right). Transmission electron micrograph of symbiotic* Frankia *within an alder root nodule. Symbiotic vesicles (v) are compartmentalized by septa (s) and attached to the hyphae (h) by a stalk (st) located at the constricted basal portion of the vesicle. Vesicles are surrounded by a so-called void space (\*), which most likely results from extraction of the lipid containing vesicle envelope during sample preparation. Hyphae are shown in both transverse and longitudinal sections. The large arrow points to the capsule, which is a barrier contiguous with the plant cell wall (cw) and consists of pectin and some fibrils. Other plant cell features include: mitochondrion (m), vacuoles (va), plastids (pl), which are largely devoid of starch, cytoplasm (cy), and cell membrane (small arrows). Bar = 1 μm. Photo courtesy of P.-Å. Vikman, P.-O. Lundquist, and K. Huss-Danell; published previously by Gallon (1992).*

their senescence follows a seasonal pattern of growth within the perennial nodules.

Ineffective nodules that do not fix nitrogen occasionally occur. They are typically small, contain relatively little *Frankia* (Berry and Sunnell 1990), and normally do not contain vesicles (Mian et al. 1976; Hahn et al. 1988).

Although virtually all *Frankia* strains can be induced to form sporangia in pure culture, this is not necessarily the case in nodules (Schwintzer 1990). *Frankia* strains observed within nodules form either many sporangia (sp$^+$) or none to very few sporangia (sp$^-$). Sporangia form as the *Frankia* strain ages, generally developing after vesicles have matured. The spores held within the sporangia presumably serve as a resting stage, thereby allowing *Frankia* to survive nodule senescence and ultimately to germinate after nodule decay. Whether nodules are sp$^+$ or sp$^-$ seems to be a function of the *Frankia* genotype, although the host can influence whether sporulation is expressed (Schwintzer 1990). No sp$^+$ *Frankia* strains exist in pure culture, which complicates research in this area.

**Nodule physiology.** Carbon, nitrogen, and hydrogen metabolism are important aspects of nodule physiology (Tjepkema et al. 1986; Huss-Danell 1990). Alder root nodules and consequently symbiotic *Frankia* obtain their carbon from photosynthate. The plant metabolites used by *Frankia* in nodules are not known with certainty, but both simple sugars and carboxylic acids have been shown to stimulate respiration of symbiotic *Frankia* (Huss-Danell 1990).

Nitrogen metabolism in nodules includes both nitrogen fixation and assimilation of fixed nitrogen (Huss-Danell 1990). *Frankia* fixes nitrogen in its vesicles, which contain the common FeMo nitrogenase. Nitrogenase activity in alders follows seasonal changes in photosynthesis, suggesting the close coupling of the two processes. Once fixed as $NH_3$, nitrogen is assimilated by the plant via the glutamine synthase-glutamine oxo-glutarate aminotransferase system. In alder, subsequent transamination reactions result in the production of citrulline, the major nitrogenous compound transported in the xylem.

Hydrogenase acts as a potential energy conservation mechanism within nodules; nodules with active hydrogenase are more efficient at fixing nitrogen than are nodules without this enzyme (Huss-Danell 1990). Hydrogenase activity is universally present in *Frankia* and actinorhizal nodules, with the one exception of a "local source" of *Frankia* from Sweden that is infective on alders (Sellstedt and Huss-Danell 1984; Sellstedt 1989).

## Isolation, Identification, and Physiology of *Frankia* Isolates

Since the isolation of *Frankia* by Callaham et al. (1978), numerous strains have been isolated from various actinorhizal plants and cultured for physiological, biochemical, inoculation, and genetic studies. *Frankia* associated with hosts in the families Rosaceae and Rhamnaceae have not yet been reported to be cultured. *Frankia* is unique among actinomycetes because it forms irregular shaped sporangia in liquid culture. Sporangia can be intercalary or terminal. The aseptate hyphae, 0.5 to 2.0 µm wide, are poorly branched. Aerial mycelium does not form in solid medium. Cells may be colorless or pigmented, and produce various soluble pigments depending on growth medium. Vesicles, site of nitrogen fixation under ambient oxygen concentrations (Tjepkema et al. 1980; Murry et al. 1984; Noridge and Benson 1986), form in the absence of combined nitrogen. Presence of combined nitrogen *in vitro* inhibits vesicle formation and nitrogen fixation.

**Isolation of *Frankia*.** The slow growth of *Frankia* and presence of contaminating microorganisms on the nodule surface impeded early progress in isolating *Frankia*. Several methods are now available to successfully isolate *Frankia* from actinorhizal plants: serial dilution (Berry and Torrey 1979), micro-dissection (Diem and Dommergues 1988), selective incubation (Quispel and Burggraaf 1981), filtration (Benson 1982), and sucrose density gradient centrifugation (Baker and O'Keefe 1984). Li (unpub.) has modified Benson's (1982) filtration techniques to routinely isolate *Frankia* from root nodules of Alnus species as follows:

*Isolation medium.* A defined, modified liquid BAP medium (Murry et al. 1984) in test tubes consists of the following in mM in distilled water: $KH_2PO_4$ $7H_2O$, 3.4; sodium propionate, 5; $MgSO_4$, 0.1; $CaCl_2$, 0.07; FeNaEDTA, 10 mg/l; biotin, 450 µg/l; and trace

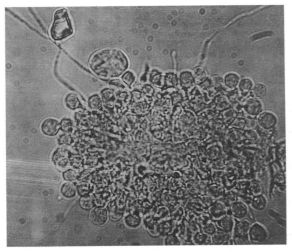
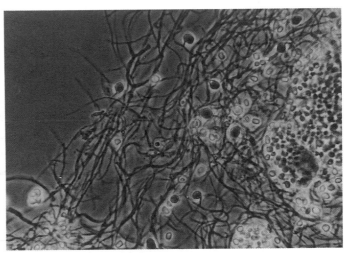

*Figures 4 and 5. Vesicles and hyphae of* Frankia.

*Figure 4 (left). Vesicle clusters of* Frankia *isolated from root nodules red alder (1100x).*

*Figure 5 (right). Vesicles formed on hyphae in liquid culture when* Frankia *is grown without a nitrogen source (1100x).*

elements and vitamins according to Tjepkema et al. (1981). The pH of the medium is adjusted to 6.7.

*Nodule preparation.* Wash root nodules in a stream of water to remove loose soil. Separate nodule lobes and place them in a beaker of 2.5-percent sodium hypochlorite solution plus one drop of Tween 20. The surface is sterilized by continuous agitation with a magnetic stir in the solution for 20 min. Then rinse the lobes thoroughly in sterile distilled water and homogenize them in 5 ml sterile distilled water in a tissue grinder to release the vesicle clusters. Filter the homogenate through autoclaved nylon screens with mesh openings of 50 and 20 μm. Then thoroughly wash vesicle clusters (Fig. 4) retained on the 20-μm screen with sterile distilled water. Collect vesicle clusters with a Pasteur pipette, place in tubes of modified BAP medium, and incubate at 30°C with occasional gentle shaking. *Frankia* filament mats appear within two to three weeks and can be transferred to new tubes of BAP containing 5 mM $NH_4Cl$ to enhance growth. *Frankia* has not been isolated from soil under alder, although Baker and O'Keefe (1984) isolated a strain from the rhizosphere of *Cercocarpus montanus*.

**Confirmation of *Frankia* isolation.** Confirmation of *Frankia* isolates is based on presence of characteristic sporangia and vesicles, branching hyphae, lack of aerial mycelium, and ability to colonize and form nodules with a host plant. Some isolates, however, lack the ability to colonize host plants even though they can fix nitrogen in culture (Diem et al. 1982; Torrey 1990). Acetylene is not reduced without vesicle formation (Tjepkema et al. 1980, 1981). Vesicles are surrounded by multi-laminate envelopes of lipid that limit diffusion of oxygen into the vesicle cytoplasm, thereby protecting nitrogenase (Parsons et al. 1987; Lamont et al. 1988; Berry et al. 1991; Harriott et al. 1991). Vesicles can differentiate into hyphae for vegetative growth (Schultz and Benson 1989). Some *Frankia* strains sporulate in nodules of host plants after inoculation (Racette and Torrey 1991).

***In vitro* growth of *Frankia* isolates.** Most physiological studies of *Frankia* use homogenized cultures so that numerous hyphal fragments are available to initiate active growth. Regrowth from hyphal fragments is optimized when homogenization is performed during the exponential stage of growth. Thus, only healthy cultures that have been homogenized are transferred to fresh medium for physiological studies (Benson and Schultz 1990).

*Carbon use.* Propionate and acetate are universal carbon sources for growing *Frankia* (Stowers 1987). Some *Frankia* strains use sugars such as glucose, fructose, mannitol, trehalose, maltose, and sucrose (Burggraaf and Shipton 1983; Lechevalier et al. 1983; Tsai et al. 1983; Lopez and Torrey 1985).

Other strains, particularly from *Alnus, Comptonia,* or *Myrica* root nodules, prefer short-chain fatty acids or organic acids, such as succinate, fumarate, malate, and pyruvate (Shipton and Burggraaf 1982; Lopez and Torrey 1985).

*Nitrogen use.* Ammonium chloride is most widely used as a nitrogen source. Some isolates can use amino acids, such as aspartate and glutamate, and inorganic sources, such as ammonium nitrate and potassium nitrate (Stowers 1987). *Frankia* also grows in the absence of combined nitrogen and under these conditions forms vesicles (Fig. 5).

*Growth determination. Frankia* grows slowly and is typically measured by correlating biomass to total protein (Burggraaf and Shipton 1983). To determine total protein, the *Frankia* culture is harvested by centrifugation and washed with 0.01 M potassium phosphate buffer, pH 6.8. The washed culture is homogenized in a tissue grinder and digested with 0.5 ml of 1.0 M NaOH in a boiling water bath for 10 minutes. The amount of protein released is determined by the methods of Markwell et al. (1978).

### Soil Ecology of Frankia

Much of what is known or surmised about the ecology of *Frankia* in soil has been learned indirectly by studying the ecology of the symbiosis. Many studies have focused on either the nitrogen-fixing ability of particular host-strain combinations or on the sp+/sp- nature of the nodules. The outcome of such studies will be discussed first, followed by a review of studies that have followed populations of *Frankia* in soil.

**Studies based on symbiotic characteristics.** Ineffective nodules often occur because of intergeneric or interspecific incompatibilities (i.e., *Frankia* strains isolated from one host genus may form only ineffective nodules when tested on a different actinorhizal host genus) (Weber et al. 1987; van Dijk et al. 1988). Some *Frankia* isolates form ineffective nodules even when inoculated on the same host species from which they were isolated (Hahn et al. 1988).

The nitrogen-fixing ability of various *Frankia* strain-plant combinations often differs significantly in laboratory studies (Carpenter et al. 1984; Hooker and Wheeler 1987; Domenach et al. 1988; Sheppard et al. 1988; Weber et al. 1989; Kurdali et al. 1990).

For example, sp- and sp+ nodules can differ in both absolute and relative nitrogenase efficiency (Normand and LaLonde 1982; Wheeler et al. 1986). Presumably the differences in effectiveness among *Frankia* strains influence their competitiveness and fitness in nature, but this has not been studied.

The spore type of *Alnus* nodules displays interesting and complex ecological relations (Schwintzer 1990). Reasons for these ecological patterns include the following: degree of host selection (e.g., sp+ strains form effective nodules on *A. incana* subsp. *incana* but not on *A. glutinosa*) (Domenach et al. 1988; van Dijk et al. 1988; Kurdali et al. 1990); sp- strains often dominate on newly disturbed sites and may be more saprophytically competent than sp+ strains, which often are found in established actinorhizal stands and may be maintained primarily through nodule turnover (van Dijk 1984; Weber 1986; Holman and Schwintzer 1987; Smolander and Sundman 1987); and low soil pH tends to favor sp+ strains (Holman and Schwintzer 1987; Smolander et al. 1988). *Alnus rubra* forms both sp+ and sp- nodules, but their ecological significance has not been studied.

**Soil *Frankia* populations.** Nearly all the ecological information about *Frankia* populations in soil has been collected through a plant bioassay system (Myrold 1993). Surveys of forest sites in Finland (Smolander and Sundman 1987; van Dijk et al. 1988; Smolander 1990) and the Pacific Northwest (Hilger and Myrold, unpub. data) have shown *Frankia* populations to range from 0 to 4600 infective units g⁻¹ soil. An infective unit is the amount of *Frankia* necessary to form one nodule. This variation in *Frankia* populations is caused partly by differences in plant species present and by soil conditions. Often higher numbers are found under nonactinorhizal plants (e.g., birch) than actinorhizal plants (Smolander and Sundman 1987; van Dijk et al. 1988; Smolander 1990).

In a greenhouse bioassay, Miller et al. (1992) determined the nodulation potential of red alder seedlings when grown in soils of six forest sites in the Oregon Coast Range: young alder, old alder, conifer clearcut, conifer plantation, rotation-age conifer, and old-growth conifer. They reported that nodules formed on red alder grown in all soils, but highest nodule numbers and rates of nodule formation occurred on seedlings grown in the alder and conifer plantation soils. Because they found that levels of nitrate and mineralizable nitrogen in the bioassayed soils were highest in the alder and conifer plantation sites, they suggested that a nitrogen-priming effect might be involved in the nodulation rates observed in these soils.

Laboratory studies of soil and rhizosphere effects on *Frankia* populations mostly have confirmed field observations. *Frankia* populations are inversely correlated with soil pH (Smolander and Sundman 1987). Survival of added *Frankia* was greater in limed than in unlimed soil (Hilger and Myrold, unpub. data; Smolander et al. 1988; Smolander and Sarsa 1990); however, rhizosphere effects are quite variable (Hilger and Myrold, unpub. data; Smolander and Sarsa 1990). Van Dijk and Sluimer-Stolk (1990) found populations of ineffective *Frankia* present at much higher levels than effective *Frankia* in sand dunes soils planted with alders.

Hilger and Myrold (unpub. data) have developed a method of estimating *Frankia* biomass in soil based on *Frankia*-specific DNA sequences. Their method determines the proportion of *Frankia* biomass to total soil biomass, not just those *Frankia* that are infective under the conditions of a seedling bioassay. This technique provides a useful tool for studying *Frankia* populations in soil and their response to environmental conditions and forest management practices.

**Diversity of *Frankia* strains.** Initially, *Frankia* strains were classified by their host, much as has been done with *Rhizobium*. It is now clear that the boundaries between strains do not fall strictly along the lines of host infectivity (Torrey 1990). Four host-specificity groups are recognized: strains that nodulate *Alnus* and *Myrica*, *Casuarina* and *Myrica*, *Elaeagnus* and *Myrica*, and only members of the Elaeagnaceae (Baker 1987). Some *Frankia* strains, however, cross these host-specificity boundaries, and there also is variation within *Frankia* strains that nodulate different *Alnus* species. For example, an effective strain isolated from one *Alnus* species may form only ineffective nodules on another species of *Alnus* (Weber et al. 1987). Different combinations of *Frankia* strain by *Alnus* species also can yield differences in tree growth and rate of nitrogen fixation (Wheeler et al. 1986).

The diversity of *Frankia* in nature typically is determined by isolating *Frankia* strains from nodules. Isolates obtained from nodules are differentiated by several means. For example, one-dimensional sodium dodecyl sulfate-polyacrylamide gel electrophoresis first was used to differentiate *Frankia* strains by Benson and Hanna (1983); they separated 43 isolates from one stand of *Alnus* into six groups, with one group being dominant and containing 35 of the isolates. They also found evidence for dual occupancy within a nodule. Another method used isozyme patterns to compare the diversity of strains isolated from a single *A. rubra* (Faure-Raynaud et al. 1990) and from individual nodules of *A. glutinosa* (Faure-Raynaud et al. 1991). Thirteen *Frankia* strains were found on a single red alder root system, but individual nodules harbored only one strain. These studies indicate a wide diversity of *Frankia* strains capable of colonizing alder.

Methods using differences in DNA or RNA sequences to differentiate *Frankia* strains eliminate the need for isolating *Frankia* from nodules. For example, Simonet et al. (1988) used a radio-labeled *Frankia* plasmid to probe nodule DNA obtained from stands of *Alnus*. This allowed them to detect a single strain in those stands. Chromosomal genes also can be used. Oligonucleotide probes based on sequence differences in the variable region of the 16S rRNA gene were used to detect a particular *Frankia* strain in root nodules of *Alnus glutinosa* (Hahn et al. 1990b), and Simonet et al. (1990) used oligonucleotide probes based on sequence differences in the *nifH* gene to detect and differentiate between two *Frankia* strains in *A. glutinosa* and *A. incana* nodules. Such techniques and others currently being developed (e.g., the use of polymerase chain reaction with *Frankia* specific primers) will provide means for performing sophisticated auteco-logical studies of *Frankia*.

**Interactions between soil organisms and *Frankia*.** Nodule initiation by *Frankia* can be depressed by low pH, combined nitrogen, or by lack of other rhizosphere microorganisms. Knowlton et al. (1979, 1980) and Knowlton and Dawson (1983) demonstrated that coinoculation of *Frankia* with certain bacteria could enhance *Frankia* colonization and nodule development of red alder; inoculation of *Frankia* strains under aseptic conditions resulted in few nodules. When plants were inoculated with the bacteria plus *Frankia,* nodulation increased even at low pH (pH 4.0 to 6.0). These bacteria grew rapidly in low pH conditions and raised the rhizosphere pH to levels conducive to *Frankia* growth and infection (Knowlton and Dawson 1983). Bacteria other than *Frankia* also can cause massive root-hair deformation, thereby promoting nodulation by *Frankia.* Studies by Mansour and Torrey (1991) found that *Frankia* spores applied in nonsterile conditions caused root-hair deformation, which led to nodule formation by germinating spores; under axenic conditions, the incidence of root-hair deformation was low and nodulation did not occur with germinating spores. Research by Perinet and LaLonde (1983) and Myrold (unpub.), however, has shown that effective nodulation occurs under aseptic or axenic conditions. Further research is needed to address the role of "helper" bacteria on nodulation.

Dual inoculation of *Frankia* with root-associated nitrogen-fixing *Azospirillum* results in significantly larger seedlings than inoculation with *Frankia* alone in *Alnus rubra* and *Casuarina cunninghamiana* (Li 1987; Li et al. 1987, p. 247; Rodriguez-Barrueco et al. 1991). This dual inoculation also produces larger sizes and greater quantities of root nodules than inoculation with *Frankia* alone. *Azospirillum* may facilitate nodule formation by promoting *Frankia* infection. Simultaneous inoculation of effective and ineffective *Frankia* strains also enhances growth and nodule formation of *A. glutinosa* (Hahn et al. 1990a). Similarly, *Alnus* species inoculated with mixtures of *Frankia* strains grew better than *Alnus* inoculated with single strains, but the number of nodules formed remained the same for inoculations of both single and mixed strains (Prat 1989). Thus, several different strains of *Frankia* may normally be present in the nodules of actinorhizal plants, as demonstrated for Casuarinaceae by Reddell and Bowen (1985) and for *Alnus* by Simonet et al. (1985). Interactions between mixed strains likely take place in natural soils and are apparent in inoculation studies. Rojas et al. (1992) demonstrated that other actinomycetes on roots, nodules, and in soil of red alder do not affect *Frankia* infection and nodule development and can, in fact, reduce red alder growth by production of allelochemicals.

**Soil and root chemistry effects on actinorhizae.** Nodulation and endosymbiont nitrogenase activity are highly sensitive to excess inorganic nitrogen in some actinorhizal plants (Hughes et al. 1968; Righetti and Munns 1981; Granhall et al. 1983; Righetti et al. 1986). In contrast, nodulated alders provided with ammonium or nitrate nitrogen can produce greater nodule and plant biomass than those without added nitrogen (Ingestad 1980). Lipp (1987) found that 100 mg nitrogen/l enhances nodulation and acetylene reduction rates over low (1 or 10 mg) or high (1000 mg) nitrogen additions. Adequate phosphorus availability is also required for nodulation and nitrogen fixation (Benoit and Berry 1990). Potassium, magnesium, and calcium are needed by some alders for maximum growth and nitrogen fixation (Pregent and Camire 1985). Salinity in soils is not a major concern for growing actinorhizal host plants because growth of *Frankia* and nodule development are not greatly affected by high salt concentrations (Dawson and Gibson 1987; Ng 1987).

Root nodules of red alder contain phenolic compounds (Li et al. 1972). Some of these compounds also exist in soil and understory species in stands of red alder (Li et al. 1970; Li 1974). Compounds such as ferulic, *o*-coumaric, *p*-coumaric, and caffeic acids inhibit the growth of *Frankia* isolates. Some phenolic compounds cause increased hyphal ramification, and others induce *Frankia* to form numerous spherical structures while not affecting vesicle formation (Perradin et al. 1983; Vogel and Dawson 1986). Vogel and Dawson (1986) suggested that plant phenolics may affect growth and development of *Frankia* in soil and in actinorhizal hosts by acting as chemical mediators in the growth regulation of *Frankia* within the root tissues from the stage of initial infection through the latter stage of its development within the root nodule (Perradin et al. 1983). Root exudates also can facilitate nodulation by stimulating spore germination and the colonization process.

## Mycorrhizal Symbioses of Red Alder
"Mycorrhiza" translates literally as "fungus root" and represents the symbiotic association between

plant roots and specialized soil fungi (mycorrhizal fungi). The fungi act as extensions of the root systems and improve host nutrition by their ability to extract nutrients and water from a volume of soil hundreds to thousands of times greater than the volume roots alone can explore. Other host benefits include protection of fine roots from pathogens, increased drought-resistance, and increased root development and longevity. In return, the mycorrhizal fungi depend on the host for carbon, primarily in the form of simple sugars, and some vitamins. The mycorrhizal symbiosis has strongly coevolved over the millennia such that each partner depends on the other for survival and fitness in natural ecosystems. Readers are referred to Allen (1992), Harley and Smith (1983), and Safir (1987) for greater detail on mycorrhizae.

Only two of the several types of mycorrhizae are relevant to alder: ectomycorrhizae (EM) and vesicular-arbuscular mycorrhizae (VAM). The former (EM) are characterized by the formation of a sheath or mantle of fungus mycelium that surrounds the fine, feeder roots and by the intercellular penetration of the fungus between epidermal or cortical cells to form a network of host-fungus contact called the Hartig net. The fungi are typically basidiomycetes or ascomycetes (occasionally zygomycetes). In the Pacific Northwest, ectomycorrhizal hosts occur in the families Betulaceae, Ericaceae (*Arbutus* and *Arctostaphylos*), Fagaceae, Pinaceae, Rosaceae, and Salicaceae.

Vesicular-arbuscular mycorrhizae are formed by zygomycetes in the order Glomales (Morton and Benny 1990). Unlike EM, VAM do not cause differentiation of the roots (swelling and branching); roots must be stained to reveal the internal structures of the fungus colonization. The VAM fungi ramify within the cortical tissue, both intercellularly and intracellularly, forming balloon-shaped vesicles (storage organs) and arbuscules (multibranched, intracellular structures that function in nutrient exchange with the host).

### Ectomycorrhizae of Red Alder

Ectomycorrhizae are by far the predominant type on red alder (Neal et al. 1968; Miller et al. 1991). Neal et al. (1968) described two EM types on red alder formed with unknown fungi; they referred to them as dark-brown clavate type and pale-brown glabrous type. Since that work, several EM fungi have been inoculated successfully onto red alder roots (Molina 1979; Miller et al. 1991) and additional suspected fungal associates have been noted from repeated occurrence of sporocarps (mushrooms and truffles) in association with red alder.

Miller et al. (1991) characterized many red alder EM from laboratory inoculations and field collected roots. They found two general morphological types: succulent (Figs. 6, 7) and flexuous (Fig. 8). The succulent EM type is short and thick, mostly determinate in growth with a rounded apex and thin or thick mantle; the Hartig net may penetrate only slightly or deeply and is well developed even near the root apex (Figs. 12, 13). The succulent type is typical of fungi host-specific to red alder. Flexuous EM typically are long and thin, seemingly indeterminate in growth, and usually have an acute root apex and thin mantle. The Hartig net commonly does not completely penetrate to the depth of one cell. Mantle and Hartig net are poorly developed at the apex. Flexuous EM are characteristic of fungi with broad host ranges.

Nonmycorrhizal short roots are rarely seen on red alders in nature. When red alder seedlings are grown in the laboratory or greenhouse, non-mycorrhizal short roots are densely covered with root hairs (Fig. 9). Red alder EM lack root hairs.

Miller et al. (1991) recognized 11 distinct EM types on red alder. The majority of fungal associates are basidiomycetes. Proven fungus associates include *Alpova diplophloeus* (Figs. 6, 7), *Lactarius obscuratus, Cortinarius bibulus, Thelephora terrestris* (Fig. 8), *Paxillus involutus,* and *Laccaria laccata.* Several species in the genera *Hebeloma, Inocybe, Naucoria,* and *Russula* are suspected EM fungus associates but have proven difficult to isolate and confirm in experimental mycorrhizal synthesis.

Of the fungi noted above, *Alpova diplophloeus* and *Lactarius obscuratus* are by far the most widespread. The former forms large, succulent types of EM on red alder that are pale brown to golden brown and darken with age (Figs. 6, 7); the tips show a distinct blue-bruising reaction (Miller et al. 1988). Miller et al. (1991) reported A. *diplophloeus* as "almost ubiquitous in the field" and particularly

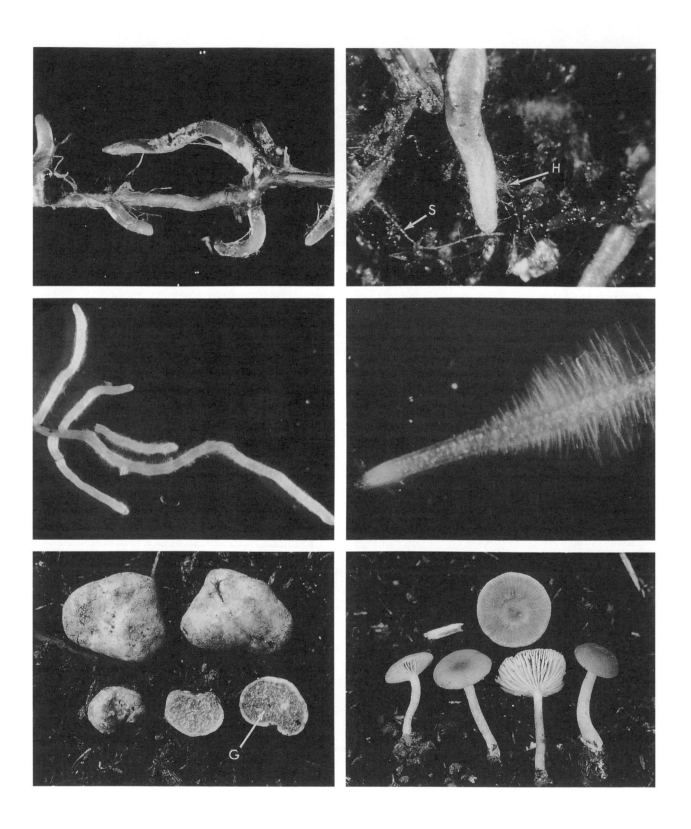

abundant in mesic areas. The "dark-brown, clavate" red alder ectomycorrhiza described by Neal et al. (1968) likely was formed by *A. diplophloeus*.

*Lactarius obscuratus* also forms large, succulent EM, but they are pale yellowish orange to orange, with a smooth mantle surface typical of EM formed by other *Lactarius* species. *Lactarius obscuratus* EM also are abundant in the field. They correspond to the "pale, brown, glabrous" ectomycorrhizae described by Neal et al. (1968) and Froidevaux (1973).

The reader is referred to Miller et al. (1991) for detailed anatomical descriptions of these and other EM types of red alder; a key to field identification of red alder EM types also is provided. Massicotte et al. (1989a) provided additional detailed ontological characterization of red alder EM formed by *Alpova diplophloeus*. They emphasized that the EM morphology depends upon the stage of lateral root elongation at the time of fungal contact. Thus, some EM types may show a gradation between flexuous and succulent morphology. See Massicotte et al. (1989a, 1989b) for detailed light and electron microscopic descriptions of red alder + *Alpova diplophloeus*.

### Host Specificity of Red Alder Ectomycorrhizal Fungi

Red alder, and alder species in general, form EM with comparatively few fungus species. For example, in contrast to Trappe's (1977) estimate of about 2000 EM fungi associated with *Pseudotsuga menziesii*, only 11 are known or suspected on red alder and less than 50 are known for the entire genus *Alnus* (Brunner et al. 1990; Miller et al. 1991). The reason is that most EM fungi of red alder are host specific to the genus. Molina (1979) attempted pure culture EM syntheses between 28 confirmed EM fungi and red alder; only four developed EM: *Alpova diplophloeus, Paxillus involutus, Scleroderma hypogeum,* and *Astraeus pteridis*. Several of the other 28 fungus species tested had typically broad host ranges when tested on eight species of Pinaceae (Molina and Trappe 1982a) yet were incompatible with red alder. In a related study, Molina (1981) found that four other *Alnus* species showed restricted fungus associates and hypothesized that the entire genus shows strong specialization regarding fungus associates.

It is important to note that both *Alpova diplophloeus* and *Lactarius obscuratus* are host specific to *Alnus*. Sporocarps of each species (Figs. 10, 11) occur exclusively with *Alnus*. Molina and Trappe (1982a, 1982b) and Massicotte and Molina (unpub. data) could not synthesize EM between *A. diplophloeus* and 10 species of Pinaceae and Ericaceae. Such close specialization of mycorrhizal association and predominance on root systems indicates a strong coevolutionary development and dependence between these symbionts (Molina et al. 1992).

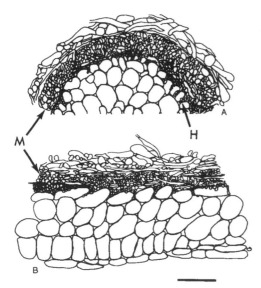

*Figure 12. Cross section (A) and longitudinal section (B) of* Alpova diplophloeus + *red alder. Bar = 100 µM; M = mantle; H = Hartig net. Drawing courtesy of Steven Miller; previously published by Miller et al. (1991).*

## Ecology and Distribution of Red Alder Ectomycorrhizae

The EM fungi of red alder are widespread in Oregon coastal forests and likely in other Pacific Northwest forest habitats wherever red alder is common on the landscape (Miller et al. 1992). Most of these fungi form mushrooms whose spores are dispersed by wind over great distances. The alder-specific *Alpova diplophloeus,* however, develops hypogeous (subterranean) sporocarps (truffles, Fig. 10) that are dispersed by small mammals. These mammals seek out the truffles by smell, dig them from the soil, consume them, and disperse the still viable spores in their fecal pellets (Maser et al. 1978). The Pacific Northwest is rich in truffle fungi (many hundreds of species), so this dispersal mechanism is common and highly successful. For example, Miller et al. (1992) noted that red alder seedlings develop abundant *A. diplophloeus* EM when grown in clearcut conifer forest soil; because *A. diplophloeus* is specific to red alder and therefore would not have been resident on previous conifer roots, the spores most likely were dispersed into the clearcut by mammals.

Miller et al. (1992) conducted a seedling greenhouse bioassay for EM fungal propagules using the same six forest soils noted previously for nodulation potential: young alder, old alder, conifer clearcut, conifer plantation, rotation-age conifer, and old-growth conifer. Red alder seedlings developed EM in all soils, with highest levels of colonization in young alder, old alder, and conifer clearcut soils; light to moderate levels of EM colonization occurred in plantation, rotation-age, and old-growth conifer soils. Development of EM also was delayed in the conifer soils, thereby indicating low initial inoculum potential. Five EM types developed during the course of the seedling bioassays. The highest diversity of EM types occurred in the clearcut conifer soil followed by the rotation-age conifer, young alder, and old alder soils; lowest EM type diversity occurred in the conifer plantation and old-growth conifer soils. *Alpova diplophloeus* was the most common and abundant type from all soils and also increased in abundance during the bioassay period, dominating the root system and displacing other EM types. This dominance and abundance by *A. diplophloeus,* especially in clearcut conifer soils, and competitive interactions with other EM fungi clearly emphasize the ecological importance of this host-specific fungus of red alder in the Pacific Northwest.

## Vesicular-Arbuscular Mycorrhizae of Red Alder

The prevalence and ecological importance of VAM on red alder in Pacific Northwest forests is unclear. Rose (1980) reported VAM development on red alder growing in coastal Oregon and northern California forest habitats: VAM colonization ranged from 50 to 90 percent in the fine roots. Miller et al. (1992), however, found no VAM on red alder seedlings grown in soils from the six coastal forest soils noted above. They suggested that their soil bioassay may have negatively affected the VAM fungus inoculum or that the VAM fungi were competitively excluded by EM fungi. Indeed, Rose (1980) was unable to develop greenhouse pot cultures of VAM fungi on red alder if EM fungi colonized first. She hypothesized that the EM present a physical barrier to VAM colonization. Development of VAM also has been reported on *Alnus incana* (Rose 1980; Arveby 1988, p. 173; Chatarpaul et al. 1989), *A. accuminata* (Russo 1989), *A. firma* (Lee 1988), *A. glutinosa* (Rose 1980; Fraga-Beddiar and Le Tacon 1990), and *A. sinuata* (Rose 1980); inoculation with *Glomus* spp. has increased growth and nutrient acquisition for some alders (see section on mycorrhiza-actinorhiza interactions).

Arveby (1988, p. 173) reported that VAM were present only during the first year of natural seedling establishment of *Alnus incana* in Sweden and that EM were dominant thereafter. Similar patterns of VAM to EM succession have been reported for other angiosperms that form both VAM and EM (Molina et al. 1992). The ecological advantage of this situation is that if either VAM or EM fungi are absent from the site, one can substitute for the other. The VAM fungi are widespread throughout Pacific Northwest forests, so they may sometimes be important in the early establishment of red alder seedlings. Thus, the natural development of VAM on red alder seedlings and mature trees, interactions with EM, and effects of VAM on plant nutrition need further study.

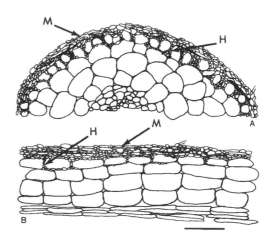

*Figure 13. Cross section (A) and longitudinal section (B) of* Lactarius obscuratus + *red alder. Bar = 100 µM; M = mantle; H = Hartig net. Drawing courtesy of Steven Miller; previously published by Miller et al. (1991).*

### Physiology and Host Benefits of Mycorrhizae

Mycorrhizae are most well known for improving phosphorus nutrition of host plants (Harley and Smith 1983). Mejstrik and Benecke (1969) found that EM of *Alnus viridis* absorb phosphorus five times as rapidly as non-mycorrhizal roots. Koo (1989) explored effects of light intensity, drought stress, and nitrogen and phosphorus fertilization on EM development of red alder by *Alpova diplophloeus* and subsequent effects on seedling growth and nutrition. Water stress decreased EM development and *Alpova diplophloeus* only enhanced red alder growth under well-watered conditions. Development of EM was greatest under strong light. Fertilization with nitrogen tended to enhance EM development and substitute for similar stimulating effects of nodulation (nitrogen fixation). Fertilization with phosphorus did not affect EM development. Overall, Koo (1989) found that EM formation by *Alpova diplophloeus* increased seedling growth and improved phosphorus status of the red alder seedlings. These mycorrhizal benefits were strongly dependent on the interaction with nodulation, however, and so are discussed in greater detail in the section on actinorhiza-mycorrhiza interactions.

The above studies by Miller et al. (1992) and Koo (1989) provide only a glimpse of the ecological and physiological contribution of mycorrhizae to the establishment and growth of red alder. Because mycorrhizae are essential to phosphorus acquisition for most plants, and because nitrogen fixation demands high phosphorus input, Trappe (1979) hypothesized that mycorrhizae of alder and other actinorhizal plants are essential to the healthy functioning of these tripartite symbioses. Koo (1989) suggested that the function of mycorrhizae for red alder may become increasingly important as stands develop and soil phosphorus levels decline. Field studies are urgently needed to address these matters.

## Interactions Between Mycorrhizae and Actinorhizae

Several studies report the existence of synergistic interactions between legumes, *Rhizobium,* and VAM (see reviews by Barea et al. 1992; Bethlenfalvay 1992). In general, dual-inoculated legumes grow better than legumes inoculated with either VAM or *Rhizobium* alone. The VAM typically enhance phosphorus uptake and *Rhizobium* enhances nitrogen nutrition. Similar results are evident in actinorhizal-VAM interactions with nonleguminous plants (Cervantes and Rodríquez-Barrueco 1992). Dual inoculation of *Ceanothus velutinus* with VAM and *Frankia* (crushed nodules) increases total dry weight of shoots and roots, number and weight of nodules, and nitrogen fixation compared to *Frankia* inoculation alone (Rose and Youngberg 1981). *Causuarina equisetifolia* inoculated with *Glomus mosseae* (VAM) and *Frankia* contains twice as much nitrogen and grows better than plants inoculated with *Frankia* alone (Diem and Gauthier 1982). *Hippophae rhamoides* displays significantly better growth, uptake of phosphorus, and nitrogenase activity in dual-inoculated versus nodulated- or mycorrhizal-only plants (Gardner et al. 1984). Berliner and Torrey (1989), however, showed that actinorhizal *Comptonia* and *Myrica* species grow as well as those with both root symbioses under greenhouse conditions.

Similar results apply to interactions among actinorhiza, mycorrhiza, and *Alnus*. *Alnus firma* seedlings inoculated with *Glomus mosseae* or with *G. mosseae + Frankia* weighed 27 and 83 percent greater than noninoculated control seedlings, respectively (Lee 1988). At low phosphorus fertilization (10 ppm) *A. acuminata* seedlings inoculated with *Glomus intraradices + Frankia* displayed acetylene reduction rates 87 percent higher than *Frankia*-only inoculated seedlings, although seedling fresh weights did not differ; at 50 ppm phosphorus, *Glomus + Frankia* increased leaf and nodule weight over *Frankia*-only inoculated seedlings (Russo 1989). Inoculation with VAM of nodulated *A. glutinosa* increased dry weight of nodules tenfold, doubled nitrogenase activity, and increased seedling growth compared to seedlings inoculated only with *Frankia* plus supplemental (50 ppm) phosphorus fertilizer (Fraga-Beddiar and Le Tacon 1990). Chatarpaul et al. (1989) examined tetrapartite interactions of EM-VAM-*Frankia* on growth of *A. incana* and found that inoculation with all three symbionts produced larger seedlings than either fungus or *Frankia* inoculation alone; inoculation with all three symbionts increased total seedling dry weight over both *Frankia* + EM and *Frankia* + VAM treatments. Both EM and VAM colonization increased in the presence of actinorhizal development.

Koo (1989) found actinorhizae and EM (inoculation with spores of *Alpova diplophloeus*) interacted significantly in affecting red alder seedling growth, nutrient status, and nitrogen-fixation rates. *Alpova diplophloeus* EM only enhanced growth and phosphorus tissue concentration of red alder seedlings when *Frankia* also was present, and then increases were typically less than 20 percent compared to *Frankia* inoculation alone. Formation of EM by *A. diplophloeus* typically did not enhance nodulation or nitrogen-fixation rates. *Frankia* strongly influenced EM development, however; Koo (1989) reported that in some experiments *Frankia* enhances EM formation by as much as six times over nonnodulated seedlings grown with supplemental nitrogen fertilization. Nonnodulated, non-nitrogen-fertilized red alder seedlings remained stunted and with only a trace of *A. diplophloeus* EM development. Koo (1989) also notes that nodulation preceded EM development in the field; in greenhouse studies, early nodule formation promoted more rapid EM development than when nodulation was delayed. Koo's data emphasize the importance of actinorhizal development and function in influencing red alder growth and EM development; actinorhizae appear the dominant root symbiosis in the tripartite symbiosis during the seedling stage because of the seedling's high nitrogen demand. Development of EM and physiological function likely follow once these nitrogen demands are met.

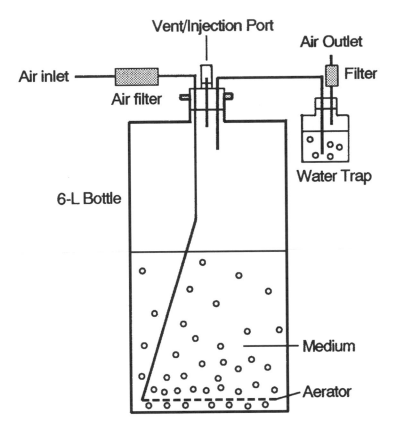

Air inlet

Air filter

Vent/Injection Port

Air Outlet

Filter

Water Trap

6-L Bottle

Medium

Aerator

*Figure 14. Batch culture apparatus for large scale production of* Frankia, *suitable for production of inoculum for field use. The stopper in the 10-L bottle is held in place with two slotted steel plates, one below the bottle lip and one above the stopper, connected by four bolts and wing-nuts. The air inlet, air outlet, and vent are made from 1/8-inch stainless steel tubing and are inserted through the stopper. The air entering the bottle is passed through a filter consisting of foam plugs inside two 30-mL syringe barrels fitted onto a 20-mL syringe barrel and sealed with epoxy. This filter is autoclaved with capped needles at both ends to maintain sterility. The filter is attached to the air inlet via a Luer-lock fitting sealed to the end of the tubing. The other end of the air inlet tube is formed into an aerator by crimping the end and making a series of holes on one side with a triangular file and bending the tubing at a right angle. The air outlet is fitted with a small flask to catch condensate and an air filter to prevent contamination by back-flow. The top of the vent tube is inserted through the septum of a vacutainer tube so that it can be sealed after autoclaving. Under the vacutainer tube, a small length of glass tube, melted closed at one end, covers the end of the vent. This inner cap maintains sterility and allows for flame sterilization so that solutions may be injected through this port. BAP medium (6L) without phosphate buffer and propionate solutions are autoclaved separately in sealed serum bottles and later added through the vent.* Frankia *inoculum is also added through the vent, allowed to settle, and later siphoned from the media via the air inlet tube. The cells are concentrated from the remaining media by centrifugation.*

## Management Implications

### Nursery Inoculations with Frankia

*Frankia* inoculations have been developed primarily for experimental purposes, but with some modifications, these techniques can be used for nursery practices. What follows is practical advice on preparation of crushed nodule inoculum, an example of how production of pure culture *Frankia* can be scaled-up, guidelines for seedling inoculation, and a summary of the results of previous greenhouse and nursery inoculation trials.

**Crushed nodule preparation.** Healthy alder nodules are collected and used fresh, although storage at 4°C for a few weeks is acceptable; storage longer than a few weeks may result in nodule decay and growth of unwanted microorganisms. Nodules are cleansed of soil. Young nodule lobes are selected and homogenized in water in a blender (Martin et al. 1991; Wheeler et al. 1991), although more elaborate preparation procedures may be used to reduce possible adverse effects of released plant phenolics (Perinet et al. 1985). Inoculum amounts in the range of 1 to 6 g nodule material per 1000 seedlings produce good inoculation success in the greenhouse (Perinet et al. 1985) and field (Wheeler et al. 1991; Martin et al. 1992); amounts at the high end of this range are used for inoculating nursery beds. Crushed nodule inoculum can remain viable for at least one year when stored frozen at -20°C (Akkermans and Houwers 1979).

**Large volume culturing methods.** Scaling up from flasks to multiliter bottles is not particularly difficult and works well for producing large quantities of *Frankia*. Diem and Dommergues (1990) presented one such approach using a batch culture method. Diem et al. (1988) used a chemostat to develop mass inoculum.

Myrold et al. (unpub. data) used a batch culture approach similar to that of Diem and Dommergues (1990). Large batch culture of *Frankia* is achieved in a closed bottle with slow agitation by bubbles (Fig. 14). In this system, 1 liter of medium produces about 1 ml of packed cell volume (pcv—the volume of cells in a 1.5-ml microfuge tube that has been spun at 16,000 x *g* for 10 min.) in about one week and provides inoculum for 1000 to 10,000 seedlings.

**Inoculation protocols.** Although crushed nodules are effective inocula (Sheppard et al. 1988; Wheeler et al. 1991), many studies report pure culture inoculum to outperform crushed nodules (Perinet et al. 1985; Hooker and Wheeler 1987; Martin et al. 1991). Direct comparison between pure culture and crushed nodule inoculations is difficult, however, because of the difficulty in equating the two sources of inoculum. The selection of inoculum source must consider the relative availability of the two sources and the potential adverse affects of using crushed nodules. The latter may produce less uniform nodulation than pure cultures (Perinet et al. 1985) and also may be a source of pathogens (Stowers and Smith 1985).

Inoculum is best applied when seed is sown (Stowers and Smith 1985; Wheeler et al. 1991). Rates of 0.1 to 1.0 µl pcv per seedling prove adequate to nodulate seedlings in both the greenhouse (Berry and Torrey 1985; Stowers and Smith 1985) and field (Martin et al. 1991). Inoculum can be surface applied as a liquid suspension or injected into the soil. Although surface application has succeeded in nursery beds (Wheeler et al. 1991), Martin et al. (1991) found improved inoculation success when pure culture *Frankia* inoculum was mixed with peat moss. Using an inoculum stabilizer such as peat reduces the total amount of inoculum needed.

**Effect of *Frankia* inoculation on seedling production.** In all reported studies, inoculation improved seedling size and quality relative to noninoculated controls. This is likely always to be true in the greenhouse when sterilized potting mixes are used and in fumigated nursery beds. In nursery situations, a major benefit of inoculation is the early formation and functioning of nodules (Martin et al. 1991; Wheeler et al. 1991); noninoculated controls often become nodulated later in the growing season as inoculum invades from outside sources or as roots grow into soil not effectively fumigated. Inoculation of alders produces larger seedlings than noninoculated seedlings supplied with supplemental nitrogen (Hilger et al. 1991; Wheeler et al. 1991).

The performance of inoculated versus non-inoculated seedlings at outplanting has received little study. McNeill et al. (1989) measured better growth of inoculated seedlings versus noninoculated seedlings over the first few seasons after outplanting, particularly on nutrient-poor soils. More studies of this sort are needed, especially to determine how long the apparent benefit of seedling inoculation lasts in nature where indigenous *Frankia* will begin to colonize the root system and interact with the introduced strain.

*Nursery Inoculations with Mycorrhizal Fungi*
Unlike the work cited above for *Frankia,* inoculation of red alder seedlings in nurseries with EM fungi has not received attention. Inoculation is possible, however, particularly with *Alpova diplophloeus.* Koo (1989) found that spores of this fungus were effective in inoculating red alder seedlings grown in containers in the greenhouse or growth chamber. He prepared slurry spore suspensions of homogenized sporocarps in water and added about 10 million spores in 5 ml water to each seedling; mycorrhizal colonization ranged from 35 to 90 percent of total short roots when inoculated concurrently with *Frankia.* Castellano and Molina (1989) reviewed the technology for operational mycorrhizal inoculation of container seedlings and reported high success of inoculating several species of Pinaceae with spores of EM fungi in the genus *Rhizopogon,* close truffle relatives of *A. diplophloeus.* In the most advanced inoculation system, spores of *Rhizopogon* are introduced through the irrigation mist system. It is likely that such a technology could work for introduction of *A. diplophloeus* spores onto red alder seedlings and could even be done in conjunction with introduction of *Frankia* cultures.

Much more research is needed to address the benefits of mycorrhizal fungi to red alder seedling nutrition, growth, and plantation performance before recommendations on the need for mycorrhizal inoculation can be made. Growth response of red alder seedlings to mycorrhizal inoculation is limited to data on only one symbiont (*Alpova diplophloeus*) and then only under growth chamber and greenhouse conditions in artificial soil mixes (Koo 1989). Trappe (1977) and Castellano and Molina (1989) outlined detailed research programs for selecting mycorrhizal fungi for inoculating seedlings based on nutritional benefits, ecological

conditions of planting site, and ease of manipulating the candidate fungi. Although such research programs have been conducted for seedlings of Pinaceae in the United States and elsewhere, they are lacking for the most part on broad-leaf EM trees in North America. Given the positive benefits seen in EM inoculation programs elsewhere, it is likely that with adequate research, EM inoculation of red alder with selected, beneficial strains of EM fungi for use under prescribed ecological conditions could prove effective.

Two major considerations are important in deciding the need for mycorrhizal inoculation. First, does the fungus improve seedling growth or reduce cull percentage in the nursery? This can be answered only through a series of experimental inoculations in various nursery settings. Techniques that work in bare-root nurseries may not work in container nurseries. Secondly, does inoculation improve outplanting performance? Outplanting performance often is affected by the history of site disturbance and residual populations of mycorrhizal fungi. Because EM fungi are widespread for red alder (Miller et al. 1992), planted seedlings may quickly develop EM after out-planting. Thus, if adequate growth is obtained in the nursery with *Frankia* inoculation only, and seedlings quickly develop EM following outplanting, then inoculation with EM fungi may not be necessary. Inoculation with EM fungi may be necessary and beneficial, however, if the red alder seedlings are planted on severely disturbed sites or on sites far from natural red alder stands (e.g., plantations in agricultural settings) where natural populations of alder EM fungi are low or absent. Such need will require experimental validation of the low EM fungus inoculum potential of the sites and then development of the proper technology to inoculate seedlings with appropriate EM fungus plus *Frankia* symbionts.

*Effects of Site Disturbance on Red Alder Root Symbioses*
Few studies have examined direct effects of soil/forest disturbance on the presence and function of the root symbionts. The soil bioassay by Miller et al. (1992) clearly showed the widespread nature of both EM fungi and *Frankia* in a variety of forest

types in the Oregon Coast Range. They found that red alder forms as many nodules in a conifer plantation soil as in a young alder forest soil. Similarly, EM development was as great in conifer clearcut soil as in young alder soil as was the diversity of EM types. Development of EM and actinorhizae were delayed on seedlings grown in mature forest soil, but the symbioses did develop. The widespread dispersal characteristics of both root symbionts likely contribute to their quick development and physiological function for benefiting red alder establishment. Indeed, this pervasiveness likely contributes to the pioneering ability of red alder on disturbed sites.

Specific impacts of soil disturbances such as fire (e.g., hot, localized burns) or soil compaction have not been examined for effects on development and function of these symbioses. If such sites are planned for rehabilitation with red alder, then further study is warranted to decide the need for symbiont inoculation or amelioration of the disturbed site to allow the effective development and function of the symbioses.

## Conclusions and Recommendations

Natural red alder seedlings typically develop both actinorhizae and mycorrhizae soon after seed germination in the field. Seedling studies indicate that red alder growth is highly responsive to nodule formation and subsequent nitrogen fixation. Mycorrhizal development likewise improves growth and phosphorus status, but during the first year of seedling growth, increases in growth due to ectomycorrhizal development are small compared to the actinorhizal benefit alone. It must be emphasized, however, that the benefits of ectomycorrhizae to red alder have barely been explored, and what is known is limited to experimental greenhouse studies with seedlings. The two symbioses also interact, the most notable result being the strong stimulation in mycorrhizal development as a result of early nodule formation.

Both fungal and actinomycete endophytes are widespread in forest soils throughout the Pacific Northwest wherever red alder is common on the landscape. This pervasiveness likely contributes to the pioneer colonizing ability of red alder. It also indicates the high survival and dispersal capabilities of the root symbionts.

Previous reviews of alder symbioses called for optimizing the use of root symbionts through development of technologies to inoculate seedlings with selected, beneficial strains. Such techniques are now available, particularly for inoculation with specific strains of *Frankia*. At this point, we recommend at least the inoculation of red alder seedlings in nurseries with *Frankia*, preferably with pure cultures of proven, effective isolates. Such an inoculation scheme will improve seedling growth in the nursery, and reduce both the percentage of culls and need for nitrogen fertilizer to compensate for the absence of *Frankia*. To optimize such a *Frankia* inoculation scheme, more work is needed on selecting strains based on ecological characteristics that will benefit seedling survival and growth on specific planting sites (i.e., matching *Frankia* strains to outplanting locations and environments).

The immediate need for inoculating with specific mycorrhizal fungi is less clear, primarily because of the strong effect of nodulation on first-year growth. Active research is needed similar to that conducted for conifers. Mycorrhizal fungi of red alder are widespread, even on recently cut over forest sites, so adequate mycorrhizal development likely will take place soon after seedling planting. Mycorrhizal inoculation may prove beneficial if seedlings are planted on sites distant from natural stands of red alder. Techniques to inoculate red alder with the alder-specific fungus *Alpova diplophloeus* are available.

The next logical step in developing these management tools is the installation of carefully designed nursery inoculation studies using a variety of *Frankia* and mycorrhizal fungus isolates and then outplanting these inoculated seedlings over a variety of planting sites.

## Acknowledgments

We thank Drs. James Trappe and David Hibbs for their technical review of this manuscript. We also thank Drs. Steven Miller and Kerstin Hauss-Danell for allowing us to use some of their excellent photographs and drawings.

# Literature Cited

Akkermans, A. D. L., and A. Houwers. 1979. Symbiotic nitrogen-fixers available for use in temperate forestry. *In* Symbiotic nitrogen fixation in the management of temperate forests. *Edited by* J. C. Gordon, C. T. Wheeler, and D. A. Perry. Oregon State Univ. Press, Corvallis. 23-26.

Allen, M. F., *ed.* 1992. Mycorrhizal functioning: an integrative plant-fungal process. Chapman and Hall, New York.

Arveby, A. S. 1988. Occurrence and successions of mycorrhizas in *Alnus incana*. *In* Ectomycorrhiza and acid rain. *Edited by* A. E. Jensen, J. Dighton, and A. H. M. Bresser. Bilthoven, The Netherlands.

Baker, D. 1987. Relationships among pure cultured strains of *Frankia* based on host specificity. Physiol. Plant. 70:245-248.

———. 1988. Opportunities for autecological studies of *Frankia,* a symbiotic actinomycete. *In* Biology of actinomycetes. *Edited by* Y. Okami, T. Beppu, and H. Ogawara. Japan Scientific Societies Press, Tokyo. 271-276.

———. 1989. Methods for the isolation, culture, and characterization of the Frankiaceae: soil actinomycetes and symbionts of actinorhizal plants. *In* Isolation of biotechnological organisms from nature. *Edited by* D. P. Labeda. McGraw-Hill, New York. 213-236.

Baker, D., and D. O'Keefe. 1984. A modified sucrose fractionation procedure for the isolation of *Frankia* from actinorhizal root nodules and soil samples. Plant and Soil 78:23-28.

Baker, D., and E. Seling. 1984. *Frankia*: new light on an actinomycete symbiont. *In* Biological, biochemical and biomedical aspects of actinomycetes. *Edited by* L. Ortiz-Ortiz, L. F. Bojalil, and V. Yakoleff. Academic Press, New York. 563-574.

Barea, J. M., R. Azcón, and C. Azcón-Aguilar. 1992. Vesicular-arbuscular mycorrhizal fungi in nitrogen-fixing systems. *In* Methods in microbiology, vol. 24, Techniques for the study of mycorrhiza. *Edited by* J. R. Norris, D. J. Read, and A. K. Varma. Academic Press, London. 391-416.

Benoit, L. F., and A. M. Berry. 1990. Methods for production and use of actinorhizal plants in forestry, low-maintenance landscapes, revegetation. *In* The biology of *Frankia* and actinorhizal plants. *Edited by* C. R. Schwintzer and J. D. Tjepkema. Academic Press, New York. 281-297.

Benson, D. R. 1982. Isolation of *Frankia* strains from alder actinorhizal root nodules. Appl. Environ. Microbiol. 44:461-465.

Benson, D. R., and D. G. Hanna. 1983. *Frankia* diversity in an alder stand as estimated by sodium dodecyl sulfate-polyacrylamide gel electrophoresis of whole-cell proteins. Can. J. Bot. 61:2919-2923.

Benson, D. R., and N. A. Schultz. 1990. Physiology and biochemistry of *Frankia* in culture. *In* The biology of *Frankia* and actinorhizal plants. *Edited by* C. R. Schwintzer and J. D. Tjepkema. Academic Press, New York. 107-127.

Berliner, R., and J. G. Torrey. 1989. On tripartite *Frankia*-mycorrhizal associations in the Myricaceae. Can. J. Bot. 67:1708-1712.

Berry, A. M., M. Moreau, R. A. Jones, and A. Daniel. 1991. Bacteriohopanetetrol: abundant lipid in *Frankia* cells and in nitrogen-fixing nodule tissue. Plant Physiol. 95:111-115.

Berry, A. M., and L. A. Sunell. 1990. The infection process and nodule development. *In* The biology of *Frankia* and actinorhizal plants. *Edited by* C. R. Schwintzer and J. D. Tjepkema. Academic Press, New York. 61-81.

Berry, A. M., and J. G. Torrey. 1979. Isolation and characterization *in vivo* and *in vitro* of an actinomycetous endophyte from *Alnus rubra* Bong. *In* Symbiotic nitrogen fixation in the management of temperate forests. *Edited by* J. C. Gordon, C. T. Wheeler, and D. A. Perry. Oregon State Univ. Press, Corvallis. 69-83.

———. 1985. Seed germination, seedling inoculation and establishment of *Alnus* spp. in containers in greenhouse trials. Plant and Soil 87:161-173.

Bethlenfalvay, G. J. 1992. Vesicular-arbuscular mycorrhizal fungi in nitrogen-fixing legumes: problems and prospects. *In* Methods in microbiology, vol. 24, Techniques for the study of mycorrhiza. *Edited by* J. R. Norris, D. J. Read, and A. K. Varma. Academic Press, London. 375-389.

Bond, G., W. W. Fletcher, and T. P. Ferguson. 1954. The development and function of the root nodules of *Alnus, Myrica* and *Hippophae*. Plant and Soil 5:309-323.

Burggraaf, A. J. P., and W. A. Shipton. 1983. Studies on the growth of *Frankia* isolates in relation to infectivity and nitrogen fixation (acetylene reduction). Can. J. Bot. 61:2774-2782.

Brunchorst, J. 1886-88. Über einige Wurzelanschwellungen, besonders diejenigen von *Alnus* und den *Elaeanaceen*. Unters. Bot. Inst. Tubingen 2:150-177.

Brunner, I. L., F. Brunner, and O. K. Miller, Jr. 1990. Ectomycorrhizal synthesis with Alaskan *Alnus tenuifolia*. Can. J. Bot. 68:761-767.

Callaham, D., P. Del Tredici, and J. G. Torrey. 1978. Isolation and cultivation *in vitro* of the actinomycete causing root nodulation in *Comptonia*. Science 199:899-902.

Carpenter, C. V., L. R. Robertson, J. C. Gordon, and D. A. Perry. 1984. The effect of four new *Frankia* isolates on growth and nitrogenase activity in clones of *Alnus rubra* and *Alnus sinuata*. Can. J. For. Res. 14:701-706.

Castellano, M. A., and R. Molina. 1989. Mycorrhizae. *In* The container tree nursery manual, vol. 5. *Edited by* T. D. Landis, R. W. Tinus, S. E. McDonald, and J. P. Barnett. USDA For. Serv., Agric. Handb. 674. Washington, D.C. 101-167.

Cervantes, E., and C. Rodríquez-Barrueco. 1992. Relationships between the mycorrhizal and actinorhizal symbioses in non-legumes. *In* Methods in microbiology, vol. 24, Techniques for the study of mycorrhiza. *Edited by* J. R. Norris, D. J. Read, and A. K. Varma. Academic Press, London. 417-432.

Chatarpaul, L., P. Chakravarty, and P. Subramaniam. 1989. Studies in tetrapartite symbioses. I. Role of ecto- and endomycorrhizal fungi and *Frankia* on the growth performance of *Alnus incana*. Plant and Soil 118:145-150.

Dawson, J. O., and A. H. Gibson. 1987. Sensitivity of selected *Frankia* isolates from *Casurina, Allocasurina* and North American host plants to sodium chloride. Physiol. Plantarum 70:272-278.

Diem, H. G., and Y. R. Dommergues. 1988. Isolation, characterization and cultivation of *Frankia*. *In* Biological nitrogen fixation, recent developments. *Edited by* N. S. Subba Rao. Gordon and Breach Science Publishers, New York. 227-254.

————. 1990. Current and potential uses and management of Casuarinaceae in the tropics and subtropics. *In* The biology of *Frankia* and actinorhizal plants. *Edited by* C. R. Schwintzer and J. D. Tjepkema. Academic Press, New York. 317-342.

Diem, H. G., E. Duhoux, P. Simonet, and Y. R. Dommergues. 1988. Actinorhizal symbiosis biotechnology: the present and future. Proc. 8th Int. Biotechnol. Symp. 2:984-995.

Diem, H. G., and D. Gauthier. 1982. Effet de l'infection endomycorrhizienne (*Glomus mosseae*) sur la nodulation et la croissance de *Casuarina equisetifolia*. C. R. hebd. Seances Acad. Sci., Ser. C, 294:215-218.

Diem, H. G., D. Gauthier, and Y. R. Dommergues. 1982. Isolation of *Frankia* from nodules of *Casuarina equisetifolia*. Can. J. Microbiol. 28:526-530.

Domenach, A. M., F. Kurdali, C. Daniere, and R. Bardin. 1988. Determination de l'identite isotopique de l'azote fixe par le *Frankia* associe au genre *Alnus*. Can. J. Bot. 66:1241-1247.

Faure-Raynaud, M., C. Daniere, A. Moiroud, and A. Capellano. 1991. Diversity of *Frankia* strains isolated from single nodules of *Alnus glutinosa*. Plant and Soil 132:207-211.

Faure-Raynaud, M., M. A. Bonnefoy-Poirier, and A. Moiroud. 1990. Diversity of *Frankia* strains isolated from actinorhizae of a single *Alnus rubra* cultivated in nursery. Symbiosis 8:147-160.

Fraga-Beddiar, A., and F. Le Tacon. 1990. Interactions between a VA mycorrhizal fungus and *Frankia* associated with alder (*Alnus glutinosa* {L.} Gaetn.). Symbiosis 9:247-258.

Froidevaux, L. 1973. The ectomycorrhizal association, *Alnus rubra + Lactarius obscuratus*. Can. J. For. Res. 3:601-603.

Gallon, J. R. 1992. Tansley review No. 44. Reconciling the incompatible: $N_2$ fixation and $O_2$. New Phytol. 122: 571-609.

Gardner, I. C., D. M. Clelland, and A. Scott. 1984. Mycorrhizal improvement in non-leguminous nitrogen-fixing associations with particular reference to *Hippophae*. Plant and Soil 78:189-199.

Granhall, U., T. Ericsson, and M. Clarholm. 1983. Dinitrogen fixation and nodulation by *Frankia* in *Alnus incana* as affected by inorganic nitrogen in pot experiments with peat. Can. J. Bot. 61:2956-2963.

Hahn, D., M. J. C. Starrenburg, and A. D. L. Akkermans. 1988. Variable compatibility of cloned *Alnus glutinosa* ecotypes against ineffective *Frankia* strains. Plant and Soil 107:233-243.

————. 1990a. Growth increment of *Alnus glutinosa* upon dual inoculation with effective and ineffective *Frankia* strains. Plant and Soil 122:121-127.

————. 1990b. Oligonucleotide probes against rRNA as a tool to study *Frankia* strains in root nodules. Appl. Environ. Microbiol. 56:1324-1346.

Hall, R. B., H. S. McNabb, Jr., C. A. Maynard, and T. L. Green. 1979. Toward development of optimal *Alnus glutinosa* symbioses. Bot. Gaz. (Suppl.) 140:120-126.

Harley, J. L., and S. E. Smith. 1983. Mycorrhizal symbiosis. Academic Press, London.

Harriott, O. T., L. Khairallah, and D. R. Benson. 1991. Isolation and structure of the lipid envelopes from the nitrogen-fixing vesicles of *Frankia* sp. strain Cp11. J. Bacteriol. 173:2061-2067.

Hilger, A. B., Y. Tanaka, and D. A. Myrold. 1991. Inoculation of fumigated nursery soil increases nodulation and yield of bare-root red alder (*Alnus rubra* Bong.). New For. 5:35-42.

Hiltner, L. 1895. Über die Bedeutung der Wurzelknöllchen von *Alnus glutinosa* für die Stickstoffernährung dieser Pflanze. Die Landwirtschaftlichen Versuchs-Stationen 46:153-161.

Holman, R. M., and C. R. Schwintzer. 1987. Distribution of spore-positive and spore-negative nodules of *Alnus incana* spp. *rugosa* in Maine, USA. Plant and Soil 104:103-111.

Hooker, J. E., and C. T. Wheeler. 1987. The effectivity of *Frankia* for nodulation and nitrogen fixation in *Alnus rubra* and *A. glutinosa*. Physiol. Plant. 70:333-341.

Hughes, D. R., S. P. Gessel, and R. B. Walker. 1968. Red alder deficiency symptoms and fertilizer trials. *In* Biology of alder. *Edited by* J. M. Trappe, J. F. Franklin, R. F. Tarrant, and G. M. Hansen. USDA For. Serv., PNW For. Range Exp. Sta., Portland, OR. 225-237.

Huss-Danell, K. 1990. The physiology of actinorhizal nodules. *In* The biology of *Frankia* and actinorhizal plants. *Edited by* C. R. Schwintzer and J. D. Tjepkema. Academic Press, New York. 129-156.

Huss-Danell, K., and A.-K. Frey. 1986. Distribution of *Frankia* in soil from forest and afforestation sites in northern Sweden. Plant and Soil 90:407-418.

Ingestad, T. 1980. Growth, nutrition, and nitrogen fixation in grey alder at varied rate of nitrogen addition. Physiol. Plant. 50:353-364.

Knowlton, S., A. M. Berry, and J. G. Torrey. 1979. The role of rhizosphere microorganisms in nodule formation in *Alnus rubra* Bong. *In* Symbiotic nitrogen fixation in the management of temperate forests. *Edited by* J. C. Gordon, C. T. Wheeler, and D. A. Perry. Oregon State Univ. Press, Corvallis. 479-480.

———. 1980. Evidence that associated soil bacteria may influence root hair infection of actinorhizal plants by *Frankia*. Can. J. Microbiol. 26:971-977.

Knowlton, S., and J. O. Dawson. 1983. Effects of *Pseudomonas cepacia* and cultural factors on the nodulation of *Alnus rubra* roots by *Frankia*. Can. J. Bot. 61:2877-2882.

Koo, C. D. 1989. Water stress, fertilization, and light effects on the growth of nodulated mycorrhizal red alder seedlings. Ph.D. thesis, Oregon State Univ., Corvallis.

Kurdali, F., F. Rinaudo, A. Moiroud, and A. M. Domenach. 1990. Competition for nodulation and $^{15}N_2$-fixation between a Sp$^+$ and a Sp$^-$ *Frankia* strain in *Alnus incana*. Soil Biol. Biochem. 22:57-64.

Lamont, H. C., W. B. Silvester, and J. G. Torrey. 1988. Nile red fluorescence demonstrates lipid in the envelope of vesicles from $N_2$-fixing cultures of *Frankia*. Can. J. Bot. 34:656-660.

Lechevalier, M. P., D. Baker, and F. Horriere. 1983. Physiology, chemistry, serology and infectivity of two *Frankia* isolates from *Alnus incana* subsp. *rugosa*. Can. J. Bot. 61:2826-2833.

Lee, K. J. 1988. Growth stimulation of *Alnus frima* and *Robinia pseudoacacia* by dual inoculation with VA mycorrhizal fungi and nitrogen-fixing bacteria and their synergistic effect. J. Korean For. Soc. 77:229-234.

Li, C. Y. 1974. Phenolic compounds in understory species of alder, conifer, and mixed alder-conifer stands of coastal Oregon. Lloydia 37:603-607.

———. 1987. Association of nitrogen-fixing bacteria with mycorrhizal fungi and feces of forest-dwelling mammals in relation to forest productivity. *In* Forest productivity and site evaluation. *Edited by* T. Kiang, J. C. Yang, and Y. P. Kao. Council of Agriculture, Taipei, Taiwan. 125-130.

Li, C. Y., K. C. Lu, J. M. Trappe, and W. B. Bollen. 1970. Separation of phenolic compounds in alkali hydrolysates of a forest soil by thin-layer chromatography. Can. J. Soil Sci. 50:458-460.

———. 1972. *Poria weirii*-inhibiting and other phenolic compounds in roots of red alder and Douglas-fir. Microbios 5:65-68.

Li, C. Y., K. V. B. R. Tilak, and R. Molina. 1987. Growth enhancement of red alder by inoculation with *Frankia-Azospirillum* combination. 1987 American Society of Microbiology Annual Meeting.

Lipp, C. C. 1987. The effect of nitrogen and phosphorus on growth, carbon allocation and nitrogen fixation of red alder seedlings. M.S. thesis, Oregon State Univ., Corvallis.

Lopez, M. F., and J. G. Torrey. 1985. Enzymes of glucose metabolism in *Frankia* sp. J. Bacteriol. 162:110-116.

Mansour, S., and J. G. Torrey. 1991. *Frankia* spores of strain HFPCg14 as inoculum for seedlings of *Casuarina glauca*. Can. J. Bot. 69:1251-1256.

Markwell, M. A. K., S. M. Haas, L. L. Bieber, and N. E. Tolbert. 1978. A modification of the Lowry procedure to simplify protein determination in membrane and lipoprotein samples. Anal. Biochem. 87:206-210.

Martin, K. J., Y. Tanaka, and D. D. Myrold. 1991. Peat carrier increases inoculation success with *Frankia* on red alder (*Alnus rubra* Bong.) in fumigated nursery beds. New For. 5:43-50.

Maser, C., J. M. Trappe, and R. A. Nussbaum. 1978. Fungal-small mammal interrelationships with emphasis on Oregon coniferous forests. Ecology 59:799-809.

Massicotte, H. B., R. L. Peterson, and L. H. Melville. 1989a. Ontogeny of *Alnus rubra*—*Alpova diplophloeus* ectomycorrhizae. I. Light microscopy and scanning electron microscopy. Can. J. Bot. 67:191-200.

Massicotte, H. B., D. A. Ackerley, and R. L. Peterson. 1989b. Ontogeny of *Alnus rubra*—*Alpova diplophloeus* ectomycorrhizae. II. Transmission and electron microscopy. Can. J. Bot. 67:201-210.

McNeill, J. D., M. K. Hollingsworth, W. L. Mason, A. J. Moffat, L. J. Sheppard, and C. T. Wheeler, 1989. Inoculation of *Alnus rubra* seedlings to improve seedling growth and forest performance. Great Britain For. Comm. Res. Div., Res. Inf. Note 144.

Mejstrik, V., and U. Benecke. 1969. The ectotrophic mycorrhizas of *Alnus viridis* (Chaix.) D.C. and their significance in respect to phosphorus uptake. New Phytol. 68:141-149.

Mian, S., G. Bong, and C. Rodriquez-Barrueco. 1976. Effective and ineffective root-nodules in *Myrica faya*. Proc. R. Soc. Lond. 194:285-293.

Miller, S. L., C. D. Koo, and R. Molina. 1988. An oxidative blue-bruising reaction in *Alpova diplophloeus* (Basidiomycetes, Rhizopogonaceae) + *Alnus rubra* ectomycorrhizae. Mycologia 80:576-581.

———. 1991. Characterization of red alder ectomycorrhizae: a preface to monitoring belowground ecological responses. Can. J. Bot. 69:516-531.

———. 1992. Early colonization of red alder and Douglas-fir by ectomycorrhizal fungi and *Frankia* in soils from the Oregon Coast Range. Mycorrhiza 2:53-61.

Molina, R. 1979. Pure culture synthesis and host specificity of red alder mycorrhizae. Can. J. Bot. 57(11):1223-1228.

———. Ectomycorrhizal specificity in the genus *Alnus.* Can. J. Bot. 59:325-334.

Molina, R., H. Massicotte, and J. M. Trappe. 1992. Specificity phenomena in mycorrhizal symbioses: community-ecological consequences and practical implications. *In* Mycorrhizal functioning: an integrative plant-fungal process. *Edited by* M. F. Allen. Chapman and Hall, New York. 357-423.

Molina, R., and J. M. Trappe. 1982a. Patterns of ectomycorrhizal host specificity and potential among Pacific Northwest conifers and fungi. For. Sci. 28:423-458.

———. 1982b. Lack of mycorrhizal specificity by the ericaceous hosts *Arbutus menziesii* and *Arctostaphylos uva-ursi.* New Phytol. 90:495-509.

Morton, J. B., and G. L. Benny. 1990. Revised classification of arbuscular mycorrhizal fungi (Zygomycetes): a new order, Glomales, two new suborders, Glomineae and Gigasporineae, and two new families, Acaulosporaceae and Gigasporaceae, with an emendation of Glomaceae. Mycotaxon 37:471-491.

Murry, M. A., M. S. Fontaine, and J. G. Torrey. 1984. Growth kinetics and nitrogenase induction in *Frankia* sp. HFPAr13 grown in batch culture. Plant and Soil 78:61-78.

Myrold, D. D. 1993. *Frankia* and the actinorhizal symbiosis. *In* Methods of soil analysis: part 3—microbiological properties. 3rd ed. *Edited by* R. W. Weaver, J. S. Angle, and P. J. Bottomley. American Society of Agronomy, Madison, WI. (In press.)

Neal, J. L., Jr., J. M. Trappe, K. C. Lu, and W. B. Bollen. 1968. Some ectotrophic mycorrhizae of *Alnus rubra. In* Biology of alder. *Edited by* J. M. Trappe, J. F. Franklin, R. F. Tarrant, and G. M. Hansen. USDA For. Serv., PNW For. Range Exp. Sta., Portland, OR. 179-184.

Newcomb, W., and S. M. Wood. 1987. Morphogenesis and fine structure of *Frankia* (Actinomycetales): the microsymbiont of nitrogen-fixing actinorhizal root nodules. Int. Rev. Cytol. 109:1-88.

Ng, B. H. 1987. The effects of salinity on growth, nodulation and nitrogen fixation of *Casuarina equisetifolia.* Plant and Soil 103:123-125.

Noridge N. A., and D. R. Benson. 1986. Isolation and nitrogen-fixing activity of *Frankia* sp. strain Cp11 vesicles. J. Bacteriology 166:301-305.

Normand, P., and M. LaLonde. 1982. Evaluation of *Frankia* strains isolated from provenances of two *Alnus* species. Can. J. Microbiol. 28:1133-1142.

Parsons, R., W. B. Silvester, S. Harris, W. T. M. Gruijters, and S. Bullivant. 1987. *Frankia* vesicles provide inducible and absolute protection for nitrogenase. Plant Physiol. 83:723-731.

Perinet, P., J. G. Brouillette, J. A. Fortin, and M. LaLonde. 1985. Large scale inoculation of actinorhizal plants with *Frankia.* Plant and Soil 87:175-183.

Perinet, P., and M. LaLonde. 1983. Axenic nodulation of *in vitro* propagated *Alnus glutinosa* plantlets by *Frankia* strains. Can J. Bot. 61:2883-2888.

Perradin, Y., M. J. Mottet, and M. LaLonde. 1983. Influence of phenolics on *in vitro* growth of *Frankia* strains. Can. J. Bot. 61:2807-2814.

Pommer, E. H. 1959. Über die Isolierung des Endophyten aus den Wurzelknöllchen von *Alnus glutinosa* Gaertn. und über Erfolgreiche Reinfektionsversuche. Ber. Dtsch. Bot. Ges. 72:138-150.

Prat, D. 1989. Effects of some pure and mixed *Frankia* strains on seedling growth in different *Alnus* species. Plant and Soil 113:31-38.

Pregent, G., and C. Camire. 1985. Mineral nutrition, dinitrogen fixation, and growth of *Alnus crispa* and *Alnus glutinosa.* Can. J. For. Res. 15:855-861.

Program in Forest Microbiology. 1988. *Frankia* bibliography, 1977-1987. Yale Univ., CT.

Quispel, A. 1990. Discoveries, discussions, and trends in research on actinorhizal root nodule symbioses before 1978. *In* The biology of *Frankia* and actinorhizal plants. *Edited by* C. R. Schwintzer and J. D. Tjepkema. Academic Press, New York. 15-33.

Quispel, A., and A. J. P. Burggraaf. 1981. *Frankia,* the diazotrophic endophyte from actinorhizas. *In* Current perspectives in nitrogen fixation. *Edited by* A. Gibson and W. Newton. Australian Academy of Science, Canberra. 229-236.

Racette, S., and J. G. Torrey. 1991. Sporulation in root nodules of actinorhizal plants inoculated with pure cultured strains of *Frankia.* Can. J. Bot. 69:1471-1476.

Reddell, P., and G. D. Bowen. 1985. Do single nodules of Casuarinaceae contain more than one *Frankia* strain? Plant and Soil 88:275-279.

Righetti, T., C. H. Chard, and R. A. Backhaus. 1986. Soil and environmental factor related to nodulation in *Cowania* and *Purshia.* Plant and Soil 91:147-160.

Righetti, T. L., and D. N. Munns. 1981. Soil factors limiting nodulation and nitrogen fixation in *Purshia. In* Genetic engineering of symbiotic nitrogen fixation and conservation of fixed nitrogen. *Edited by* J. M. Lyons, R. C. Valentine, D. A. Phillips, D. W. Rains, and R. C. Juffaker. Plenum Press, New York. 395-407.

Rodríguez-Barrueco, C., E. Cervantes, N. S. Subba Rao, and E. Rodriguez-Caceres. 1991. Growth promoting effect of *Azospirillum brasilense* on *Casuriana cunninghamiana* Miq. seedlings. Plant and Soil 135:121-124.

Rojas, N. S., D. A. Perry, C. Y. Li, and J. Friedman. 1992. Influence of actinomycetes on *Frankia* infection, nitrogenase activity, and seedling growth of red alder. Soil Biol. Biochem. 24:1043-1049.

Rose. S. L. 1980. Mycorrhizal associations of some actinomycete nodulated nitrogen-fixing plants. Can. J. Bot. 58:1449-1454.

Rose, S. L., and C. T. Youngberg. 1981. Tripartite associations in snowbrush (*Ceanothus velutinus*): effect of vesicular-arbuscular mycorrhizae on growth, nodulation, and nitrogen fixation. Can. J. Bot. 59:34-39.

Russo, R. O. 1989. Evaluating alder-endophyte (*Alnus acuminata-Frankia* mycorrhizae) interactions. I. Acetylene reduction in seedlings inoculated with *Frankia* strain Ar13 and *Glomus* intra-radices, under three phosphorus levels. Plant and Soil 118:151-155.

Safir, G. R. 1987. Ecophysiology of VA mycorrhizal plants. CRC Press, Boca Raton, FL.

Schaede, R. 1933. Über die Symbionten in den Knöllchen der Erle und des Sanddornes und die cytologischen Verhältnisse in ihnen. Planta 19:389-416.

Schultz, N., and D. R. Benson. 1989. Developmental potential of *Frankia* vesicles. J. Bacteriology 171:6873-6877.

Schwintzer, C. R. 1990. Spore-positive and spore-negative nodules. *In* The biology of *Frankia* and actinorhizal plants. *Edited by* C. R. Schwintzer and J. D. Tjepkema. Academic Press, New York. 177-193.

Schwintzer, C. R., and J. D. Tjepkema, *eds*. 1990. The biology of *Frankia* and actinorhizal plants. Academic Press, New York.

Sellstedt, A. 1989. Occurrence and activity of hydrogenase in symbiotic *Frankia* from field-collected *Alnus incana*. Physiol. Plant. 75:304-308.

Sellstedt, A., and K. Huss-Danell. 1984. Growth, nitrogen fixation and relative efficiency of nitrogenase in *Alnus incana* grown in different cultivation systems. Plant and Soil 78:147-158.

Sheppard, L. J., J. E. Hooker, C. T. Wheeler, and R. I. Smith. 1988. Glasshouse evaluation of the growth of *Alnus rubra* and *Alnus glutinosa* on peat and acid brown earth soils when inoculated with four sources of *Frankia*. Plant and Soil 110:187-198.

Shipton, W. A., and A. J. P. Burggraaf. 1982. A comparison of the requirements for various carbon and nitrogen sources and vitamins in some *Frankia* isolates. Plant and Soil 69:149-161.

Silvester, W. B. 1977. Dinitrogen fixation by plant associations excluding legumes. *In* A treatise on dinitrogen fixation, section IV: agronomy and ecology. *Edited by* R. W. F. Hardy and A. H. Gibson. Wiley, New York. 141-190.

Silvester, W. B., S. L. Harris, and J. D. Tjepkema. 1990. Oxygen regulation and hemoglobin. *In* The biology of *Frankia* and actinorhizal plants. *Edited by* C. R. Schwintzer and J. D. Tjepkema. Academic Press, New York. 157-176.

Simonet, P., N. T. Le, E. T. du Cros, and R. Bardin. 1988. Identification of *Frankia* strains by direct DNA hybridization of crushed nodules. Appl. Environ. Microbiol. 54:2500-2503.

Simonet, P., P. Normand, A. Moiroud, and R. Bardin. 1990. Identification of *Frankia* strains in nodules by hybridization of polymerase chain reaction products with strain-specific oligonucleotide probes. Arch. Microbiol. 153:235-240.

Simonet, P., P. Normand, A. Moiroud, and M. LaLonde. 1985. Restriction enzyme digestion patterns of *Frankia* plasmids. Plant and Soil 87:49-60.

Smolander, A. 1990. *Frankia* populations in soils under different tree species—with special emphasis on soils under *Betula pendula*. Plant and Soil 121:1-10.

Smolander, A., and M.-L. Sarsa. 1990. *Frankia* strains of soil under *Betula pendula*: behaviour in soil and in pure culture. Plant and Soil 122:129-136.

Smolander, A., and V. Sundman. 1987. *Frankia* in acid soils of forests devoid of actinorhizal plants. Physiol. Plant. 70:297-303.

Smolander, A., C. van Dijk, and V. Sundman. 1988. Survival of *Frankia* strains introduced into soil. Plant and Soil 106:65-72.

Stowers, M. D. 1987. Collection, isolation, cultivation, and maintenance of *Frankia*. *In* Symbiotic nitrogen fixation technology. *Edited by* G. H. Elkan. Marcel Dekker, New York. 29-53.

Stowers, M. D., and J. E. Smith. 1985. Inoculation and production of container-grown red alder seedlings. Plant and Soil 87:153-160.

Tarrant, R. F., and J. M. Trappe. 1971. The role of *Alnus* in improving the forest environment. Plant and Soil (spec. vol.) 1971:335-348.

Tjepkema, J. D., W. Ormerod, and J. G. Torrey. 1980. Vesicle formation and acetylene reduction activity in *Frankia* sp. CP11 cultured in defined nutrient media. Nature 287:633-635.

———. Factors affecting vesicle formation and acetylene reduction (nitrogenase activity) in *Frankia* sp. Cp11. Can. J. Microbiol. 27:815-823.

Tjepkema, J. D., C. R. Schwintzer, and D. R. Benson. 1986. Physiology of actinorhizal nodules. Annu. Rev. Plant Physiol. 37:209-232.

Tjepkema, J. D., and J. G. Torrey. 1979. Symbiotic nitrogen fixation in actinomycete-nodulated plants. Preface. Bot. Gaz. (Suppl.) 140:i-ii.

Torrey, J. G. 1990. Cross-inoculation groups within *Frankia* and host-endosymbiont associations. *In* The biology of *Frankia* and actinorhizal plants. *Edited by* C. R. Schwintzer and J. D. Tjepkema. Academic Press, New York. 83-106.

Trappe, J. M. 1977. Selection of fungi for ectomycorrhizal inoculation in nurseries. Ann. Rev. Phytopathol. 15:203-222.

————. 1979. Mycorrhiza-nodule-host interrelationships in symbiotic nitrogen fixation: a quest in need of questers. *In* Symbiotic nitrogen fixation in the management of temperate forests. *Edited by* J. C. Gordon, C. T. Wheeler, and D. A. Perry. Oregon State Univ. Press, Corvallis. 276-286.

Tsai, Y.-L., L. M. McBride, and J. C. Ensign. 1983. Studies of growth and morphology of *Frankia* strains EAN1pec, EuI1c, CPI1, and ACN1$^{AG}$. Can. J. Bot. 61:2768-2773.

van Dijk, C. 1984. Ecological aspects of spore formation in the *Frankia-Alnus* symbiosis. Ph.D. thesis, State Univ., Leiden, The Netherlands.

van Dijk, C., A. Sluimer, and A. Weber. 1988. Host range differentiation of spore-positive and spore-negative strain types of *Frankia* in stands of *Alnus glutinosa* and *Alnus incana* in Finland. Physiol. Plant. 72:349-358.

van Dijk, C., and A. Sluimer-Stolk. 1990. An ineffective strain type of *Frankia* in the soil of natural stands of *Alnus glutinosa* (L.) Gaertner. Plant and Soil 127:107-121.

Virtanen, A. I., Y. Moisio, R. M. Allison, and R. H. Burris. 1954. Fixation of molecular nitrogen by excised nodules of alder. Acta Chem. Scand. 8:1730-1731.

Vogel, C. S., and J. O. Dawson. 1986. *In vitro* growth of five *Frankia* isolates in the presence of four phenolic acids and juglone. Soil Biol. Biochem. 18:227-231.

Weber, A. 1986. Distribution of spore-positive and spore-negative nodules in stands of *Alnus glutinosa* and *Alnus incana* in Finland. Plant and Soil 96:205-213.

Weber, A., E. L. Nurmiaho-Lassila, and V. Sundman. 1987. Features of intrageneric *Alnus-Frankia* specificity. Physiol. Plant. 70:289-296.

Weber, A., M.-L. Sarsa, and V. Sundman. 1989. *Frankia-Alnus incana* symbiosis: effect of endophyte on nitrogen fixation and biomass production. Plant and Soil 120:291-297.

Wheeler, C. T., J. E. Hooker, A. Crowe, and A. M. M. Berrie. 1986. The improvement and utilization in forestry of nitrogen fixation by actinorhizal plants with special reference to *Alnus* in Scotland. Plant and Soil 90:393-406

Wheeler, C. T., M. K. Hollingsworth, J. E. Hooker, J. D. McNeill, W. L. Mason, A. J. Moffat, and L. J. Sheppard. 1991. The effect of inoculation with either cultured *Frankia* or crushed nodules on nodulation and growth of *Alnus rubra* and *Alnus glutinosa* seedlings in forest nurseries. For. Ecol. Manage. 43:153-166.

# 3

# Influences of Red Alder on Soils and Long-Term Ecosystem Productivity

B. T. BORMANN, K. CROMACK, JR., & W. O. RUSSELL, III

Red alder (*Alnus rubra* Bong.) is widely perceived as enhancing the long-term ecosystem productivity of coastal forests of western North America. Alders are the primary symbiotic fixers of atmospheric nitrogen ($N_2$) in a region thought to be generally deficient in nitrogen (Gessel et al. 1973) and are likely to be responsible for the large accumulations of nitrogen observed in many coastal forest soils. Nitrogen accumulations in Pacific Northwest soils are difficult to explain on the basis of nitrogen in precipitation alone, especially when nitrogen losses from combustion, drainage, and denitrification are considered. Alders certainly played an important role in revegetation following Wisconsin-era glacial retreat; alder pollen dominated sediments deposited during the first several thousand years after the retreat of continental glaciers (Heusser 1960). Alder presence has recently expanded as human disturbances have increased (Davis 1973).

Sitka alder (*Alnus sinuata* {Regal} Rydb.) greatly facilitated primary succession and improved long-term ecosystem productivity after the more recent glacial retreat in Glacier Bay, Alaska (Bormann and Sidle 1990). Such improvement at this and similar sites presumably is due to the substantial accumulations of nitrogen and organic matter in the soil, and the increased availability of phosphorus, exchangeable calcium, and magnesium (Crocker and Major 1955; Ugolini 1968; Bormann and Sidle 1990). Facilitation by alder is demonstrated also by changes that occur after alder had been shaded out by Sitka spruce (*Picea sitchensis* Bong.) in Glacier Bay. In only 50 years after alder decreased as a successional component, aboveground net primary production of magnesium decreased from about 8 to 5 ha$^{-1}$yr$^{-1}$ (Bormann and Sidle 1990). Associated with this decline were many mineral soil changes including decreased soil pH (5.85 to 5.60), available phosphorus (14 to 3 μg g$^{-1}$), and exchangeable calcium (0.71 to 0.25 mg g$^{-1}$), and increased exchangeable iron (93 to 167 μg g$^{-1}$).

Evaluations of secondary succession have also suggested that, in some circumstances, alder can improve the growth of associated species. In perhaps the best available example, Douglas-fir (*Pseudotsuga menziesii* {Mirb.} Franco) growing in a mixed stand with red alder, planted as a vegetative fire break, was compared with adjacent pure stands of Douglas-fir (Miller and Murray 1978). In 48 years, mixed stands had accumulated 159 m$^3$ ha$^{-1}$ of wood (88 and 71 m$^3$ ha$^{-1}$ in fir and alder, respectively) as compared to 82 m$^3$ ha$^{-1}$ in the adjacent pure Douglas-fir stands. Site index of Douglas-fir (50-year basis) was increased on average by 6.4 m. Improved soil fertility is in large part responsible for the economic viability of both mixed and crop rotation culture of alder and Douglas-fir (Atkinson et al. 1979; Tarrant et al. 1983).

Various ecosystem processes or mechanisms have been proposed and studied in an attempt to explain how alder can bring about these changes. The mechanism most studied is symbiotic fixation of atmospheric nitrogen although some attention has been given to nutrient cycling and organic-matter decomposition. Increased nitrification has been proposed as a mechanism to reduce pH and increase cation leaching, especially for subsequent crops of alder (Van Miegroet and Cole 1984; Cole et al. 1989). Predicting effects of alder on long-term ecosystem productivity for a specific site is difficult because the generality of mechanisms has not been evaluated on a wide range of environments within the region, most likely other mechanisms have received little or no recognition, and research has not been integrative, typically focusing on mechanisms in isolation of one another.

Synthesis of mechanisms and their interactions, and development of this knowledge to improve management strategies, is clearly difficult. It is an important task, however, because land-management decisions cannot wait for the assembly of a large body of empirical growth and yield data that apply directly to many regional environments. New directives within the USDA Forest Service to implement ecosystem management—based in large part on ecosystem sustainability (Overbay 1992; Robertson 1992)—make this task crucially important.

In this chapter we review observed effects and related mechanisms to see how alder and its associates can influence soil properties. In addition, we evaluate the interaction of mechanisms, propose potentially important unstudied mechanisms, and discuss the implications of using alder to manage for sustained long-term ecosystem productivity.

## Observed Effects and Mechanisms

### Biological Fixation of Atmospheric Nitrogen

Red alder and other alders to a similar or lesser extent are thought to have among the highest rates of symbiotic fixation of atmospheric nitrogen in temperate-forest ecosystems, especially under unfertilized and unirrigated conditions (Cromack et al. 1979; Binkley et al. 1993; Binkley et al., Chapter 4). Red alder has shown both hydrogenase and nitrogenase enzyme activity in nodules (Dixon and Wheeler 1983; Molina et al., Chapter 2). Peak rates of fixation of atmospheric nitrogen in thinleaf alder (*Alnus tenuifolia* Nutt.) are greater than 300 kg ha$^{-1}$ yr$^{-1}$ (Van Cleve et al. 1971). Fixation of atmospheric nitrogen leads to large accumulations of nitrogen in forest floor and mineral soil in alder stands ranging from 500 to 7500 kg ha$^{-1}$ in 50 years (Table 1). Nitrogen is thought to be the most commonly limiting nutrient in this region (Gessel et al. 1973). From this perspective, red alder—as the principal fixer of atmospheric nitrogen—is likely to be important to the long-term ecosystem productivity of the region.

### Nutrient Cycling

Early interest in alder was based on the high nutrient concentrations found in its litter as compared to the litter of other hardwoods or conifers (Tarrant and Chandler 1951). Nitrogen was the most abundant nutrient component of red alder litter, although $Ca^{2+}$, $Mg^{2+}$, $K^+$, and P also were high. Red alder also produces more aboveground litterfall than do associated conifers (Waring and Schlesinger 1985). Together, the higher nutrient concentrations and greater litterfall of red alder produce significantly higher annual litter returns of N, $Ca^{2+}$, $Mg^{2+}$, $K^+$, and P than does adjacent Douglas-fir (Gessel and Turner 1974; Cole et al. 1978). Red alder's deciduous growth habit and higher demand for phosphorus (Sprent 1988) and cations drive this process. Once on the ground, alder litter, especially woody components, decomposes more rapidly than Douglas-fir litter (Neal et al. 1965). Clearly, alder cycles many nutrients at an accelerated rate as compared with conifers, which is likely to lead to enhanced availability and redistribution of nutrients from within the soil profile. This, in turn, is likely to improve ecosystem productivity as long as plant uptake can prevent leaching losses. The abundant growth of understory plants under alder as compared to conifers may be due to enhanced nutrient availability as well as increased sunlight.

### Soil Organic Matter Decomposition and Accumulation

Alders are known to speed soil development during primary succession (Crocker and Major 1955; Lawrence 1958; Ugolini 1968). By producing large quantities of nutrient-rich litter biomass annually, alder is able to rapidly increase soil organic matter in as little as five years (Van Cleve et al. 1972). While accumulation of nitrogen is generally lower in mixed alder stands relative to pure alder stands, accumulation of organic matter is quite constant (Table 1). The litter of other trees is more rapidly decomposed and incorporated into soil organic matter when mixed with nutrient-rich alder litter (Bormann and Sidle 1990).

Alders may contribute to increased organic matter in the soil through the production of polyphenol-rich litter, since both leaves and woody tissues contain polyphenols (Kurth and Becker 1953; Horhammer and Scherm 1955; Uvarova et al. 1970; Karchesy 1974). Red alder bark was found to contain 9.8 percent tannin, for example (Peacock 1900); and Karchesy (1974) found condensed tannin in red alder bark to polymerize with the red staining phenol oregonin. Decomposition of red alder leaf litter slows after the first year of decay in the forest floor; its decay rate is similar to that of conifer species such as Douglas-fir after two years in the field (Edmonds 1980); this effect may be due in part to the potential role of phenol polymerization in creating decay-resistant complexes (Stevenson 1982). Phenol polymer-ization through oxidative reactions of phenols leached from litter components could lead to increased soil organic matter through the mechanism of phenol precipitation from soil solution as polymerization proceeds, and thus lead to formation of humus-like materials (Stevenson 1982). Incorporation of organic nitrogen into polymerized phenols and humus could occur from reactions with amino acids in litter leachates and root exudates under alder. The rapid increase of organic matter under alder containing greater concentrations of organic nitrogen in different density fractions of soil organic matter (Sollins et al. 1984) implies that such a mechanism occurs. Incorporation of organic nitrogen into a more decay-resistant carbon complex would conserve the added nitrogen in forms slower to mineralize and thus act as a mechanism slowing release of nitrogen available for plant uptake (Stevenson 1986).

Alder rapidly increases organic matter in soil by producing large quantities of rapidly decomposable biomass annually. One result of this increase is aggregation and decreased bulk density (Wild 1988). In existing alder examples, bulk density was either unaffected or dramatically reduced (Table 1). Soil structure and organic matter are thought to be major drivers of ecosystem productivity (Powers 1991). Increased moisture-holding capacity is also an important effect of increased soil organic matter. Crocker and Dickson (1957) found a doubling in moisture-holding capacity in the first 50 years of post-glacial Sitka alder chronosequence near Juneau, Alaska.

Litter of alder and other species that fix atmospheric nitrogen is a preferred food of earthworms (Voigt and Steucek 1969; Brown and Harrison 1982; Graham and Wood 1991). Certain species of them feed on the soil surface and transport litter into the mineral soil (Darwin 1881), a mechanism likely to occur under alder wherever litter-feeding earthworms occur (Brown and Harrison 1982). Earthworm activity may explain much of the increased organic matter incorporation and aggregation observed under alder.

Alder produces more litter than do conifers (Zavitkovski and Newton 1971) and thus more energy flowing into detrital foodchains. More detritivores may mean that alder can support a greater biomass of detrital insectivores.

Fixed-depth sampling was used in all studies we reviewed (Table 1). Increasing organic matter, aggregation, and porosity causes soil to expand upward (Wild 1988; Bormann et al., in prep.). As organic matter accumulates in alder soils, a progressively deeper depth would have to be sampled to maintain the same mass of mineral material as an equivalent depth in an adjacent stand without alder or in the same stand through time. The values in Table 1, therefore, are conservative for accretion of organic matter and nitrogen under alder and may also overestimate changes in pH and cation exchange capacity. This is a minor problem for values summed over a deep soil profile.

## Soil Organic Matter and Nitrification

The large inputs of nitrogen and organic matter into mineral soil beneath alder are likely to be interrelated to other observed mechanisms. Nitrification is likely to occur where processes such as incorporation of nitrogen into soil organic matter, or nitrogen uptake and immobilization fall behind fixation of atmospheric nitrogen. The downward movement of dissolved and fine-particulate organic matter and nitrogen in soil solution is important in bringing these materials into contact with mineral soil particles and is one of the key processes by which these materials are stored in the soil profile (Qualls and Haines 1992).

Many initial reactions probably occur during solution transport between compounds such as reactive polyphenols and amino acids. More

Table 1. Effects of alder on soil, based on a review of studies that measured a wide range of soil changes. Note a 50-year basis was used for all variables and sites. This represents an extrapolation for some studies.

| Study basis | Location | Soil description | Mineral soil depth (cm) | Mineral soil + forest floor | | Change in mineral soil | | | Reference |
|---|---|---|---|---|---|---|---|---|---|
| | | | | Total N (kg/ha/50yrs) | Soil organic matter (mg/ha/50 yrs) | pH (unit/50 yrs) | Bulk density (g/cc/50yrs) | Cation exchange capacity (meq/1/50 yrs) | |
| *Pure alder stands* | | | | | | | | | |
| Red alder chrono-sequence (0-45 yrs) | Capitol Forest, WA | Old, weathered basalt PM | 0-20 | 2540 | 41.0 | -0.65 | -0.23 | — | Bormann and DeBell 1981 |
| Sitka alder chrono-sequence (0-17 yrs) | Glacier Bay, AK | Very young glacial till | 0-25 | 911 | 42.0 | -0.68 | -0.21 | 3.50 | Ugolini 1968 |
| Thinleaf alder chrono-sequence (0-15 yrs) | Tanana River, AK | Very young alluvium | 0-70 | 7561 | 42.0 | -1.00 | -0.91 | 15.79 | Van Cleve et al. 1971, Van Cleve and Viereck 1972 |
| Pure red alder 1984 vs. adjacent pure Douglas-fir (50 yrs old) | Cedar River, WA | Wisconsin-era glacial till | 0-15 | 1610 | — | -0.18 | -0.15 | -0.75 | Van Miegroet and Cole |
| Pure red alder 1988 vs. adjacent pure Douglas-fir | Cedar River, WA | Wisconsin-era glacial till | 0-15 | 1837 | 62.0 | -80.0 | 0.03 | 21.37 | Van Miegroet and Cole |

| Study basis | Location | Soil description | Mineral soil depth (cm) | Total N (kg/ha/50yrs) | Soil organic matter (mg/ha/50 yrs) | pH (unit/50 yrs) | Bulk density (g/cc/50yrs) | Cation exchange capacity (meq/1/50 yrs) | Reference |
|---|---|---|---|---|---|---|---|---|---|
| | | | | | | | Change in mineral soil | | |
| | | | Mineral soil + forest floor | | | | | | |
| *Mixed alder stands* | | | | | | | | | |
| Red alder/ Douglas-fir vs. adjacent Douglas-fir (30 yrs old) | Wind River, WA | Colluvium from basalt | 0-30.5 | 724 | 50.5 | — | -0.08 | — | Tarrant and Miller 1963 |
| Red alder/ Douglas-fir vs. adjacent Douglas-fir (30 yrs old) | Wind River, WA | Colluvium from basalt | 0-40 | — | — | -0.23 | -0.03 | — | Binkley and Sollins 1990 |
| Red alder/ Douglas-fir vs. adjacent Douglas-fir (40 yrs old) | Cascade Head, OR | Old residuum from sandstone | 0-30 | 2213 | 26.0 | -0.45 | 0.01 | — | Franklin et al. 1968, Binkley et al. (in press) |
| Red alder/ Douglas-fir vs. adjacent Douglas-fir (23 yrs old) | Skykomish, WA | Glacial lake sediments | 0-20 | 540 | 53.0 | -0.56 | -0.05 | — | Binkley 1983 |
| Red Alder/ Douglas-fir vs. adjacent Douglas-fir (23 yrs old) | Mt. Benson, Vancouver Is., B.C., Canada | Wisconsin-era glacial till | 0-20 | 2475 | 30.0 | -0.17 | 0.02 | — | Binkley 1983 |

complex materials form during the soil mineral adsorption phase. Although some of these materials are decomposed during solution transport, Qualls and Haines (1992) hypothesized that most carbon mineralization and nitrogen release occurs at mineral surfaces by bacteria and fungi associated with saprophytic or mycorrhizal activities. Losses of nitrogen through nitrification, denitrification, or solution transport occur where the physical and chemical reactive capacity of the soil organo-mineral storage mechanisms is exceeded by nitrogen input rates or where nitrogen mineralization is exceeded by uptake capacity of vegetation.

Cation exchange capacity generally increases with accumulation of organic matter (Russell 1973). This occurred on three of four alder sites (Table 1). Long-term mineralization of organic matter and utilization of accreted soil organic nitrogen would re-expose mineral soil surfaces (Qualls and Haines 1992). Decreased reserves of organic matter in soil could decrease cation exchange capacity and pH. The hydrogen ion ($H^+$) effects of nitrification could be minimized through sustained uptake of nitrogen by trees that do not fix atmospheric nitrogen and by increased production of woody tissues with a high ratio of carbon to nitrogen (Sollins et al. 1980). Thus, cycles of carbon and nitrogen accretion under alder could alternate with other tree species or occur in mixed stands with other vegetation, which is made more productive due to the increased nitrogen resources and improved soil structure contributed by alder. Relative to pure alder stands, mixed alder stands generally accumulate soil organic matter more rapidly than they accumulate nitrogen and are likely to have higher nitrogen uptake by roots. These factors can explain the less rapid decline of pH in mixed alder stands relative to pure alder stands (Table 1).

## Soil Acidification and Cation Leaching

Nitrification has been proposed as the principal mechanism to increase acidity and leach cations under alder (Van Miegroet and Cole 1984, 1985, 1988). Nitrification reactions produce protons that can displace exchangeable bases. Nitrate can balance cation charges and can then be leached below the rooting zone if root uptake is insufficient; it is also relatively easily displaced from anion exchange sites.

Acidification was observed in all studies we reviewed (Table 1). The reduction in pH was greatest in basic primary succession soils and least in older soils that were initially acidic. Acidification is a complicated process, and nitrification is only one of many possible causes. Plant uptake of nitrate consumes $H^+$ ions produced by nitrification (Van Breeman et al. 1983). Many other reactions also produce and consume $H^+$ ions, including incomplete organic-matter decomposition, uptake of cations, metal complexation, and sulfate oxidation. Hydrogen ions are consumed by weathering, uptake of anions, and denitrification. Change in pH is the net effect of all these reactions. Finding causal factors for reductions in pH among these many interacting reactions will require study of proton mass balance in the ecosystem (Van Breeman et al. 1983).

Evidence exists for increased cation leaching under alder at the Cedar River site in Washington (Van Miegroet and Cole 1988), presumably the net result of acidification processes. This process needs to be evaluated over a wide range of soils before any generalities can be made.

It is difficult to assess whether increased acidity is a symptom of positive or negative impacts on the long-term ecosystem productivity. Certainly, leaching of cations below the rooting zone in excess of weathering inputs could eventually degrade ecosystem productivity for that site, but perhaps increased acidity is simply a result of increased organic matter or cation uptake and redistribution. Also, we should not downplay the role of both natural and human-caused episodic disturbances. For example, those disturbances that mix upper and lower horizons are likely to reverse acidification.

It is also important to compare soil acidification under alder with that under other species. Soil development under spruce-hemlock forests in coastal southeast Alaska can reduce pH of the upper mineral soil more rapidly than can alder, from 5.0 to 3.5 in only 50 years (Bormann et al., in prep.).

## Soil Weathering

The effect of alder on primary mineral weathering has not been sufficiently studied. Several lines of evidence suggest, however, that rates of weathering may be higher in alder soils than in those of other conifers. Data from Van Cleve and Viereck (1972) indicate that up to 3000 kg ha$^{-1}$ of additional phosphorus were made available under alder in 20 years with as much as 1500 kg ha$^{-1}$ in 5 years. Most of this phosphorus would have been present in mineral form, and subsequently weathered, taken up, and redeposited as organic phosphorus as measured by perchloric acid digest (Van Cleve and Viereck 1972). Significant increases in cation leachates and marginal or insignificant exchangeable base depletion suggest increased mineral weathering rates in alder stands (Cole et al. 1989). When cation exchange capacity is increased, as occurred in most alder study sites (Table 1), and base saturation remains nearly the same, an increase in total exchangeable bases occurs. This with increased leaching and plant uptake suggests rapid primary mineral weathering.

Production of H$^+$ ions and organic acids may be greater under alder relative to conifers or other hardwoods, a possible mechanism for increased primary mineral weathering. This would be an interesting area to research in the future. Decreased pH has demonstrable effects on a wide range of biotic and abiotic soil processes, including biologically mediated nutrient cycles (Paul and Clark 1989) and primary mineral weathering (Boyle et al. 1974). Boyle et al. (1974) found that the type of acid was more important than pH in affecting primary mineral weathering rates. Compared with a hydrochloric acid control, various organic acids exhibited different effectiveness in weathering biotite. To evaluate the importance of weathering rates under alder, specific acids associated with alder need to be identified and their production rates established. This could help determine the importance of soil weathering under alder relative to comparable-aged conifer stands on similar parent material.

If alder increases primary mineral weathering and if these inputs outweigh leaching losses, alder thus contributes to increased cation supply and presumably long-term ecosystem productivity. If leaching losses outweigh weathering inputs, then that productivity will eventually be reduced. The age and chemical composition of the soil will likely determine the relative ranking of these processes. In old, deep, highly weathered residuum, fewer primary minerals are available for weathering reactions and many exchange sites are available. Thus, displacement will likely consume available protons. To the extent that nitrate is not taken up by plant roots, leaching losses are likely to exceed weathering inputs. In younger soils—especially those with easily weatherable particles such as finely ground glacial silts and unweathered alluvium—weathering is likely to greatly exceed leaching. The relative importance of leaching and weathering may influence the growth of alder. Highest alder productivity is generally found on young alluvial soils and the lowest productivity on older soil surfaces (Harrington and Courtin, Chapter 10).

## Implications for Long-term Ecosystem Productivity

Ecosystem productivity is broader than the traditional concern with the growth of trees, principally crop conifers. Our ecosystem concept of productivity focuses on net primary productivity as influenced by activity of all organisms (Gordon et al., in prep.). Frequent references to alder are contained in this chapter, but it should be noted that other species play important roles in alder-dominated ecosystems. Certainly *Frankia* is essential (Binkley et al., Chapter 4). We also recognize that alder mycorrhizae are distinct from those found under many conifers (Molina 1979; Molina et al., Chapter 2). It is also apparent that understory plants, soil animals, and free-living microbes form a distinctly different community than do conifer-dominated ecosystems. Effects of alder on long-term ecosystem productivity should consider the role of these other organisms and the role of associative rhizosphere fixation of atmospheric nitrogen by conifers (Bormann et al. 1993).

A picture emerges of how alder and its associates interact with their environment to bring about changes in soils and subsequent changes in the productive potential of the land. On young, nitrogen-poor soils with abundant weatherable minerals, a positive feedback is likely. Initially, fixation of atmospheric nitrogen increases tree growth which in turn provides the energy to extend roots and to produce organic acids capable of accelerated mineral weathering. Weathering and redistribution make cations and phosphorus more available, which in turn stimulates more fixation of atmospheric nitrogen, growth, and so on. Alders appear especially suited to primary succession soils and soils damaged by disturbance, erosion, or repeated hot fires.

On nitrogen-rich sites with deep, highly weathered substrates, a negative feedback may develop to reduce growth of pure alder stands and the potential productivity of subsequent ecosystems. Further additions of organic matter and nitrogen lead to the production of $H^+$ ions that are not countered by plant uptake or weathering. Production of nitrates leaches released cations deep into the profile. Repeated pure alder rotations on such soils are risky. Mixed culture to reduce nitrate buildup or soil scarification may be needed to maintain long-term ecosystem productivity when growing alder.

The leaching of nitrogen and cations, if higher under alder than under conifers, may be important to the productivity of riparian and aquatic ecosystems and even near-shore oceanic productivity. The ubiquity of various alders in riparian zones from California to Alaska and inland to northern Idaho suggests that alder may enhance the productivity of aquatic systems limited by nitrogen and cations. Here again, the type of system may determine whether increased productivity has a positive or negative effect on any given component or process.

Many alder stands are on soils that lie between the extremes of soil age and nitrogen content. It is difficult to speculate on the impact of alder on the long-term ecosystem productivity of such sites because of the lack of research on the interactions of these mechanisms. Certainly alder culture should be considered on nitrogen-deficient sites, and if nitrogen or cation leaching is perceived as a potential problem on a given site, then a conservative approach would be to grow alder in mixed culture or crop rotations to minimize the risks.

## Literature Cited

Atkinson, W. A., B. T. Bormann, and D. S. DeBell. 1979. Crop rotation of Douglas-fir and red alder: a preliminary biological and economic assessment. Bot. Gaz. (Suppl.) 140:102-109.

Binkley, D. 1983. Ecosystem production in Douglas-fir plantations: interaction of red alder and site fertility. For. Ecol. Manage. 5:215-227.

Binkley, D., and P. Sollins. 1990. Factors determining differences in soil pH in adjacent conifer and alder-conifer stands. Soil Sci. Soc. Am. J. 54:1427-1433.

Binkley, D., P. Sollins, R. Bell, D. Sachs, and D. Myrold. 1993. Biogeochemistry of adjacent conifer and alder/conifer stands. Ecology 73:2022-2033.

Bormann, B. T., F. H. Borman, W. B. Bowden, D. Wang, M. C. Snyder, C. Y. Li, and R. C. Ingersoll. 1993. Rapid $N_2$ fixation in pines, alder and locust: evidence from the sandbox ecosystem study. Ecology 74:583-598.

Bormann, B. T., and D. S. DeBell. 1981. Nitrogen content and other soil properties related to age of red alder stands. Soil Sci. Soc. Am. J. 45:428-432.

Bormann, B. T., and R. C. Sidle. 1990. Changes in productivity and distribution of nutrients in a chronosequence at Glacier Bay National Park, Alaska. J. Ecol. 78:561-578.

Bormann et al. Rapid changes in soils during secondary succession: implications for long-term ecosystem productivity. (In prep.)

Boyle, J. R., G. K. Voigt., and B. L. Sawhney. 1974. Chemical weathering of biotite by organic acids. Soil Sci. 117(1):42-45.

Brown, A. H. F., and A. F. Harrison. 1982. Effects of tree mixtures on earthworm populations and nitrogen and phosphorus status in Norway spruce (*Picea abies*) stands. *In* New trends in soil biology. *Edited by* P. Lebrun, H. M. Andre, A. De Medts, C. Gregoire-Wibo, and G. Wauthy. Proceedings of the VII International Colloquium of Soil Zoology, Dieu-Brichart, Belgium. 101-108.

Cole, D. W., S. P. Gessel, and J. Turner. 1978. Comparative mineral cycling in red alder and Douglas-fir. *In* Utilization and management of alder. *Compiled by* D. G. Briggs, D. S. DeBell, and W. A. Atkinson. USDA For. Serv., Gen. Tech. Rep. PNW-70. 327-336.

Cole, D. W., J. Compton., H. V. Miegroet, and P. Homann. 1989. Changes in soil properties and site productivity caused by red alder. Paper presented at IUFRO Symposium on Management of Nutrition in Forests Under Stress, 18-21 September, Freiberg, Germany.

Crocker, R. L., and J. Major. 1955. Soil development in relation to vegetation and surface age at Glacier Bay, Alaska. J. Ecol. 43:427-448.

Crocker, R. L., and B. A. Dickson. 1957. Soil development on the recessional moraines of the Herbert and Mendehall glaciers south-eastern Alaska. J. Ecol. 45:169-185.

Cromack, K., C. C. Delwiche, and D. H. McNabb. 1979. Prospects and problems of nitrogen management using symbiotic nitrogen fixers. *In* Symbiotic nitrogen fixation in the management of temperate forests. *Edited by* J. C. Gordon, C. T. Wheeler, and D. A. Perry. Oregon State Univ. Press, Corvallis. 210-223.

Darwin, D. 1881. The formation of vegetable mould through the action of worms. Bookworm Publishing Co., Ontario, CA.

Davis, M. D. 1973. Pollen evidence of changing land use around the shores of Lake Washington. Northwest Sci. 47(3):133-148.

Dixon, R. O. D., and C. T. Wheeler. 1983. Biochemical, physiological and environmental aspects of symbiotic nitrogen fixation. *In* Biological nitrogen fixation in forest ecosystems: foundations and applications. *Edited by* J. C. Gordon and C. T. Wheeler. Martinus Nijhoff/D. W. Junk, The Hague. 107-171.

Edmonds, R. L. 1980. Litter decomposition and nutrient release in Douglas-fir, red alder, western hemlock, and Pacific silver fir ecosystems in western Washington. Can. J. For. Res. 10:327-337.

Franklin, J. F., C. T. Dyrness, D. G. Moore, and R. F. Tarrant. 1968. Chemical soil properties under coastal Oregon stands of red alder and conifers. *In* Biology of alder. *Edited by* J. M. Trappe, J. F. Franklin, R. F. Tarrant, and G. M. Hansen. USDA For. Serv., PNW For. Range Exp. Sta., Portland, OR. 157-172.

Gessel, S. P., Cole, D. W., and E. C. Steinbrenner. 1973. Nitrogen balances in forest ecosystems of the Pacific Northwest. Soil Biol. Biochem. 5:19-34.

Gessel, S. P., and J. Turner. 1974. Litter production by red alder in western Washington. For. Sci. 20:325-330.

Gordon, J. C., B. T. Bormann, L. Jacobs, R. H. Waring, S. N. Little, and L. Bednar. Ecosystem sustainability: can it be measured? (In prep.)

Graham, R. C., and H. B. Wood. 1991. Morphologic development and clay redistribution in lysimeter soils under chaparral and pine. Soil Sci. Soc. Am. J. 55:1638-1646.

Heusser, C. J. 1960. Late-Pleistocene environments of North Pacific America. American Geographic Society Spec. Publ. 35.

Horhammer, L., and A. Scherm. 1955. Über das vorkommen Zyklischer Pflanzensauren bei einigen polygonaceen und betulaceen. Arkiv der Pharmazie Band 288(60):441-447.

Karchesy, J. J. 1974. Polyphenols of red alder: chemistry of the staining phenomenon. Ph.D. thesis. Oregon State Univ., Corvallis.

Kurth, E. F., and E. L. Becker. 1953. The chemical nature of extractives from red alder. Tappi 36:461-466.

Lawrence, D. B. 1958. Glaciers and vegetation in southeast Alaska. American Scientist 46:89-122.

Miller, R. E., and M. D. Murray. 1978. The effects of red alder on the growth of Douglas-fir. *In* Utilization and management of alder. *Compiled by* D. G. Briggs, D. S. DeBell, and W. A. Atkinson. USDA For. Serv., Gen. Tech. Rep. PNW-70. 283-306.

Molina, R. 1979. Pure culture synthesis and host specificity of red alder mycorrhizae. Can. J. Bot. 57(11):1223-1228.

Neal, J. L., W. B. Bollen, and K. C. Lu. 1965. Influence of particle size on decomposition of red alder and Douglas-fir sawdust in soil. Nature 205:991-993.

Overbay, J. C. 1992. Ecosystem management. Paper read at National Workshop on Taking an Ecological Approach to Management, 27 April 1992. USDA For. Serv., Salt Lake City, UT.

Paul, E. A., and F. E. Clark. 1989. Soil microbiology and biochemistry. Academic Press. San Diego, California.

Peacock, J. C. 1900. Some of the unpublished results of the investigation of tannins by the late professor Henry Trimble. Am. J. Phar. 72:429-433.

Powers, R. 1991. Are we maintaining the productivity of forest lands? Establishing guidelines through a network of long-term studies. *In* Proc. management and productivity of western-montane forest soils. *Edited by* A. E. Harvey and L. P. Neuenschwander. USDA For. Serv., Gen. Tech. Rep. INT-280. 70-81.

Qualls, R. G., and B. L. Haines. 1992. Biodegradability of dissolved organic matter in forest throughfall, soil solution, and stream water. Soil Sci. Soc. Am. J. 56:578-586.

Robertson, F. D. 1992. Ecosystem management of the national forests and grasslands. Memo to regional foresters and station directors, 4 June 1992. USDA For. Serv., Washington, D.C.

Russell, E. W. 1973. Soil conditions and plant growth. Longman, New York.

Sollins, P., C. C. Grier, F. M. McCorison, K. Cromack, Jr., R. Fogel, and R. L. Fredriksen. 1980. The internal element cycles of an old-growth Douglas-fir stand in Oregon. Ecol. Monogr. 50:261-285.

Sollins, P., G. Spycher, and C. A. Glassman. 1984. Net nitrogen mineralization from light- and heavy-fraction forest soil organic matter. Soil Biol. and Biochem. 16:31-37.

Sprent, J. 1988. The ecology of the nitrogen cycle. Cambridge Univ. Press, Cambridge, England.

Stevenson, F. J. 1982. Humus chemistry. Wiley, New York.

——. 1986. Cycles of soil carbon, nitrogen, phosphorous, sulfur, micronutrients. Wiley, New York.

Tarrant, R. F., B. T. Bormann, D. S. DeBell, and W. A. Atkinson. 1983. Managing red alder in the Douglas-fir region: some possibilities. J. For. 81(12):787-792.

Tarrant, R. F., and R. F. Chandler. 1951. Observations on litter fall and foliage nutrient content of some pacific northwest tree species. J. For. 49:914-915.

Tarrant, R. F., and R. E. Miller. 1963. Accumulation of organic matter and soil nitrogen beneath a plantation of red alder and Douglas-fir. Soil Sci. Soc. Am. Proc. 27:231-234.

Ugolini, F. C. 1968. Soil development and alder invasion in a recently deglaciated area of Glacier Bay, Alaska. *In* Utilization and management of alder. *Edited by* J. M. Trappe, J. F. Franklin, R. F. Tarrant, and G. M. Hansen. USDA For. Serv., PNW For. Range Exp. Sta., Portland, OR. 115-140.

Uvarova, N. I., G. I. Oshito, A. K. Dzizenko, and G. B. Elyakov. 1970. 1, 7-Diphenyl-3, 5-heptanediol from *Alnus fructicosa* and *Alnus manshurica*. Khim. Prir. Soedin. 6:463-464.

Van Breeman, N., J. Mulder, and C. T. Driscoll. 1983. Acidification and alkalinization of soils. Plant and Soil 75:283-308.

Van Cleve, K., and L. A. Viereck. 1972. Distribution of selected chemical elements in even-aged alder (*Alnus*) ecosystems near Fairbanks, Alaska. Arctic and Alpine Res. 4(3):239-255.

Van Cleve, K., L. A. Viereck, and R. L. Schlentner. 1971. Accumulations of nitrogen in alder (*Alnus*) ecosystems near Fairbanks, Alaska. Artic and Alpine Res. 3(2):101-114.

Van Miegroet, H., and D. W. Cole. 1984. The impact of nitrification and cation leaching in a red alder ecosystem. J. Environ. Qual. 13(4):586-590.

———. 1985. Acidification sources in red alder and douglas-fir soils—importance of nitrification. Soil Sci. Soc. Am. J. 49:1274-1279.

———. 1988. Influence of nitrogen-fixing alder on acidification and cation leaching in a forest soil. *In* Forest site evaluation and long-term productivity. *Edited by* D. W. Cole and S. P. Gessel. Univ. of Washington Press, Seattle.

Voigt, G. K., and G. L. Steucek. 1969. Nitrogen distribution and accretion in an alder ecosystem. Soil Sci. Soc. Am. Proc. 33:946-949.

Waring, R. H., and W. H. Schlesinger. 1985. Forest ecosystems: concepts and management. Academic Press, San Diego, CA.

Wild, A. 1988. Russell's soil condition and plant growth. Longman Scientific & Technical. Essex, England.

Zavitkovski, J., and M. Newton. 1971. Litterfall and litter accumulation in red alder stands in western Oregon. Plant and Soil 35:257-268.

# 4

# Nitrogen Fixation by Red Alder: Biology, Rates, and Controls

Dan Binkley, Kermit Cromack, Jr., & Dwight D. Baker

The ability of alder (*Alnus* spp.) to enrich soils has been recognized for centuries (Tarrant and Trappe 1971); growth of alders in nitrogen-free media was accomplished in the late nineteenth century (Hiltner 1896), and $^{15}$N techniques first demonstrated the ability of alders to fix atmospheric nitrogen (N$_2$) in the 1950s (Bond 1955). Since then a great deal of information has accumulated on the rates of nitrogen fixation by alders, and on the biology of alder and its nitrogen-fixing symbiont, *Frankia*. Although the factors that control nitrogen fixation by alder are well characterized in a general way, a great deal remains unknown about factors that control nitrogen-fixation rates in alder ecosystems. In this paper, we summarize the biology of nitrogen fixation by red alder (*Alnus rubra* Bong.), compile all available estimates of nitrogen-fixation rates for red alder stands, and discuss the factors that control these rates.

## Biology of Nitrogen Fixation by Red Alder

Nitrogen fixation by *Alnus rubra* is accomplished by symbiotic actinomycetes in the genus *Frankia*, which live in root nodules found on almost all naturally occurring trees (see also Molina et al., Chapter 2). *Frankia* is a filamentous bacterium that in some ways resembles a fungus but is a true prokaryote and lacks the well-developed organelles found in eukaryotic fungi. It is also phylogenetically and morphologically distinct from the rhizobia bacteria that are responsible for nitrogen fixation in legumes, such as locusts (*Robinia* spp.), acacias (*Acacia* spp.) and albizias (*Albizia* spp.).

Most nurseries and field sites have sufficient natural populations of infective (capable of forming nodules) and effective (capable of nitrogen

fixation) strains of *Frankia*. These strains have been available from pure cultures since 1978 (Callaham et al. 1978), and laboratory production and large-scale distribution of the inoculum in a nursery is relatively simple (Perinet et al. 1985; Stowers and Smith 1985). Inoculation is usually not necessary unless nursery soils have been fumigated to kill pathogens.

The symbiotic association between red alder and *Frankia* is relatively specific. *Frankia* strains that nodulate red alder will also nodulate other alder species, such as Sitka alder (*Alnus sinuata* {Regel} Rydb.) and black alder (*Alnus glutinosa* {L.} Gaertn.), but not plants from most other actinorhizal genera native to the Pacific Northwest, such as *Ceanothus*, *Purshia*, and *Shepherdia*. These strains belong to Host Specificity Group 1 (Baker 1987) and nodulate sweet fern (*Comptonia peregrina*) and many myrtle species (*Myrica* spp.). Evidence from studies conducted in Scandinavia on gray alder (*Alnus incana* {L.} Moench) indicate that alder-infecting *Frankia* can remain in soils at stable population levels even in the absence of the host plant (Smolander and Sundman 1987).

Nodulation of *Alnus* begins when *Frankia* penetrates root hairs and enters the cortical cells of young roots (Berry and Torrey 1983). The development of an incipient root nodule meristem occurs in the same way that a lateral root meristem develops. *Frankia* invades this new structure and takes up residence in enlarged cortical cells (Berry and Sunnell 1990), where it produces specialized cells (vesicles) that are the site of the nitrogen-fixing enzyme (nitrogenase) and nitrogen fixation. Alder root nodules are perennial and continue to grow in size as individual nodule lobes divide and produce a coral-like morphology.

Nitrogen fixation occurs as high-energy electrons (very low reduction potential) are added to atmospheric nitrogen, reducing it to the higher energy state of ammonia ($NH_3$):

$$N_2 + 8H^+ + 8e^- \longrightarrow 2NH_3 + H_2$$

The $H_2$ produced in this reaction may be oxidized (via the hydrogenase enzyme) to release energy (Dixon and Wheeler 1983). It is difficult to estimate the real energy cost of this reaction under normal environmental conditions, but reported rates for red alder are on the order of 5 g of carbon (12.5 g of carbohydrate) per g of nitrogen fixed (Tjepkema and Winship 1980; Schubert 1982). Once nitrogen has been fixed in the form of ammonia, it is added to an amino acid (such as glutamine, forming glutamic acid, which is then converted to citrulline [Akkermans et al. 1979]) and exported for use in roots, stems, and leaves of the alder. We know of no evidence of direct excretion of fixed nitrogen from root nodules into the soil.

## Rates of Nitrogen Fixation by Red Alder

### Methods of Assessing Rates of Nitrogen Fixation

Three approaches have been used to estimate rates of nitrogen fixation by red alder (see Silvester 1983 for a good summary of all available methods). The simplest method involves contrasting the total nitrogen content of an ecosystem at two points in time; any increase (or accretion) of nitrogen during the period is used as an estimate of nitrogen fixation. In some studies, only changes in soil nitrogen pools have been followed, and the soil accretion estimate would underrepresent the accretion rate for the total ecosystem. This approach also assumes that all nitrogen that is fixed remains within the ecosystem and results in underestimates of true nitrogen-fixation rates on sites where nitrogen leaching from the soil is high (e.g., Van Miegroet and Cole 1984, 1988; Binkley et al. 1993b; Bormann et al., Chapter 3).

A variation on the accretion method uses a time sequence of plots at different locations to approximate the accretion that would be observed in a single site through time. This approach includes the same assumption of no losses of fixed nitrogen from the ecosystem, as well as an assumption that the total nitrogen content of each site would be similar for stands of the same age. See Cole and Van Miegroet (1989) for a discussion of the chronosequence approach.

The third method uses incubations of alder nodules (typically severed from the tree with several centimeters of roots attached) in an atmosphere containing acetylene to estimate the current rate of nitrogen fixation. The nitrogenase enzyme has a high preference for acetylene over atmospheric nitrogen, and the product of acetylene reduction (ethylene) is easy to measure. The rate of acetylene reduction is converted into an estimate of nitrogen fixation by division by a factor that accounts for the greater electron requirement for nitrogen fixation per mol of atmospheric nitrogen, and this nitrogen-fixation rate may be extrapolated to an annual hectare basis by multiplying the total nodule biomass per hectare by the length of time that nodules are active. Estimates of nitrogen fixation from acetylene reduction are imprecise because of substantial variation (uncertainty) in the following: nodule activity within and among plants, nodule activity through a season, nodule biomass per hectare, the appropriate conversion factor from acetylene to nitrogen, and any errors that might arise from the effects of acetylene and ethylene on the physiology of the symbiosis (Minchin et al. 1983, 1986).

Three additional approaches for estimating nitrogen-fixation rates use the heavy, stable isotope $^{15}N$ for direct tracing of alder-fixed nitrogen. For short-term experiments, $^{15}N_2$ can be exposed to alder nodules during short incubations, and the nodules then examined for incorporation of $^{15}N$ (Silvester 1983). We know of no study that has used this method for red alder; such a study would be particularly useful for establishing the appropriate conversion ratio for acetylene to atmospheric nitrogen. For longer-term experiments, the nitrogen contained in the soil can be enriched with $^{15}N$; after one or more years, lower ratios of $^{15}N:^{14}N$ in alder plants relative to non-nitrogen-fixers can be used to calculate the proportion of the alder's nitrogen derived from the atmosphere. This method has not been used for red alder, but a recent study with Sitka alder in British Columbia found that over 95 percent of the alder's nitrogen came from the atmosphere (Mead and Preston 1992). In some situations, the natural ratio of $^{15}N$:$^{14}N$ in the soil may differ enough from that of the atmosphere to allow a calculation of the rate of nitrogen fixation by comparing this ratio in alder forests with those of adjacent conifer forests. This approach has been used in one study (with three locations) for red alder, but natural variations in the ratios of the nitrogen isotopes were too great to allow determination of nitrogen-fixation rate (Binkley et al. 1985).

The contribution of nitrogen fixation to a plant or ecosystem can be defined in several ways. The percent of a plant's nitrogen supply that is derived from nitrogen fixation is commonly referred to as "pNdfa" (percent of nitrogen derived from the atmosphere). The pNdfa indicates the importance of nitrogen fixation as a source of nitrogen for a plant and can be estimated directly from stable isotope methods (either natural abundance or $^{15}N$-enrichment methods). The absolute quantity of nitrogen fixed, "tNdfa" (total nitrogen derived from the atmosphere), refers to the magnitude of nitrogen fixation without reference to the overall nitrogen supply of the ecosystem. The tNdfa can be estimated from accretion studies, acetylene reduction assays, or calculated from isotope studies. In this review, we focus on the quantity of nitrogen fixed by red alder in terms of tNdfa rather than as a proportion of the alder's overall supply of nitrogen (pNdfa).

## Rates of Nitrogen Fixation for Pure Alder Forests

We found 15 estimates of nitrogen-fixation rates in the literature for forests dominated by red alder, with rates ranging from a low of 0 kg ha$^{-1}$ yr$^{-1}$ for a 0- to 4-year-old stand (Cole and Newton 1986) to a high of 320 kg ha$^{-1}$ yr$^{-1}$ (Newton et al. 1968). Most rates fell within the range of 100 to 200 kg ha$^{-1}$ yr$^{-1}$.

**Coast Range, Oregon.** Newton et al. (1968) examined an age sequence of 36 stands of naturally regenerated red alder. Sites were selected to insure comparable initial soil conditions, including the following: restricting all sites to the Tyee sandstone formation, and using only sites with no apparent organic matter on the soil surface after logging.

The authors were most confident that initial conditions were similar among stands less than 15 years old; for older stands, they could not be certain about the extent of organic debris present after logging of the previous forest. In each stand, two soil samples were taken at each depth interval: 0 to 0.1 m, 0.1 to 0.3 m, and 0.3 to 0.6 m. The nitrogen content of the biomass was calculated by using nitrogen concentrations for representative samples and some undescribed method of estimating biomass. A regression analysis of the total nitrogen content of the soil plus biomass nitrogen yielded an average rate of nitrogen accretion of 320 kg ha$^{-1}$ yr$^{-1}$ for 20 years ($r^2$ = 0.88 for nitrogen content with age, n = 18), with the majority of the accretion occurring in the soil. The authors concluded that they found no evidence of further accretion beyond age 30, although they had only two sites beyond this age. The older sites may have been anomalous; initial soil conditions were unknown, and they contained notably less nitrogen (about 9 to 11 Mg/ha) at age 45 and 65 than the age sequence would suggest even at age 30 (13 Mg/ha). Leaching losses of nitrogen tend to be high for alder stands in the Coast Range (Bormann et al., Chapter 3), so the actual rate of nitrogen fixation would likely be in excess of the estimated rate of accretion.

This rate of nitrogen accretion of over 300 kg ha$^{-1}$ yr$^{-1}$ across the age sequence of stands is almost twice the rate found for other studies that have used within-site accretion or acetylene reduction. It is possible that the low number of soil samples per stand or differences in initial site conditions inflated the actual rate for these sites; however, these potential sources of error should increase the overall variance in the relationship between stand age and nitrogen content rather than provide a spuriously high correlation. The Coast Range has some of the most productive temperate forests in the world, with very productive stands of red alder that may indeed have high rates of nitrogen fixation. We feel that this unusually high accretion rate should probably be accepted until other, more intensive studies in the Coast Range provide additional information.

**Cascade Head, Oregon.** Franklin et al. (1968) contrasted total soil nitrogen content in three stands on a very fertile site: pure conifer [mostly Douglas-fir (*Pseudotsuga menziesii* {Mirb.} Franco) with some western hemlock (*Tsuga heterophylla* {Raf.} Sargent) and Sitka spruce (*Picea sitchensis* {Bong.} Carr.)], conifer/alder, and pure red alder. The plots were about 0.4 ha in size, with no replication; the conifer stand was in a midslope position and the alder stand in a lower slope position. The greater nitrogen content of the alder forest relative to the conifer forest gave an average annual soil accretion for 40 years of 140 kg ha$^{-1}$ yr$^{-1}$ in the mineral soil to a depth of 0.9 m (Table 1). Only five soil pits were sampled in each stand, with only one pit sampled for bulk density, so we expect this estimate of nitrogen accretion is not precise. Total nitrogen content of the soil pits ranged from 17,700 to 19,500 kg/ha in the alder soil and from 9200 to 16,400 kg/ha in the conifer forest. To assess the likely precision of the average accretion rate of 140 kg ha$^{-1}$ yr$^{-1}$, we calculated the accretion rate represented by subtracting the lower value in the range of nitrogen contents given for the conifer soil from the upper value of the alder soil, and the upper conifer soil value from the lower alder soil value. This gave a range of 30 to 260 kg ha$^{-1}$ yr$^{-1}$. Based on more intensive studies in the alder/conifer stand (described below), we expect that leaching losses of nitrogen from the soil were likely on the order of 50 kg ha$^{-1}$ yr$^{-1}$, and that the actual rate of nitrogen fixation for this stand was likely between 100 and 200 kg ha$^{-1}$ yr$^{-1}$. This rate is substantially lower than that found in the age sequence by Newton et al. (1968), perhaps because the soils at Cascade Head are formed in a different parent material, or because the nitrogen content of the conifer stand was already in excess of 13 Mg/ha.

**Widow Creek, Oregon.** Franklin et al. (1968) also sampled adjacent 30-year-old stands of red alder and conifers (mostly Douglas-fir) at the Widow Creek site. They dug only one sample pit in each stand and estimated an annual accretion rate of soil nitrogen of 27 kg ha$^{-1}$ yr$^{-1}$ in the mineral soil to a depth of 0.9 m. Given the lack of replication of stands or subsamples within each stand, this estimate may be very different from the true rate, and we cannot speculate on why the rate might have been so low.

**Hoh River, Washington.** Luken and Fonda (1983) examined an age sequence of alder stands on terraces along the Hoh River in the Olympic National Park. They chose six stands representing three ages (14, 24 and 65 years) of river terraces. The nutrient content of biomass was estimated for a 15 x 25 m plot in each stand; only one soil pit was sampled in each stand. The original substrate of each terrace was probably not identical when the terraces were deposited; the clay content of a new terrace was about 0.2 percent, compared with 2 to 5 percent for the 14-year-old stand, 4 to 5 percent for the 24-year-old stand, and 8 to 9 percent for the 65-year-old stand. Total ecosystem nitrogen content appeared to increase from 780 kg/ha in the freshly deposited river terrace to 4660 kg/ha in the 65-year-old red alder stands, for an average annual rate of accretion of 60 kg ha$^{-1}$ yr$^{-1}$ for the ecosystems (Table 1). The rate was higher in younger stands. The rate of accretion of nitrogen in biomass was 34 kg ha$^{-1}$ yr$^{-1}$ for the first 14 years, 32 kg ha$^{-1}$ yr$^{-1}$ for the next 10 years, and 7 kg ha$^{-1}$ yr$^{-1}$ for the next 39 years. Accretion in the soil (to a depth of 0.45 m) appeared to total about 43 kg ha$^{-1}$ yr$^{-1}$, but it is not clear if this reflects the greater nitrogen content associated with the greater deposition of clay in the older stands or from nitrogen fixation (clay content was very highly [and linearly] correlated with nitrogen concentration of the soil across soil depths and stand ages [$r^2$ = 0.82]). Nitrogen losses through leaching were not examined.

**Thompson Research Site, Washington.** Cole et al. (1978) and Van Miegroet et al. (1989) reported on the accretion of nitrogen in a red alder stand relative to an adjacent stand of Douglas-fir. Both stands originated after logging of an old-growth forest; the alder regenerated naturally, whereas the Douglas-fir stand was planted. A 400-m$^2$ plot was established in each stand, and four or five soil pits were sampled to a depth of 0.45 m (Van Miegroet and Cole 1988). The rate of accretion in the ecosystem was estimated to be 85 kg ha$^{-1}$ yr$^{-1}$ for the first 38 years, and 78 kg ha$^{-1}$ yr$^{-1}$ for the first 50 years (Table 1). Nitrogen leaching losses from the red alder stand (past the 40-cm soil depth) were about 50 kg ha$^{-1}$ yr$^{-1}$ at about age 50 (Van Miegroet and Cole 1984), indicating that the nitrogen-fixation rate was likely on the order of 125 kg ha$^{-1}$ yr$^{-1}$ for 50 years.

*Table 1. In-field estimates of nitrogen-fixation rates by red alder in pure stands.*

| Location | Age, age span | Method | Rate | Reference |
|---|---|---|---|---|
| *Red alder stands* | | | | |
| Western Oregon | 0-30 | Accretion, plants, forest floor, soil to 0.6 m | 320 | Newton et al. 1968 |
| Cascade Head, OR | 0-40 | Accretion, mineral soil to 0.9 m | 140 | Franklin et al. 1968 |
| Widow Creek, OR | 0-30 | Accretion, mineral soil to 0.9 m | 27 | Franklin et al. 1968 |
| Hoh River, WA | 0-14 | Accretion, plants, forest floor, soil to 0.45 m | 164 | Luken and Fonda 1983 |
| | 15-24 | | 58 | |
| | 25-65 | | 25 | |
| Thompson Res. Site, WA | 0-38 | Accretion, plants, forest floor, soil to 0.6 m | 85 | Cole et al. 1978 |
| | 0-50 | | 78 | Van Miegroet et al. 1989 |
| Western Washington | 0-40 | Accretion, forest floor, soil to 0.2 m | 50 | Bormann and DeBell 1981 |
| Centralia, WA | 0-30 | Accretion, soil to 0.18 m | 59-65 | Heilman 1982 |
| Centralia, WA | 5 | Acetylene reduction | 150 | Heilman and Ekuan 1982 |
| Western Washington | 2-4 | Acetylene reduction | 62 | Tripp et al. 1979 |
| Lady Island, WA | 0-4 | Accretion, soil to 0.15 m | 80 | DeBell and Radwan 1979 |
| Rainier, OR | 5 | Acetylene reduction | 60-80 | Bormann and Gordon 1984 |
| *Red alder/ conifer* | | | | |
| Cascade Head, OR | 0-40 | Accretion, forest floor, soil to 0.9 m | 28 | Franklin et al. 1968 |
| | 0-55 | Accretion, plants, forest floor, soil to 0.9 m | 73 | Binkley et al. 1993 |
| | 55 | Acetylene reduction | 85 | Binkley et al. 1993 |
| Wind River, WA | 0-26 | Accretion, soil to 0.9 m | 40 | Tarrant and Miller 1963 |
| | 0-55 | Accretion, plants, forest floor, soil to 0.9 m | 54 | Binkley et al. 1993 |
| | 55 | Acetylene reduction | 75 | Binkley et al. 1993 |
| Mt. Benson, B.C. | 0-23 | Accretion, plants, forest floor, soil to 0.5 m | 65 | Binkley 1983 |
| | 23 | Acetylene reduction | 130 | Binkley 1981 |
| Skykomish, WA | 0-23 | Accretion, plants, forest floor, soil to 0.6 m | 42 | Binkley 1983 |
| Western Oregon | 0-17 | Accretion, soil to 0.15 m | 13-51 | Berg and Doerksen 1975 |
| Coast Range, OR | 0-4 | Accretion, soil to 0.15 m | 0 | Cole and Newton 1986 |
| Lennox Forest, Scotland | 0-16 | Accretion, forest floor, soil to 0.05 m | 36 | Malcolm et al. 1985 |
| *Alder/ poplar* | | | | |
| Toutle River, WA | 0-6 | Accretion, forest floor, soil to 0.3 m | 56 | Heilman 1990 |
| Lady Island, WA | 0-4 | Accretion, soil to 0.15 cm | 32 | DeBell and Radwan 1979 |

**Western Washington.** Bormann and DeBell (1981) examined nitrogen accretion in an age sequence of 13 red alder stands (and another sequence of Douglas-fir). Soil nitrogen content was very high, beginning at about 4000 kg/ha in the upper 20 cm of soil. The alder stands ranged in age up to 40 years, and the increase of nitrogen contained in the forest floor and in the 0- to 0.2-m depth mineral soil averaged 50 kg ha$^{-1}$ yr$^{-1}$ (Table 1). Total ecosystem nitrogen accretion and nitrogen fixation were not estimated, but this rate of accretion in the forest floor and upper mineral soil probably indicates a nitrogen-fixation rate of 100 to 150 kg ha$^{-1}$ yr$^{-1}$ over 40 years on this high-nitrogen site.

**Centralia, Washington.** Heilman (1982) examined nitrogen accretion under three stands of alders occupying coal-mine spoils, ranging in age from about 10 to 30 years. From 5 to 13 pits were dug with a backhoe in each site and sampled in 0.3 m increments to a depth of 1.2 m. An unspecified number of core samples were taken to determine bulk density. The rate of accretion in the soil did not differ among the three locations (and ages), with all stands showing accretion rates of about 60 to 65 kg ha$^{-1}$ yr$^{-1}$ in the upper 0.3 m (Table 1). Inclusion of the nitrogen content of biomass would probably raise the accretion rate to about 80 kg ha$^{-1}$ yr$^{-1}$, and any leaching losses of nitrogen that occurred would also raise the nitrogen-fixation estimate.

On a similar spoil site, Heilman and Ekuan (1982) estimated nitrogen-fixation rates in a 3- to 5-year-old alder stand using the acetylene reduction technique. Red alder seedlings were planted at a density of 6730/ha. The seasonal course of nitrogen fixation was based on rates measured at one or two sampling points in each of three years, and the seasonal trend found for a nearby site reported by Tripp et al. (1979). Nodule biomass was determined by excavating complete root systems in the first year, and by sampling two soil pits (0.2 m$^2$) beside each of about 20 trees. A conversion factor of 3 mol acetylene per mol N$_2$ was assumed. Heilman and Ekuan (1982) estimated the rate at age 5 to be about 150 kg ha$^{-1}$ yr$^{-1}$, with a standard error of the mean of about 50 kg ha$^{-1}$ yr$^{-1}$. This is substantially higher than the accretion estimate from nearby coal spoils (Heilman 1982), and if true would indi-

cate that substantial nitrogen leaching may occur on these sites.

A third study on two coal-mine spoil sites near Centralia, Washington, examined nitrogen-fixation rates of 2- to 4-year-old red alder seedlings using acetylene reduction (Tripp et al. 1979). Entire root systems of sample trees were excavated, shaken to remove the soil, and incubated with acetylene in large bags for 1 hour or for 24 hours (with gas sampling through a septum every 4 hours). Several trees were sampled at each of six sampling times. Nitrogen fixation commenced in early May and continued through late October. Daily peak rates of activity per g of nodule tended to be about five times the nighttime minimum rates. Using a conversion factor of 3 mol of acetylene per mol of N$_2$, the authors calculated a nitrogen-fixation rate of about 62 kg ha$^{-1}$ yr$^{-1}$ for a 4-year-old stand of red alder. This compares favorably with the accretion rates estimated for other mine spoils in the vicinity, but is less than half the rate estimated by acetylene reduction by Heilman and Ekuan (1982). The difference in these acetylene reduction estimates of nitrogen fixation derive primarily from the greater nodule biomass per tree (about 270 kg/ha in the study by Heilman and Ekuan [1982] compared with 40 kg/ha in the study by Tripp et al. [1979]). We expect the differences in these two estimates represent real site differences and are not artifacts of sampling.

**Lady Island, Washington.** DeBell and Radwan (1979) examined nitrogen accretion in 0 to 0.15 m mineral soil under plantations of red alder and red alder with black cottonwood (*Populus trichocarpa* Torr. & Gray) relative to pure stands of black cottonwood. Two replicate plots were 36 m$^2$, with densities of 13,900 trees/ha. At age 4, nine soil samples were taken in each plot for total nitrogen determination. Accretion in the pure alder stand averaged 80 kg ha$^{-1}$ yr$^{-1}$ in the upper soil; total ecosystem accretion and nitrogen-fixation rate were not estimated, but probably fell within the range of 100 to 150 kg ha$^{-1}$ yr$^{-1}$.

**Rainier, Oregon.** Bormann and Gordon (1984) examined nitrogen fixation across a range of tree densities in a 5-year-old experimental plantation of red alder. Two replicate stands of unstated size with densities of 13,400/ha, 4500/ha, and 1330/ha were assayed for acetylene reduction rates. Assays were performed monthly on 10 nodule samples in each plot taken from two trees. No diurnal trends in rates were examined, and a conversion factor of 3 mols of acetylene to 1 mol of $N_2$ was used to calculate nitrogen-fixation rate. Nodule biomass was estimated based on regression equations developed to relate nodule biomass to tree diameter at breast height and stand density ($R^2 = 0.93$). The rate of nitrogen fixation ranged from about 60 kg ha$^{-1}$ yr$^{-1}$ in the highest and lowest density stands to about 82 kg ha$^{-1}$ yr$^{-1}$ in the moderate density stand (Table 1).

### Rates of Nitrogen Fixation for Mixed Stands of Alder and Conifers

We found seven case studies that provided estimates of nitrogen fixation for stands of red alder mixed with conifers, and several of these studies used more than one approach to obtain the estimates.

**Cascade Head, Oregon.** As mentioned above, three stands were established near the Oregon Coast to examine stand productivity in a pure red alder stand of moderate density, a pure conifer stand of low density, and a mixed stand of high density (Berntsen 1961). Franklin et al. (1968) compared the soil nitrogen content (to 0.9-m depth) of the conifer stand (13,100 kg/ha) with that of the alder/conifer stand (14,200 kg/ha), and estimated that nitrogen accretion in the soil had occurred at a rate of 28 kg ha$^{-1}$ yr$^{-1}$ for 40 years. Only three soil pits were sampled in each stand, and bulk density was measured in just one pit.

Binkley et al. (1993a) resampled the conifer and alder/conifer stand more intensively in 1985, with 10 pits in each stand. They estimated that the soil in the conifer stand contained only 9800 kg/ha of nitrogen to a depth of 0.9 m, with another 720 kg/ha of nitrogen in biomass. The reason for the lower nitrogen content relative to the earlier sampling by Franklin et al. (1968) is not clear. We expect that three soil pits were too few to accurately estimate soil nitrogen content; in fact, the estimate by Binkley et al. (1993a) falls within the range of the values found for the three pits sampled by Franklin et al. (1968). The nitrogen content of the alder/conifer stand in 1985 was estimated to be 13,720 kg/

ha in the mineral soil (to 0.9 m) plus 820 kg/ha in biomass (Binkley et al. 1993a). The soil nitrogen content estimate gain was lower than that reported by Franklin et al. (1968), but well within the range they reported for their three pits. Subtracting the nitrogen content of the conifer stand from that of the alder/conifer stand gives an average annual accretion rate of 73 kg ha$^{-1}$ yr$^{-1}$ for 55 years. At age 55, the estimated rate of nitrogen leaching from the alder/conifer stand (50 kg ha$^{-1}$ yr$^{-1}$) was about 30 kg ha$^{-1}$ yr$^{-1}$ greater than that of the conifer stand (21 kg ha$^{-1}$ yr$^{-1}$), so the actual rate of nitrogen fixation was likely averaged between 100 to 125 kg ha$^{-1}$ yr$^{-1}$ for 55 years. Denitrification is another possible avenue of nitrogen loss, but it was of minor importance ($< 0.2$ kg ha$^{-1}$ yr$^{-1}$) at this site.

Nitrogen fixation was also estimated by the acetylene reduction method (Kim 1987; Binkley et al. 1993a), with a rate of 85 kg ha$^{-1}$ yr$^{-1}$ calculated with a stand nodule biomass of 250 kg/ha at age 55. The authors expected this acetylene reduction estimate was likely to be within $\pm$ 40 kg ha$^{-1}$ yr$^{-1}$ of the real rate. The acetylene reduction estimate of the current (age 55) nitrogen-fixation rate was somewhat less than the estimate based on accretion plus leaching, and may indicate that nitrogen fixation was lower in the stand at age 55 than at an earlier age. Indeed, mortality of alder in the mixed stand has been high (Binkley and Greene 1983; Binkley et al. 1993a).

**Wind River, Washington.** A 20-m wide strip of a 4-year-old Douglas-fir plantation was interplanted with 2-year-old red alder seedlings to serve as a 1-km long firebreak. Twenty-six years after planting of the alder, Tarrant and Miller (1963) sampled soils to a depth of 0.9 m. Soils were sampled within 12 plots in the Douglas-fir stand, and 12 paired plots in the adjacent alder/Douglas-fir stand. Six samples of forest floor, 0- to 0.07-m and 0.07- to 0.15-m depth soils were collected and composited within each plot, and deeper horizons were sampled from a single pit in each plot. This intensive sampling provided an nitrogen content estimate of 3170 kg N/ha in the Douglas-fir stand, and 4220 kg N/ha in the alder/Douglas-fir stand. The annual rate of accretion in the soil was 40 kg ha$^{-1}$ yr$^{-1}$ for 26 years.

Binkley et al. (1993a) resampled the stands in 1985 using 10 pits per stand. In the Douglas-fir stand, their estimate of the nitrogen content of the forest floor plus mineral soil to 0.9 m was 2070 kg/ha, with another 160 kg/ha in biomass. Comparable values for the alder/Douglas-fir stand were 4690 kg N/ha in soil, and 340 kg N/ha in the biomass. The average annual accretion rate for 52 years was 50 kg ha$^{-1}$ yr$^{-1}$ for the soil, and 54 kg ha$^{-1}$ yr$^{-1}$ for the total soil plus vegetation. The current rate of nitrogen leaching from the alder/Douglas-fir stand was about 25 kg ha$^{-1}$ yr$^{-1}$ (denitrification was negligible), which would increase estimate of nitrogen fixation to about 80 kg ha$^{-1}$ yr$^{-1}$.

The nitrogen content of the conifer stand in the 1985 sampling was substantially lower than that reported by Tarrant and Miller (1963); given the relatively intensive sampling in each study, we expect the difference derives from sampling different portions of the alder/Douglas-fir strip. The nitrogen content of the alder/Douglas-fir stand in the 1985 sampling was about 810 kg/ha greater than that from the 1959 sampling.

Nitrogen fixation estimated by acetylene reduction assays in 1985 was 75 kg ha$^{-1}$ yr$^{-1}$ for 325 kg/ha of nodule biomass (Binkley et al. 1993a), which matches the estimate derived from accretion plus leaching. The precision of this estimate was thought to be within ± 40 kg ha$^{-1}$ yr$^{-1}$.

**Mt. Benson, British Columbia.** Binkley (1981, 1983) estimated nitrogen fixation in a 23-year-old stand of red alder and Douglas-fir adjacent to pure Douglas-fir stand. Ten soil pits were sampled in each stand to a depth of 0.5 m, and biomass regression equations were used to estimate stand biomass. The total nitrogen content of the Douglas-fir stand was 1730 kg/ha, compared with 3220 in the alder/Douglas-fir stand, giving an annual accretion rate for the ecosystem of 65 kg ha$^{-1}$ yr$^{-1}$. Leaching losses were not measured; we expect they might have increased the nitrogen fixation estimate to about 75 to 95 kg ha$^{-1}$ yr$^{-1}$. The acetylene reduction estimate of current nitrogen fixation was 130 kg ha$^{-1}$ yr$^{-1}$ for 390 kg/ha of nodules, with a precision of probably ± 60 kg ha$^{-1}$ yr$^{-1}$.

**Skykomish River, Washington.** Binkley (1983) also estimated nitrogen accretion for another pair of 23-year-old stands of Douglas-fir and red alder mixed with Douglas-fir. Ten soil pits per stand were sampled for the 0- to 0.15-m depth, and five per stand for deeper horizons to 0.5 m. Regression equations were used to estimate stand biomass. The total nitrogen content of the Douglas-fir stand was 5370 kg/ha, compared with 6340 kg/ha in the mixed stand, giving an annual accretion rate of 42 kg ha$^{-1}$ yr$^{-1}$ for the ecosystem. Leaching losses were not measured, but would probably increase the estimate of nitrogen fixation to 65 to 85 kg ha$^{-1}$ yr$^{-1}$.

**Columbia County, Oregon.** Berg and Doerksen (1975) examined changes in soil nitrogen content that resulted from colonization of a heavily thinned stand of Douglas-fir. Seventeen years after thinning, as the invading alders were dying under the closing conifer canopy, samples of 0- to 0.15-m depth mineral soils were collected from locations with major alder influence, from areas with light alder influence, and from the adjacent lightly thinned (no alder) stands. Each level of alder influence (heavy, light, and none) was represented by three composited samples, with the number of samples in each composite not specified. The nitrogen content of the 0- to 0.15-m depth soils was 1840 kg/ha without alder influence, 2060 kg/ha with light alder influence, and 2710 kg/ha with heavy alder influence. The 17-year accretion rate in soils for light alder influence was 13 kg ha$^{-1}$ yr$^{-1}$, compared with 51 kg ha$^{-1}$ yr$^{-1}$ for areas with heavy alder influence. Given the low sampling intensity, the apparent difference between the no-alder and light-alder areas may not have been significant (no statistical analyses were performed), but the high levels under the heavy influence of alder (a 50-percent increase) were probably real.

**Coast Range, Oregon.** Cole and Newton (1986) designed a study to evaluate (among other things) competition between red alder and Douglas-fir using a Nelder design that maintained a constant (1:1) proportion of the species while allowing total density to change. Within each of three locations, 48 soil samples (0 to 0.15 m) were collected under Douglas-fir, and under the mixed plots of alder and Douglas-fir. After four growing seasons, no evidence of nitrogen accretion in the topsoil was found (the alders were well nodulated). The authors concluded that any fixation that had occurred was probably not great and had failed to increase soil nitrogen capital. The sampling intensity in this study was greater than that used in all other studies, so we have high confidence in the precision of the accretion rate. We expect that nitrogen fixation was occurring (no cases have been reported with alder nodules failing to fix nitrogen), but several factors may have limited any accumulation in the soil. Much of the fixed nitrogen was retained in alder biomass, and much of the nitrogen cycled in aboveground litterfall may have blown out of the relatively small plots. In addition, nitrate leaching may have been high under the influence of alder, and counterbalanced any addition from the alders.

**Lennox Forest, Scotland.** Malcolm et al. (1985) examined nitrogen accretion in a mixed stand of red alder and Sitka spruce. Three replicate plots were sampled (0- to 0.05-m depth only) representing each combination of species and site treatments: soil drainage or no drainage, and spruce mixed with alder or without alder. Across both drainage treatments, accretion averaged 585 kg/ha, for an annual rate of 37 kg ha$^{-1}$ yr$^{-1}$ for 16 years.

### Rates of Nitrogen Fixation for Mixed Stands of Red Alder and Black Cottonwood

We found two studies that examined nitrogen fixation in mixed stands of red alder and black cottonwood designed for production of biomass fuels. DeBell and Radwan (1979) included a mixed plantation of red alder and black poplar in the study mentioned above, and calculated an annual accretion rate (0- to 0.15-m soil) of 32 kg ha$^{-1}$ yr$^{-1}$. Heilman (1990) established plantations of black cottonwood (3200/ha) and hybrid poplar (black cottonwood x eastern cottonwood [*Populus deltoides* Bartr. ex Marsh.]) on mudflows along the Toutle River that resulted from the 1980 eruption of Mt. St. Helens, and red alder seedlings established naturally (14,600/ha) within the plantations. The alder quickly outgrew the cottonwood, and after six years alder trees accounted for about 93 percent of the basal area of the stands. The annual rate of nitrogen accretion in forest floor plus 0 to 0.3 m mineral soil was 56 kg ha$^{-1}$ yr$^{-1}$ through age 6.

### Controls on Rates of Nitrogen Fixation by Red Alder

Surprisingly little research has focused on factors that control the rate of nitrogen fixation by red alder. In a general way, nitrogen fixation is favored by environmental factors that promote alder growth: full sunshine, adequate soil moisture, warm temperatures, and adequate nutrition (Dixon and Wheeler 1983). The role of genetics (of both the alder host and the *Frankia* endophyte) is also poorly explored (see Ager and Stettler, Chapter 6).

### Sunlight

With the exception of black cottonwood, red alder requires more light than any of the tree species found in association with it (Harrington 1990). It is very sensitive to competition; alder canopies must remain in the upper canopy of a stand for alder to survive. We know of no direct experimentation on the effect of shading on nitrogen-fixation rates, but competition studies have shown that nitrogen fixation declines as competition increases. For example, Heilman and Stettler (1983) found that red alder trees with intermediate (partially shaded from above) canopies had half the nodule biomass of trees with codominant or dominant (full sunshine) canopies, and the lower nodule biomass was accompanied by a lower rate of acetylene reduction per unit of nodule weight. Heilman and Stettler (1985) found that nitrogen fixation by red alder declined in the fourth growing season in mixtures with black cottonwood; increased shading by the taller cottonwoods probably impaired nitrogen fixation by alder.

### Soil Moisture

Only one study has experimentally examined the effects of soil moisture on acetylene reduction rates in red alder. Heilman and Stettler (1985) examined nodule activity on root systems of red alder that received trickle irrigation compared with nodules from the other (nonirrigated) side of the same trees. They found that acetylene reduction rates were about 40 percent on the irrigated side of the trees. Other studies have documented the importance of soil moisture by following trends in acetylene reduction rates during dry summer periods when conditions are otherwise favorable for nitrogen fixation (Tripp et al. 1979; Binkley 1981; Bormann and Gordon 1984; Binkley et al. 1993a). D. Hibbs (Oregon State University, pers. com.) has noted yellowish alder foliage on wet, poorly drained sites and suspects that nitrogen fixation may be limited in such cases. It might be possible to estimate annual water balances for sites with estimates of nitrogen-fixation rates, but we expect that other factors (such as competition and phosphorus nutrition) may be too important to allow any strong relationship between site water balance and nitrogen fixation across sites. We tentatively conclude that seasonal dynamics in soil moisture likely affect current nitrogen-fixation activity, but that differences in water balances across sites do not produce any clear effect on cumulative nitrogen fixation.

## Temperature

The best evidence of the role of temperature in affecting nitrogen fixation derives from seasonal patterns of acetylene reduction activity (Tripp et al. 1979; Binkley 1981; Bormann and Gordon 1984; Binkley et al. 1993a); rates rise as soils warm in spring and summer, and decline into autumn. It is not clear whether this pattern of nitrogen-fixation activity derives from direct effects of temperature on the nitrogen-fixation process, or simply from the indirect effects mediated by temperature influences on plant phenology and physiology. The tight connection between nitrogen-fixation rate and plant physiology is well illustrated by the large diurnal fluctuations in acetylene reduction activity noted by Tripp et al. (1979). Cooler, high elevation sites (such as Wind River in Washington) show shorter seasons of nitrogen-fixation activity than do warmer, low elevation sites (such as Cascade Head, OR; Binkley et al. 1993a).

## Phosphorus Nutrition

Phosphorus is a key nutrient for nitrogen-fixing species, which commonly have a greater phosphorus requirement than do non-nitrogen-fixing plants (Sprent 1988). The complex biogeochemical cycle of phosphorus appears to differ markedly in stands of red alder compared to conifers, including both increases and decreases in phosphorus availability reported for different sites (see Bormann et al., Chapter 3; see also Cole et al. 1990; Zou 1992). No fertilization studies have examined the response of nitrogen fixation by alder to phosphorus fertilization in the field. In a bioassay of soils from a stand of red alder and Douglas-fir on Mt. Benson, British Columbia, Binkley (1986, p. 144) showed that

addition of phosphorus and sulfur allowed red alder seedlings to double in biomass and increase acetylene reduction activity per seedling by fivefold (additions of molybdenum had no effect). Experimental fertilization across a range of sites (followed by assessment of response in growth and nitrogen fixation) is clearly warranted; J. Compton and D. Cole (pers. com.) at the University of Washington have installed some fertilization experiments at the Thompson (Washington) site.

## Nitrogen Nutrition

It seems reasonable that high levels of available ammonium and nitrate in soils would lead to low rates of nitrogen fixation as plants optimize energy expenditures by taking advantage of these lower-cost forms of nitrogen. Indeed, evidence has accumulated for over 100 years that nitrogen fixation by leguminous crop species declines sharply as nitrogen availability in the soil increases (Dixon and Wheeler 1983), but the evidence is less clear for alder species. Some laboratory/greenhouse studies have concluded that nitrogen fixation by alder declines as nitrogen availability increases, whereas other studies have concluded the opposite.

In one of the earliest studies, Stewart and Bond (1961) showed that nitrogen fixation by black alder was greater when grown in solutions containing 10 mg/L of ammonium-nitrogen than in nitrogen-free solutions; higher concentrations depressed nitrogen fixation. We have never heard of a forest soil solution exceeding even 1 mg N/L as ammonium, so this study would suggest that nitrogen fixation might increase on sites with increasing ammonium availability.

Zavitkovski and Newton (1968) showed that nodulation of red alder seedlings improved with the addition of nitrate fertilizer up to a rate of 30 mg N/kg of soil in pots, declining at higher levels. We know of no examples where the concentration of extractable nitrate-nitrogen in alder soils exceeded 30 mg/kg (e.g., Binkley and Matson 1983; Binkley et al. 1993b). Nitrogen fixation by red alder seedlings decreased when 60 to 240 mg of urea nitrogen were added per kg of soil in the pots; however, any effect of the urea may have derived from effects other than nitrogen supply, such as direct toxicity of such high rates or drastic rise in soil pH.

Monaco et al. (1981) found that the effects of adding 20 mg-N/L as ammonium nitrate varied among genotypes of red alder seedlings, but no overall effects were apparent on either nodule biomass per plant or acetylene reduction activity per plant. Lipp (1987) examined the effects of increasing solution concentrations of nitrogen (as ammonium + nitrate) and phosphorus on red alder seedlings, and found increasing nitrogen from 1 mg-N/L to 10 mg-N/L decreased acetylene reduction activity/plant by about 25 percent. The levels of phosphorus she used were, however, about three orders of magnitude above normal values for soil solutions. Ingestad (1980, 1981) demonstrated that nitrogen fixation by black alder increased as the supply of nitrate was increased, with increasing nitrate supplies provided by more frequent applications at low, realistic concentrations.

Alder species in general are more sensitive to the inhibitory effects of available soil nitrogen than are other actinorhizal genera (Kohls and Baker 1989). Significant reductions in numbers and weight of nodules occurred when nitrate concentrations increased to 20 mg-N/L. Ammonium inhibited nodule formation more than nitrate did, but these authors could not separate the effects of available nitrogen from those of pH. The effect of high concentrations of ammonium or nitrate derived from inhibition of the formation of root hairs, which prevented infection of roots (Kohls and Baker 1989).

The relevance of these laboratory studies to field conditions is not clear. Typical nitrate concentrations in soil solutions under alder stands are in the range of 1 to 8 mg-N/L, and ammonium levels are typically between 0.02 and 0.2 mg-N/L (Binkley et al. 1982; Van Miegroet and Cole 1988; Binkley et al. 1993a). Studies that use unrealistically high concentrations of nitrate or ammonium may be useful for detailed examination of processes, but probably produce unrealistic responses relative to field conditions.

The few laboratory studies that used realistic concentrations of ammonium or nitrate generally indicated that availability of ammonium and nitrate at rates commonly observed in the field should not inhibit alder nitrogen fixation, and this is consistent with studies that have documented high rates of nitrogen fixation on all sites, even those where

leaching losses of nitrogen exceed 50 kg-N ha$^{-1}$ yr$^{-1}$ (Cole et al. 1978; Van Miegroet and Cole 1984; Binkley et al. 1993a). We know of no cases where healthy stands of alder failed to fix substantial quantities of nitrogen.

All this evidence begs the question, why does red alder fix nitrogen on sites where it is abundantly available as ammonium and nitrate in the soil? Fixation of 100 kg of nitrogen costs between 1000 and 2000 kg of carbohydrate that could otherwise be used for growth or reproduction. We have no answer for this question, but likely avenues of inquiry would involve examination of the total costs of acquisition (including root growth) and assimilation of each source of nitrogen, as well as the degree of control the host plant exerts over the activity of endophytes in nodules.

## Stand Density

A key issue in the management of red alder is the rate of nitrogen fixation as a function of alder density. Only one study has directly examined the relationship between alder density and nitrogen fixation. As noted earlier, Bormann and Gordon (1984) examined nitrogen fixation at three densities (from 1330 to 13400/ha) in a 5-year-old plantation. Despite an order-of-magnitude difference in density, rates of nitrogen fixation ranged only between 60 kg ha$^{-1}$ yr$^{-1}$ to 82 kg ha$^{-1}$ yr$^{-1}$.

Two indirect estimates have also been attempted. Miller and Murray (1978, 1979) noted that Douglas-fir within 10 m of an alder/Douglas-fir stand at Wind River, Washington, were larger than Douglas-fir farther from the stand boundary. From this, they estimated that 20 to 100 dominant red alder per hectare might be sufficient to supply 20 to 50 kg-N/ha for use by interplanted Douglas-fir.

Based on all available studies, Binkley (1986, pp. 156-157) noted that the annual rate of nitrogen fixation typically equalled 2 to 4 percent of alder leaf biomass. Using regression equations for estimating leaf biomass from stem diameters, he calculated the number of alders needed of a given diameter to yield a desired rate of nitrogen fixation. For example, an annual rate of nitrogen fixation of 20 kg/ha might require 300 alder/ha in a 10-cm diameter class, 150 alder/ha in a 15-cm diameter class, or 120 alder in a 20-cm diameter class.

## Symbioses and Rates of Nitrogen Fixation

Rates of nitrogen fixation by alder seedlings have been shown to vary among genotypes of the alder hosts and of the *Frankia* endophyte (see also Molina et al., Chapter 2). Monaco et al. (1981) showed that acetylene reduction rates (per plant) varied by more than fivefold in a greenhouse experiment on seedlings from different parent trees. Carpenter et al. (1984) examined the influence of two strains of *Frankia* from red alder and two strains from Sitka alder on nitrogen fixation by two clones of red alder seedlings. Total acetylene reduction activity per plant depended strongly on the strain of *Frankia*, with a strong interaction between host and endophyte. For the red alder clone from Vancouver Island, the acetylene reduction rate was more than 10 times greater with an Oregon *Frankia* strain than with a Vancouver Island strain. In contrast, an Oregon alder clone showed equal acetylene reduction activity between the two *Frankia* strains. Wheeler et al. (1991) examined acetylene reduction rates in red alder seedlings inoculated with a strain of *Frankia* isolated from nodules from a plantation in Scotland, with suspensions from crushed nodules from the same field site, and with a strain supplied from the United States. The best rate of nitrogen fixation (as gauged by seedling nitrogen content) was found for the strain cultivated from the Scottish nodules, and the poorest rate occurred for seedlings inoculated using crushed nodules. The mechanisms responsible for the observed differences in effectiveness of *Frankia* strains, and of symbioses with *Alnus* strains, remain largely unknown.

Symbioses are critically important for red alder, which may be colonized by three forms of symbionts: *Frankia*, ectomycorrhizae, and vesicular-arbuscular mycorrhizae (VAM) (Trappe 1979; Molina et al., Chapter 2). The importance of these tripartite or quadripartite associations in nitrogen fixation and overall nutrition of red alder remains poorly understood. Alders appear to have relatively few mycorrhizal species, perhaps as few as 30 ectomycorrhizal species for the genus worldwide (Molina 1979; Miller et al. 1991). In an extensive survey of a wide range of habitats, Miller et al. (1991) found only 11 types of ectomycorrhizae on red alder; some of these, such as *Alpova diplo-phloeus*, form associations only with alder, and others, such as *Laccaria laccata*, were generalists in forming mycorrhizal associations with many species. In contrast, Douglas-fir (*Pseudotsuga menziesii* {Mirb.} Franco) probably forms mycorrhizal associations with over 2000 mycorrhizal fungi (Trappe 1977; Allen 1991).

There is evidence that alder mycorrhizae can increase phosphorus uptake, particularly on infertile sites (Mejstrik and Benecke 1969); fine roots of green alder (*Alnus viridis* Vill.) from a poor site in New Zealand that was low in available phosphorus had 29 percent greater mycorrhizal infection than those from a more fertile site. Uptake of $^{32}P$ by excised roots from the infertile site was twice as fast as rates for roots from the fertile site, and was directly proportional to the degree of mycorrhizal colonization (Mejstrik and Benecke 1969). Greenhouse studies with red alder seedlings have documented greater growth rates for mycorrhizal seedlings, and higher concentrations of phosphorus in foliage (Koo 1989).

Field observations of red alder seedlings on disturbed sites have found that infection by *Frankia* can precede mycorrhizal colonization (Koo 1989), and experimental evidence suggests that nodulation is more critical to early seedling development than is mycorrhizal colonization. Ectomycorrhizal colonization of red alder does not appear to increase nodulation and nitrogen fixation as much as found for colonization by VAM of another actinorhizal species, *Ceanothus velutinus* Dougl. (Rose and Youngberg 1981). Critical experiments need to be performed that examine seedlings in soils of high and low phosphorus availability, with and without nodules, ectomycorrhizae and VAM to determine the importance of each symbiont in nitrogen fixation and overall plant nutrition.

## Conclusions and Synthesis

All evidence indicates that healthy alder trees fix large amounts of nitrogen, typically ranging from 50 to 100 kg ha$^{-1}$ yr$^{-1}$ in mixed species stands, and 100 to 200 kg ha$^{-1}$ yr$^{-1}$ in pure stands. The factors that determine nitrogen-fixation rate are probably the same as those that regulate overall plant vigor. Based on current knowledge, opportunities for managing red alder for high rates of nitrogen fixation appear to include the following:

- Any treatment that improves the vigor of the alder (such as controlling competition with coniferous weed species)

- Genetic selection of hosts, endophytes, and combinations of these for a range of site conditions

- Management of phosphorus nutrition of the alder

Maximum rates of nitrogen fixation may not be a desirable goal in two cases: where nitrate leaching is pronounced (with attendant risks of soil acidification, base cation depletion, and stream pollution), and where dominant, vigorous alders impair the growth of conifer crop trees.

A very wide range of research would substantially increase the state-of-knowledge on nitrogen fixation by red alder. We know the general ballpark of these rates, but have only a limited mechanistic understanding of the environmental and genetic constraints on them. Most of the process-oriented research with red alder has been limited to seedlings or to young trees; we know little about the role of water stress, nutrient stress, and mycorrhizal/nutrient interactions in older stands of red alder.

We have three general recommendations for research. We suggest that an ecosystem-level simulation model could be developed that includes the key factors thought to influence nitrogen fixation by alder, and that insights from this model would help optimize the design of experiments and collection of data from existing experiments. Several well-designed experimental plantations have been established in Oregon and Washington (R. Miller, D. Hibbs, and M. Newton, pers. com.), and we suggest that long-term, collaborative research efforts be designed to take greatest advantage of the opportunities offered by these plantations. We also suggest that a new series of plantations be established to examine the interactions of species (red alder and conifers) and nutrition (fertilization with phosphorus and nitrogen) on nitrogen-fixation rates and other ecosystem processes. This new set of plantations would require only moderate resources to establish, and collaboration among a wide range of scientists and forest managers could yield great insights.

## Acknowledgments

This chapter was made possible by the wide array of scientists and funding agencies that have contributed to the understanding of nitrogen fixation by red alder. The chapter was also improved substantially by comments from Paul Heilman, Helga Van Miegroet, and Dave Hibbs.

## Literature Cited

Akkermans, A. D. L., W. Roelofsen, and J. Blom. 1979. Dinitrogen fixation and ammonia assimilation in actinomycetous root nodules of *Alnus glutinosa*. *In* Symbiotic nitrogen fixation in the management of temperate forests. *Edited by* J. C. Gordon, C. T. Wheeler, and D. A. Perry. Oregon State Univ. Press, Corvallis. 160-174.

Allen, M. F. 1991. The ecology of mycorrhizae. Cambridge Univ. Press, Cambridge.

Baker, D. D. 1987. Relationships among pure cultured strains of *Frankia* based on host specificity. Physiol. Plant. 70:245-248.

Berg, A., and A. Doerksen. 1975. Natural fertilization of a heavily thinned Douglas-fir stand by understory red alder. Forest Research Laboratory, Oregon State Univ., Corvallis, Res. Note 56.

Berntsen, C. M. 1961. Growth and development of red alder compared with conifers in 30-year-old stands. USDA For. Serv., Res. Pap. PNW-38.

Berry, A. M., and L. A. Sunnell. 1990. The infection process and nodule development. *In* The biology of *Frankia* and actinorhizal plants. *Edited by* C. R. Schwintzer and J. D. Tjepkema. Academic Press, New York. 61-81.

Berry, A. M., and J. G. Torrey. 1983. Root hair deformation in the infection process of *Alnus rubra*. Can. J. Bot. 61:2863-2876.

Binkley, D. 1981. Nodule biomass and acetylene reduction rates of red alder and Sitka alder on Vancouver Island, B.C. Can. J. For. Res. 11:281-286.

———. 1983. Ecosystem production in Douglas-fir plantations: interaction of red alder and site fertility. For. Ecol. Manage. 5:215-227.

———. 1986. Forest Nutrition Management. Wiley, New York.

Binkley, D., and S. Greene. 1983. Production in mixtures of conifers and red alder: the importance of site fertility and stand age. *In* IUFRO Symposium on Forest Site and Continuous Productivity. *Edited by* R. Ballard and S. P. Gessel. USDA For. Serv., Gen. Tech. Rep. PNW-163. Portland, OR. 112-117.

Binkley, D., and P. Matson. 1983. Ion exchange resin bag method for assessing forest soil nitrogen availability. Soil Sci. Soc. Am. J. 47:1050—1052.

Binkley, D., J. P. Kimmins, and M. C. Feller. 1982. Water chemistry profiles in an early- and a mid-successional forest in coastal British Columbia. Can. J. For. Res. 12:240-248.

Binkley, D., P. Sollins, and W. B. McGill. 1985. Natural abundance of nitrogen$^{15}$ as a tool for tracing alder-fixed nitrogen. Soil Sci. Soc. Am. J. 49:444-447.

Binkley, D., P. Sollins, R. Bell, D. Sachs, and D. Myrold. 1993a. Biogeochemistry of adjacent conifer and alder/conifer stands. Ecology 73:2022-2033.

Binkley, D., R. Bell, and P. Sollins. 1993b. Soil nitrogen transformations in adjacent conifer and alder/conifer forests. Can. J. For. Res. (In press.)

Bond, G. 1955. An isotopic study of the fixation of nitrogen associated with nodulated plants of *Alnus, Myrica,* and *Hippophae*. J. Exp. Bot. 6:303-311.

Bormann, B. T., and D. S. DeBell. 1981. Nitrogen content and other soil properties related to age of red alder stands. Soil Sci. Soc. Am. J. 45:428-432.

Bormann, B. T., and J. C. Gordon. 1984. Stand density effects in young red alder plantations: productivity, photosynthate partitioning, and nitrogen fixation. Ecology 65:394-402.

Callaham, D., P. Del Tredici, and J. G. Torrey. 1978. Isolation and cultivation *in vitro* of the actinomycete causing root nodulation in *Comptonia*. Science 199:899-902.

Carpenter, C. V., L. R. Robertson, J. C. Gordon, and D. A. Perry. 1984. The effect of four new *Frankia* isolates on growth and nitrogenase activity in clones of *Alnus rubra* and *Alnus sinuata*. Can. J. For. Res. 14:701-706.

Cole, D. W., and H. Van Miegroet. 1989. Chronosequences: a technique to assess ecosystem dynamics. *In* Research strategies for long-term site productivity. *Edited by* W. Dyck and C. Mees. For. Res. Inst. Bull. 152, Rotorua, New Zealand. 5-24.

Cole, D. W., S. P. Gessel, and J. Turner. 1978. Comparative mineral cycling in red alder and Douglas-fir. *In* Utilization and management of alder. *Compiled by* D. G. Briggs, D. S. DeBell, and W. A. Atkinson. USDA For. Serv., Gen. Tech. Rep. PNW-70. 327-336.

Cole, D. W., J. Compton, H. Van Miegroet, and P. Homann. 1990. Changes in soil properties and site productivity caused by red alder. Water, Air, Soil Pollut. 54:231-246.

Cole, E. C., and M. Newton. 1986. Nutrient, moisture, and light relations in 5-year-old Douglas-fir plantations under variable competition. Can. J. For. Res. 16:727-732.

DeBell, D. S., and M. A. Radwan. 1979. Growth and nitrogen relations of coppiced black cottonwood and red alder in pure and mixed plantings. Bot. Gaz. (Suppl.) 140:97-101.

Dixon, R. O. D., and C. T. Wheeler. 1983. Biochemical, physiological and environmental aspects of symbiotic nitrogen fixation. *In* Biological nitrogen fixation in forest ecosystems: foundations and applications. *Edited by* J. C. Gordon and C. T. Wheeler. Martinus Nijhoff/D. W. Junk, The Hague. 108-183.

Franklin, J., C. Dyrness, D. Moore, and R. Tarrant. 1968. Chemical soil properties under coastal Oregon stands of alder and conifers. *In* Biology of alder. *Edited by* J. M. Trappe, J. F. Franklin, R. F. Tarrant, and G. M. Hansen. USDA For. Serv., PNW For. Range Exp. Sta., Portland, OR. 157-172.

Harrington, C. A. 1990. *Alnus rubra* Bong.—red alder. *In* Silvics of North America, vol. 2, Hardwoods. *Technically coordinated by* R. M. Burns and B. H. Honkala. USDA For. Serv., Agric. Handb. 654. Washington, D.C. 116-123.

Heilman, P. 1982. Nitrogen and organic-matter accumulation in coal mine spoils supporting red alder stands. Can. J. For. Res. 12:809-813.

———. 1990. Growth and N status of *Populus* in mixture with red alder on recent volcanic mudflow from Mount St. Helens. Can. J. For. Res. 20:84-90.

Heilman, P., and G. Ekuan. 1982. Nodulation and nitrogen fixation by red alder and Sitka alder on coal mine spoils. Can. J. For. Res. 12:922-997.

Heilman, P., and R. F. Stettler. 1983. Phytomass production in young mixed plantations of *Alnus rubra* (Bong.) and cottonwood in Western Washington. Can. J. Microbiol. 29:1007-1013.

———. 1985. Mixed, short-rotation culture of red alder and black cottonwood: growth, coppicing, nitrogen fixation, and allelopathy. For. Sci. 31:607-616.

Hiltner, L. 1896. Über die Bedeutung der Wurzelknöllchen von *Alnus glutinosa* für die Stickstoffernährung dieser Pflanze. Die Landwirtschaftlichen Versuchs-Stationen 46:153-161.

Ingestad, T. 1980. Growth, nutrition, and nitrogen fixation in grey alder at varied rate of nitrogen addition. Physiol. Plant. 50:353-364.

———. 1981. Nutrition and growth of birch and grey alder seedlings in low conductivity solutions and at varied relative rates of nutrient addition. Physiol. Plant. 52:454-466.

Kim, D. Y. 1987. Seasonal estimates of nitrogen fixation by *Alnus rubra* and *Ceanothus* species in western Oregon forest ecosystems. M.S. thesis, Oregon State Univ., Corvallis.

Kohls, S. J., and D. Baker. 1989. Effects of substrate nitrate concentration on symbiotic nodule formation in actinorhizal plants. Plant and Soil 118:171-179.

Koo, C. D. 1989. Water stress, fertilization, and light effects on the growth of nodulated mycorrhizal red alder seedlings. Ph.D. thesis, Oregon State Univ., Corvallis.

Lipp, C. C. 1987. The effect of nitrogen and phosphorus on growth, carbon allocation and nitrogen fixation of red alder seedlings. M.S. thesis, Oregon State Univ., Corvallis.

Luken, J. O., and R. W. Fonda. 1983. Nitrogen accumulation in a chronosequence of red alder communities along the Hoh River, Olympic National Park, Washington. Can. J. For. Res. 13:1223-1237.

Malcolm, D., J. E. Hooker, and C. T. Wheeler. 1985. *Frankia* symbiosis as a source of nitrogen in forestry: a case study of symbiotic nitrogen-fixation in a mixed *Alnus-Picea* plantation in Scotland. Proceedings of the Royal Society of Edinburgh 85B:263-282.

Mead, D. J., and C. M. Preston. 1992. Nitrogen fixation in Sitka alder by $^{15}$N isotope dilution after eight growing seasons in a lodgepole pine forest. Can. J. For. Res. (In press.)

Mejstrik, V., and U. Benecke. 1969. The ectotrophic mycorrhizas of *Alnus viridis* (Chaix) C.C. and their significance in respect to phosphorus uptake. New Phytol. 68:141-149.

Miller, R. E., and M. D. Murray. 1978. The effects of red alder on the growth of Douglas-fir. *In* Utilization and management of alder. *Compiled by* D. G. Briggs, D. S. DeBell, and W. A. Atkinson. USDA For. Serv., Gen. Tech. Rep. PNW-70. 283-306.

Miller, R. E., and M. D. Murray. 1979. Fertilizer versus red alder for adding nitrogen to Douglas-fir forests of the Pacific Northwest. *In* Symbiotic nitrogen fixation in the management of temperate forests. *Edited by* J. Gordon, C. Wheeler, and D. Perry. Oregon State Univ. Press, Corvallis. 356-373.

Miller, S. L., C. D. Koo, and R. Molina. 1991. Characterization of red alder ectomycorrhizae: a preface to monitoring belowground ecological responses. Can. J. Bot. 69:516-531.

Minchin, F. R., J. E. Sheehy, and J. E. Witty. 1986. Further errors in the acetylene reduction assay: effects of plant disturbance. J. Exp. Bot. 37:1581-1591.

Minchin, F. R., J. F. Witty, J. E. Sheehy, and M. Muller. 1983. A major error in the acetylene reduction assay: decrease in nodular nitrogenase activity under assay conditions. J. Exp. Bot. 34:641-649.

Molina, R. 1979. Pure culture synthesis and host specificity of red alder mycorrhizae. Can. J. Bot. 57(11):1223-1228.

Monaco, P. A., T. M. Ching, and K. K. Ching. 1981. Variation of *Alnus rubra* for nitrogen fixation capacity and biomass production. Silv. Genet. 30:46-50.

Newton, M., B. A. El Hassan, and J. Zavitkovski. 1968. Role of red alder in western Oregon forest succession. *In* Biology of alder. *Edited by* J. M. Trappe, J. F. Franklin, R. F. Tarrant, and G. M. Hansen. USDA For. Serv., PNW For. Range Exp. Sta., Portland, OR. 73-83.

Perinet, P., J. G. Brouillette, J. A. Fortin, and M. LaLonde. 1985. Large scale inoculations of actinorhizal plants with *Frankia*. Plant and Soil 87:175-183.

Rose, S. L., and C. T. Youngberg. 1981. Tripartite associations in snowbrush (*Ceanothus velutinus*): effect of vesicular-arbuscular mycorrhizae on growth, nodulation, and nitrogen fixation. Can. J. Bot. 59:34-39.

Schubert, K. 1982. The energetics of biological nitrogen fixation. American Society of Plant Physiologists, Rockville, MD.

Silvester, W. B. 1983. Analysis of nitrogen fixation. *In* Biological nitrogen fixation in forest ecosystems: foundations and applications. *Edited by* J. C. Gordon and C. T. Wheeler. Martinus Nijhoff/D. W. Junk, The Hague. 173-212.

Smolander, A., and V. Sundman. 1987. *Frankia* in acid soils of forests devoid of actinorhizal plants. Physiol. Plant. 70:297-303.

Sprent, J. 1988. The ecology of the nitrogen cycle. Cambridge Univ. Press, Cambridge.

Stewart, W. D. P., and G. Bond. 1961. The effect of ammonium nitrogen on fixation of elemental nitrogen in *Alnus* and *Myrica*. Plant and Soil 14:347-359.

Stowers, M. D., and J. E. Smith. 1985. Inoculation and production of container-grown red alder seedlings. Plant and Soil 87:153-160.

Tarrant, R. F., and R. E. Miller. 1963. Accumulation of organic matter and soil nitrogen beneath a plantation of red alder and Douglas-fir. Soil Sci. Soc. Am. Proc. 27:231-234.

Tarrant, R. F., and J. M. Trappe. 1971. The role of *Alnus* in improving the forest environment. Plant and Soil (spec. vol.) 1971:335-348.

Tjepkema, J. D., and L. J. Winship. 1980. Energy requirement for nitrogen fixation in actinorhizal and legume root nodules. Science 209:279-281.

Trappe, J. M. 1977. Selection of fungi for ectomycorrhizal inoculation in nurseries. Ann. Rev. Phytopathol. 15:203-222.

————. 1979. Mycorrhizae-nodule-host interrelationships in symbiotic nitrogen fixation: a quest in need of questers. *In* Symbiotic nitrogen fixation in the management of temperate forests. *Edited by* J. C. Gordon, C. T. Wheeler, and D. A. Perry. Oregon State Univ. Press, Corvallis. 276-286.

Tripp, L. N., D. F. Bezdicek, and P. E. Heilman. 1979. Seasonal and diurnal patterns and rates of nitrogen fixation by young red alder. For. Sci. 25:371-380.

Van Miegroet, H., and D. W. Cole. 1984. The impact of nitrification on soil acidification and cation leaching in a red alder ecosystem. J. Environ. Qual. 13:586-590.

————. 1988. Influence of nitrogen-fixing alder on acidification and cation leaching in a forest soil. *In* Forest site evaluation and long-term productivity. *Edited by* D. Cole and S. P. Gessel. Univ. of Washington Press, Seattle. 113-124.

Van Miegroet, H., D. Cole, D. Binkley, and P. Sollins. 1989. The effect of nitrogen accumulation on soil chemical properties in alder forests. *In* Effects of air pollution on Western forests. *Edited by* R. Olson and A. LeFohn. Air and Waste Management Association, Pittsburgh, PA. 515-528.

Wheeler, C. T., M. K. Hollingsworth, J. E. Hooker, J. D. McNeill, W. L. Mason, A. J. Moffat, and L. J. Sheppard. 1991. The effect of inoculation with either cultured *Frankia* or crushed nodules on nodulation and growth of *Alnus rubra* and *Alnus glutinosa* seedlings in forest nurseries. For. Ecol. Manage. 43:153-166.

Zavitkovski, J., and M. Newton. 1968. Effect of organic matter and combined nitrogen on nodulation and nitrogen fixation in red alder. *In* Biology of alder. *Edited by* J. M. Trappe, J. F. Franklin, R. F. Tarrant, and G. M. Hansen. USDA For. Serv., PNW For. Range Exp. Sta., Portland, OR. 209-223.

Zou, X. 1992. Phosphorus transformations in soils: a new method and tree species effects. Ph.D. thesis, Colorado State Univ., Ft. Collins.

# 5

# Physiological Characteristics of Red Alder: Water Relations and Photosynthesis

L. J. SHAINSKY, B. J. YODER, T. B. HARRINGTON & S. S. N. CHAN

Red alder (*Alnus rubra* Bong.) has been classified as extremely shade intolerant (Harrington 1990), and water is critically important in its ecology. Historically, red alder grew almost exclusively in coastal and riparian habitats. Today, its range has expanded upslope, perhaps as a result of human-induced disturbance. Red alder is an early seral species with rapid juvenile growth rates, aggressive competitive abilities, and the capacity to fix nitrogen. As a result of these and other traits, red alder plays a key role in the succession of Pacific Northwest forest ecosystems. A growing interest in managing mixed-species stands to enhance productivity and growth through nitrogen enrichment has stimulated research to examine the mechanisms of red alder growth and those of associate species, such as black cottonwood (*Populus trichocarpa*) and Douglas-fir (*Pseudotsuga menziesii* var. *menziesii*). Because of the importance of this species, research is also underway to examine effects of climate change on red alder and their consequences to the role of red alder in future forest ecosystems.

Key processes mediating the ecology of red alder are physiological, as are some of the underlying mechanisms of productivity in managed and unmanaged red alder stands. Physiological capacities, such as maximum stomatal conductance and photosynthesis, may put upper limits on the growth and range of red alder, and determine a particular site's potential for supporting a productive red alder stand. Physiological plasticity and responses to environmental conditions also shape responses of red alder to management practices. Understanding the physiology of red alder may help us predict red alder performance under previously untested conditions through mechanistic modeling.

The objectives of this chapter are to examine how physiological processes of red alder respond to environmental conditions and resource limitations, and to explore how these processes may mediate red alder ecology. We first examine key autecological aspects of red alder physiology—leaf water potential, stomatal conductance, and assimilation rate. Drawing on experiments in both the field and the laboratory, we examine how environmental factors affect these traits. Then, we examine how physiological phenomena may be used to interpret underlying mechanisms of productivity and ecology of red alder. We explore some of the implications of stress physiology to the management of red alder and provide a physiological foundation for interpreting other chapters in this book. Although nitrogen fixation is a critically important feature of red alder physiology and biology, its discussion is left to a chapter devoted solely to this topic (Binkley et al., Chapter 4).

Our goal is to review comprehensively the available literature within the scope of our objectives. Much of the available information on red alder physiology is unpublished, and we have drawn heavily on these unpublished data sets. A brief overview of unpublished studies is provided in the Appendix. Almost all the data in the literature and in unpublished data sets come from young trees (< 10 years old); the need for observations on more mature red alder is obvious.

*In memory of Jimmy Dukes, who helped illuminate red alder's physiological character.*

## Units and Conventions

The units that are used to describe water relations and gas exchange of plants vary widely throughout the literature. Wherever possible, we adhere to most current conventions, although we present some data in the units in which they were originally reported.

Plant water potential ($\Psi$) is always negative, and is expressed in units of MPa in which 1 MPa = 10 bars = 9.87 atm. Plants that are well hydrated will have $\Psi$ near to, but less than, zero and are said to have a "high" water potential. Water-stressed plants will have a more negative $\Psi$, or a "low" water potential.

Exchange of water vapor and carbon dioxide ($CO_2$) is mediated by gradients in the concentrations of these gases, in addition to all the factors that regulate these gradients throughout the soil-plant-air continuum. Differences in water potential at the leaf-air interface are determined largely by the water vapor pressure gradient (VPG)—the difference in vapor pressure between the air at the leaf surface and the substomatal cavity. This gradient is the driving force for water loss from plants. Because quantification of VPG requires temperature measurements of both the ambient air and the leaf, VPG is sometimes approximated by the vapor pressure deficit (VPD), which is the difference between the actual and saturation vapor pressure of the air at a particular temperature. When leaf temperature equals air temperature, VPD and VPG are equivalent.

The assimilation rate of $CO_2$ ($A_s$) is best described on a molar basis per unit leaf area; here we use the units $\mu mol\ m^{-2}s^{-1}$ for $A_s$. In much of the older literature, $A_s$ is given in terms of the weight (commonly $mg\ m^{-2}s^{-1}$) rather than the moles of $CO_2$ exchanged. Conversion is simple: $1\ mg\ CO_2\ m^{-2}s^{-1} = 22.7\ \mu mol\ CO_2\ m^{-2}s^{-1}$.

The conventions for stomatal conductance of water vapor ($g_{wv}$) have shifted more drastically. Most literature sources (and this chapter) use units based on concentration gradients to describe conductance ($cm\ s^{-1}$) or its inverse, resistance ($s\ cm^{-1}$). A conversion to molar units is preferred, because these values are independent of temperature and pressure. It is also more straightforward to compare $A_s$ and $g_{wv}$ when both are reported in molar units. The relationship between the two types of units is $g_{wv}$ ($mol\ m^{-2}s^{-1}$) = $g_{wv}$ ($cm\ s^{-1}$) * 0.446 (273/(T + 273))(P/101.3), in which T is air temperature in °C, and P is atmospheric pressure in kPa.

## Water

The availability of water in both the soil and the atmosphere is probably the most important abiotic factor limiting the range of red alder. Thus, we can learn much about the ecology of red alder by examining its response to changes in water availability. Most plants respond to long-term water stress through modification of a suite of morphological and physiological characteristics. Allocation patterns may change so that leaf area is reduced, root biomass or length is increased, and leaf-level physiological well-being is conserved. Although the focus of this section is on leaf-level physiology, we also consider some morphological aspects of water relations.

### Relationships Between $\Psi_l$, $g_{wv}$ and $A_s$ as Soil Water Varies

Leaf water potential ($\Psi_l$) is related to $g_{wv}$ and $A_s$ in many ways. Before dawn, $\Psi_l$ is usually considered to be in equilibrium with the average $\Psi_{soil}$ in the rhizosphere—the soil in the immediate vicinity of the roots. A number of studies have demonstrated the expected decrease in predawn $\Psi_l$ as soil water decreases for red alder (e.g., Pezeshki and Hinckley 1988; Shainsky and Radosevich 1992; S. S. N. Chan, unpub. data; T. B. Harrington et al., unpub. data). Research on several tree species has shown that the maximum $g_{wv}$ attained during any day is related to the predawn $\Psi_l$, although this relationship is also affected by atmospheric humidity (Zobel 1974; Running 1976; Waring et al. 1981). The same appears to be true for red alder: Lu (1989) found a high positive correlation ($R^2 = 0.80$) between predawn $\Psi_l$ and $g_{wv}$ at noon. Thus, most data support the notion that soil water mediates maximum stomatal conductance of red alder through its effects on $\Psi_l$.

During daylight hours, $\Psi_l$ of a transpiring plant generally decreases as the flux of water out of leaves exceeds the flux of water from the soil to the roots, and from the roots to the leaves. The diurnal course of $\Psi_1$ depends on $g_{wv}$, in addition to the supply of water to roots, the leaf-to-air vapor pressure gradient, and the conductance of water through the xylem (e.g., Hsiao 1973; Nobel 1974). The diurnal course of $g_{wv}$, in turn, is dependent in part on concurrent $\Psi_1$, as well as on a variety of other environmental factors. Thus, no simple cause-and-effect relationship exists between $\Psi_1$ and $g_{wv}$ or $A_s$.

Nonetheless, threshold levels of $\Psi_1$ can be identified from boundary-line analysis of plots relating gas exchange parameters to $\Psi_1$. Boundary-line analysis of $A_s$ versus $\Psi_1$ of data from B. J. Yoder et al. (unpub. data) indicated a threshold of approximately -0.8 MPa, beyond which $\Psi_1$ began to limit $A_s$ (Fig. 1). When $\Psi_1$ reached -1.5 MPa, there was essentially no net carbon fixation. (The curious phenomenon of declining $A_s$ with increasing $\Psi_1$ when $\Psi_1$ is high [Fig. 1] will be discussed later in this chapter.) Two other unpublished studies, a laboratory study in which variation in $A_s$ and $\Psi_1$ was produced by manipulating watering and $CO_2$ regimes (S. S. N. Chan et al., unpub. data), and a field study in which variation in $A_s$ and $\Psi_1$ was produced by manipulating density (T. B. Harrington and B. J. Yoder, unpub. data), showed a similar threshold phenomenon: $A_s$ declined when $\Psi_1$ fell below -0.8 MPa. Lu (1989) found a higher $\Psi_1$ threshold for $g_{wv}$: stomata began to close when $\Psi_1$ reached -1.2 MPa. Interestingly, in these and other studies, $\Psi_1$ measurements below -1.5 MPa are rare. This value of $\Psi_1$ is termed "the wilting point." Red alder appears to be very sensitive to wilting, which most likely leads to the leaf abscission and death of young trees observed frequently in many of the field studies discussed in this chapter.

## Interactive Effects of Soil Water and Atmospheric Humidity on Leaf Physiology

Complicating the interaction between $\Psi_1$ and $g_{wv}$ is the fact that stomata generally respond directly to atmospheric humidity and independently from the effects of humidity on $\Psi_1$ (e.g., Tibbits 1979).

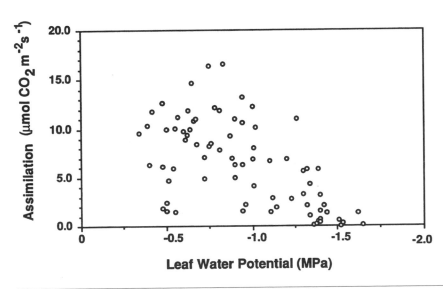

Figure 1. Relationship between $CO_2$ assimilation and leaf water potential, measured concurrently in the morning. Variation in both physiological parameters was partly in response to direct manipulations of water supply and humidity (From: B. J. Yoder et al., unpub.).

To investigate the separate effects of soil and atmospheric water on $\Psi_l$ and $g_{wv}$, Pezeshki and Hinckley (1988) measured physiological responses to different VPD conditions when soil water conditions were similar. They found that stomata of red alder seedlings closed earlier in the day under high VPD than under low VPD conditions. The stomatal closure at high VPD was great enough to "overcompensate" for the increased evaporative demand, so that $\Psi_l$ was actually higher than that observed under low VPD conditions. This is because water flux to the roots caught up with or exceeded the flux of water from the leaves to the atmosphere, causing $\Psi_l$ to rise. When VPD was similar (with low evaporative demand) and soil moisture varied, $g_{wv}$ was reduced only slightly under the drier soil conditions. As a consequence, $\Psi_l$ responded to changes in the soil moisture supply. $\Psi_l$ dropped to nearly -1.5 MPa on a droughty day, but only reached -1.0 MPa when soil moisture was high.

As is true for many other tree species (Waring and Schlesinger 1985), soil water supply affects the sensitivity of red alder stomata to VPD (Pezeshki and Hinckley 1982, 1988; B. J. Yoder et al., unpub. data). Pezeshki and Hinckley (1982) found that $g_{wv}$ of well-watered red alder seedlings followed a steep, linear response to changes in vapor pressure. As the leaf-to-air vapor pressure gradient increased from 0 to 2.5 kPa, $g_{wv}$ dropped from 0.75 cm s$^{-1}$ to 0.20 cm s$^{-1}$. The rate of decline was even steeper when plants were moderately water-stressed: at very low evaporative demand $g_{wv}$ was about 0.55 cm s$^{-1}$ and stomata were closed by the time VPD reached 2.0 kPa. Very stressed seedlings had much lower conductance even at high humidity; stomata of these trees were closed when the vapor pressure gradient was less than 1.0 kPa. Thus, stomatal closure occurs at lower VPD under drier conditions than under conditions of adequate soil moisture.

Interactions between soil water supply and humidity as they affect $\Psi_l$ and $g_{wv}$ were explored further by B. J. Yoder et al. (unpub. data). As expected, the general trend was for $\Psi_l$, $g_{wv}$, and $A_s$ to decrease as the amount of applied water was reduced (Fig. 2). The interactive effects of varying VPD and soil water, however, were more complex and reinforce the findings of Pezeshki and Hinckley (1982). At the highest rate of water supply, neither $A_s$ nor $g_{wv}$ responded to variation in VPD. At slightly lower water supply rates, the relationship between VPD and both $A_s$ and $g_{wv}$ was nearly linear: stomata closed as evaporative demand increased. Under these high VPD conditions, the decreased $g_{wv}$ appeared to have balanced the increased evaporative demand, because $\Psi_l$ remained constant (Fig. 2A). At still lower water supply rates, stomata did not close enough to compensate for increases in evaporative demand, and $\Psi_l$ dropped as VPD increased. An exception to this trend occurred at the highest VPD, in which case stomata closed so tightly that values of $\Psi_l$ were higher than those at other levels of VPD.

Pezeshki and Hinckley's (1988) data, as well as that of B. J. Yoder et al. (unpub. data) (Fig. 2) showed that when red alder seedlings were subjected to very low atmospheric humidity, $g_{wv}$ was reduced enough to "overcompensate" for the increased evaporative demand. This combination of low $g_{wv}$ and high $\Psi_l$ may explain part of the data in Figure 1. The points on the left side of the figure represent data primarily from trees in a very dry atmosphere. Stomatal closure in this environment limited $A_s$, but also reduced transpiration to the point that $\Psi_l$ remained high.

## Comparisons with Other Species

Compared with conifers, red alder generally has a much higher maximum $g_{wv}$. In addition, maximum $g_{wv}$ declines at a much higher predawn $\Psi_l$. The maximum $g_{wv}$ reported for red alder varies between 0.8 and 1.0 cm s$^{-1}$ (Pezeshki and Hinckley 1982, 1988; Lu 1989; T. B. Harrington and B. J. Yoder, unpub. data; B. J. Yoder et al., unpub. data), although Lu (1989) reported some measurements as high as 3.0 cm s$^{-1}$. In contrast, the maximum $g_{wv}$ for most conifers ranges between 0.2 and 0.4 cm s$^{-1}$ (Running 1976). No rigorous tests with red alder have been conducted to evaluate precise relationships between predawn $\Psi_l$ and maximum $g_{wv}$, but it appears that as predawn $\Psi_l$ values fall from -0.2 to -0.4 MPa, maximum conductance achieved during the day is reduced. When predawn $\Psi_l$ values reach from -0.6 to -1.0 MPa, maximum $g_{wv}$ is nearly zero, which suggests that stomata are almost completely closed (inferred from Pezeshki and Hinckley 1982; Lu 1989). In contrast, Waring

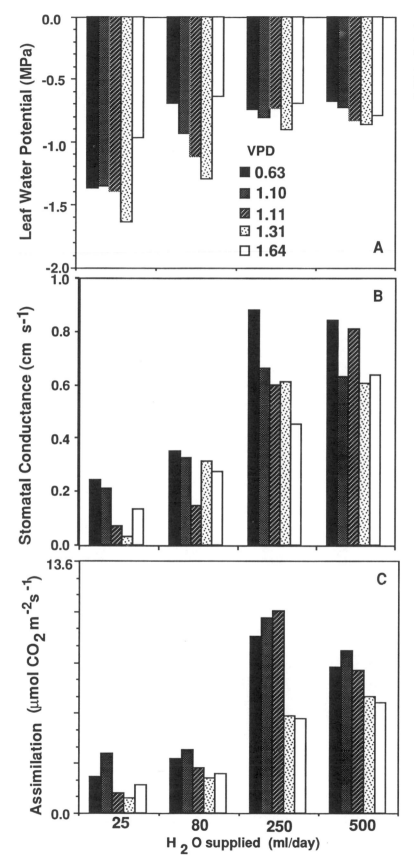

*Figure 2. (A) Leaf water potential, (B) stomatal conductance, and (C) $CO_2$ assimilation rates of red alder in response to soil water supply and vapor pressure deficit (From: B. J. Yoder et al., unpub.).*

et al. (1981) reported that $g_{wv}$ of Douglas-fir was about half of maximum when predawn $\Psi_l$ was -1.0 MPa, and complete stomatal closure did not occur until predawn $\Psi_l$ reached about -1.5 MPa.

Several experiments have demonstrated that when red alder is grown alongside other species, predawn and midday $\Psi_l$ values often differ significantly between species. For example, T. B. Harrington et al. (unpub. data) demonstrated that predawn and midday $\Psi_l$ values of red alder seedlings were almost always higher that those of Douglas-fir of the same age when they were grown together on three different sites (Fig. 3). The differences were most pronounced in August at the most droughty site (the Belfair, WA, site had rocky soil with low water-holding capacity). The $\Psi_l$ of the two species appeared to track environmental conditions similarly over the season measured; when $\Psi_l$ of red alder increased, so did that of Douglas-fir, and when alder's $\Psi_l$ recovered, so did that of Douglas-fir.

Cole and Newton (1986) also examined red alder and Douglas-fir across a variety of sites in the Pacific Northwest and consistently demonstrated that red alder maintained values of higher $\Psi_l$ than did Douglas-fir. Shainsky and Radosevich (1992) and S. S. N. Chan (unpub. data; Figs. 4A and 4B) showed that red alder $\Psi_l$ was always higher than

that of Douglas-fir, whether stress was induced artificially by withholding water directly or through competition for water. Cole and Newton (1987) hypothesized that differences in predawn $\Psi_l$ occurred between red alder and Douglas-fir seedlings because of the ability of red alder (1) to preempt soil moisture in the common root space, thus making it unavailable to Douglas-fir, and/or (2) to exploit parts of the soil not exploited by Douglas-fir. Current indications are that both hypotheses are true. For example, Shainsky (1988) showed that 3-year-old red alder consumed more soil moisture deeper in the soil and earlier in the growing season than did 3-year-old Douglas-fir companions.

Differences in allocation were also found in an analysis of Douglas-fir and red alder seedlings that were grown with and without shade, and with and without irrigation (S. S. N. Chan, unpub. data). Chan's results indicated that shaded, unirrigated red alder had a significantly higher proportion of biomass in roots compared with seedlings under other treatments. Further, Chan's data suggest that red alder may maintain a more narrow range in predawn $\Psi_l$ than Douglas-fir's by shifting allocation between shoots and roots in response to resource limitations. Even if red alder and Douglas-fir allocated the same proportion of their resources below ground, it is likely that the alder is able to

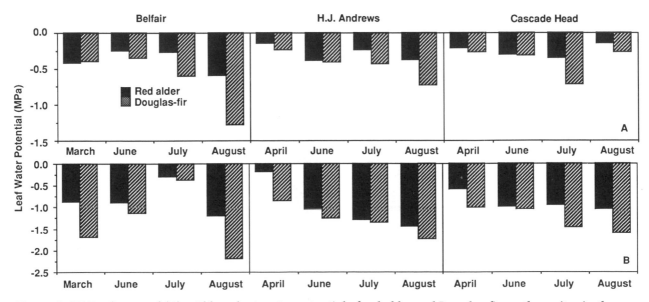

Figure 3. (A) Predawn and (B) midday plant water potential of red alder and Douglas-fir on three sites in the Pacific Northwest (From: T. B. Harrington et al., unpub.).

exploit a different soil volume. While Yoder et al. (1988) found no difference in ratios of shoot to root for red alder and Douglas-fir seedlings at the three sites used in the study of T. B. Harrington et al. (unpub. data), they also observed that fine roots of red alder seedlings tended to occur in much deeper soil strata than did those of same-aged or even same-sized Douglas-fir grown at the same site.

The differences between the $\Psi_l$ values of red alder and Douglas-fir seedlings at midday can be explained by the same mechanisms discussed for the differences in predawn $\Psi_l$. In addition, stomatal closure and leaf abscission play a role in species differences, reducing transpirational water loss at much higher $\Psi_l$ values for alder than for Douglas-fir. This greater sensitivity is characteristic of a drought avoider and may allow $\Psi_l$ to remain higher during the day, but with the potential for sharp decreases in the capacity to fix carbon.

Another probable reason for the high $Y_l$ of red alder seedlings compared with Douglas-fir is that alder seedlings die at levels of soil drought that Douglas-fir can endure. Predawn $Y_l$ for red alder in the field has rarely been reported below -0.5 MPa. Although values were measured between -0.5 and -1.0 MPa at Belfair, WA (Fig. 3A), subsequent mortality was extremely high (T. B. Harrington et al., unpub. data). Preliminary analysis of the data of T. B. Harrington et al. (unpub. data) suggests that red alder mortality was high where predawn $Y_l$ of Douglas-fir seedlings fell below -1.0 MPa. It is possible that unsuitable planting sites for red alder may be diagnosed by assessing whether or not minimum predawn $Y_l$ of Douglas-fir falls below this value. Further study is needed to verify this relationship.

Black cottonwood is another common associate that has been compared with red alder. When Pezeshki and Hinckley (1988) grew red alder and black cottonwood seedlings under identical conditions of soil moisture and VPD, the alder tended to have a slightly lower predawn $\Psi_l$. The difference between species may be the result of differing rooting behavior: cottonwood seedlings may have a more extensive rooting system and/or have roots in contact with soil of higher water potential. Pezeshki and Hinckley (1988) also observed that cottonwood $g_{wv}$ and $\Psi_l$ generally followed similar trends to the response of red alder to the environ-

ment. Cottonwood, however, maintained a higher $g_{wv}$ and $\Psi_l$ than did red alder under all conditions of VPD and soil water availability examined. Greater $g_{wv}$ in cottonwood coupled with higher $\Psi_l$ suggests that cottonwood has the ability to supply more water to its leaves. Many different mechanisms for this are possible. For example, cottonwood, a riparian species, may have relatively larger xylem vessels, which reduce resistance to water flow from roots to leaves. Data (Pezeshki and Hinckley 1988) suggest that both species exhibit physiological characteristics of drought avoiders. Relative to cottonwood, however, which tended to avoid low $\Psi_l$, red alder endured lower water potentials. Pezeshki and Hinckley (1988) showed that the cost of such endurance by red alder is premature leaf abscission.

## Radiation and Temperature

If no other factors are limiting, $A_s$ and $g_{wv}$ of red alder appear to vary with the diurnal cycle of photosynthetically active radiation (PAR) (Krueger and Ruth 1969; Webb et al. 1974; Pezeshki and Hinckley 1982). Pezeshki and Hinckley (1982) used boundary-line analysis to interpret the responses of $g_{wv}$ to PAR. In the absence of limitations imposed by VPD or soil water stress, $g_{wv}$ of red alder seedlings increased steadily with increasing PAR before noon. Stomata began to open after PAR reached 80 $\mu$mol m$^{-2}$s$^{-1}$; opening was rapid thereafter. A maximum $g_{wv}$ of 0.44 cm s$^{-1}$ was observed when PAR was approximately 400 $\mu$mol m$^{-2}$s$^{-1}$.

Webb et al. (1974) demonstrated how light and temperature can interact to influence $A_s$ of red alder. Their data differ from that of most other studies, because they measured exchange rates of whole plants, rather than of single leaves. Webb et al. (1974) found that increasing PAR (1) caused an increase in maximum $A_s$ at a given temperature, (2) increased the temperature at which the maximum $CO_2$ exchange occurred, and (3) altered the shape of the overall temperature response curve. As expected, the optimum temperature for photosynthesis increased as PAR increased: under low light the optimum temperature was approximately 12°C and under high light the optimum temperature increased to about 22°C. This is in part the result of increased photorespiration due to increased light and increased temperature, although light-tem-

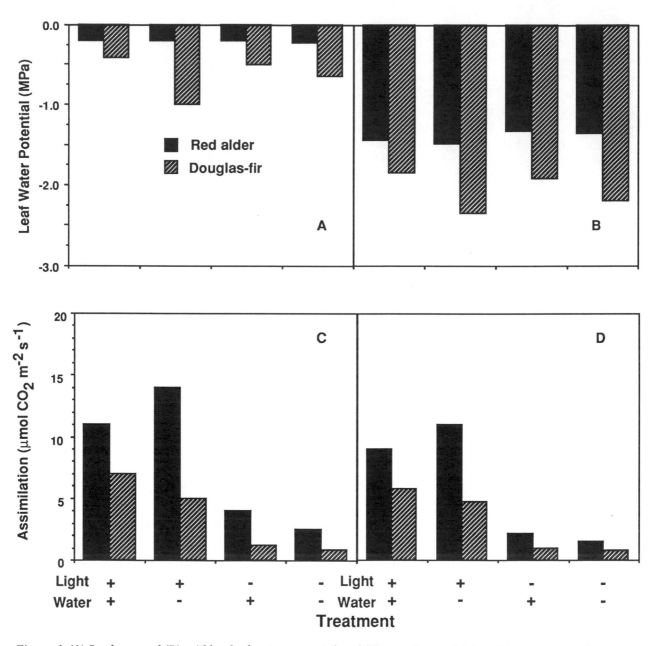

*Figure 4. (A) Predawn and (B) midday leaf water potential and (C) morning and (D) midday CO$_2$ assimilation rates of red alder and Douglas-fir in response to light and water in the field (From: S. S. N. Chan, unpub.). "+" indicates the resource was provided, "+ light" = full sun. "-" indicates the resource was withheld, "+ water" = irrigated.*

and increased temperature, although light-temperature interaction varies greatly with species and degree of acclimation (Berry and Bjorkman 1980). The broad range in temperature optima observed by Webb et al. (1974) illustrates that care must be taken to account for interactive effects of environmental factors in models of physiological activity.

Light and temperature affect many plant functions in addition to $g_{wv}$ and $A_s$. We have observed field plantings of red alder seedlings damaged by temperature extremes that do not harm associated conifers. One common symptom of very high temperatures is blistering of the cambium just above the root collar. The analysis of critical temperatures leading to mortality, and the associated physiology, is a fruitful area for future research.

### Interactions of Radiation and Temperature with Water

Recently, S. S. N. Chan (unpub. data) examined the interaction between light and water limitations on red alder growth and physiology by manipulating resource levels artificially in the field. Generally, the $\Psi_l$ of red alder did not respond to differences in watering regime or light (Fig. 4A). Mid-morning and afternoon $A_s$ rates were affected primarily by light and to a lesser degree by water (Figs. 4C, 4D). While physiological indicators varied little, biomass totals and aboveground and below-ground allocation were strongly affected by water stress. Red alder seedlings in this experiment appeared to maintain a constant $\Psi_l$ and to reduce variability in $A_s$ and $g_{wv}$ across the different watering regimes through regulation of their ratios of root to shoot biomass.

Data from J. H. Dukes (unpub. data) and S. S. N. Chan (unpub. data) indicate that an optimum PAR exists for $A_s$ (Figs. 5A and 5B), and that $A_s$ declines at very high PAR. For PAR < 400 µmol $m^{-2}s^{-1}$, $A_s$ increased with increasing PAR (Fig. 5A). No interactive effects between light and watering regime were apparent at low light. In contrast, $A_s$ declined as PAR increased to values between 800 and 1200 µmol $m^{-2}s^{-1}$. Curiously, at these high light levels unwatered trees had generally higher $A_s$ than did watered ones. The optimum PAR, inferred from the whole data set, was approximately 800 to 1000 µmol $m^{-2}s^{-1}$. Unfortunately, a gap in data between the shaded and unshaded treatments prevents a clear interpretation of an apparent parabolic response of $A_s$ to PAR. Laboratory observations on cut shoots by J. H. Dukes (unpub. data; Fig. 5B) for the two most contrasting treatments (i.e., high light/watered and low light/unwatered) in S. S. N. Chan's study (unpub. data) also showed a parabolic response of $A_s$ to PAR for leaf samples from shaded/unwatered treatments. The decline at high light is likely the result of photoinhibition, to which the shade-treated and water-stressed leaves would be particularly susceptible. Further, foliage from water-stressed treatments may have had lower $g_{wv}$, and in the absence of evaporative cooling, leaf temperature would have increased to levels that limited $A_s$.

The apparent value for PAR saturation in Pezeshki and Hinckley's (1982) study (approximately 400 µmol $m^{-2}s^{-1}$) was much lower than the apparent saturation levels (about 800 to 1000 µmol $m^{-2}s^{-1}$) found by S. S. N. Chan (unpub. data) and J. H. Dukes (unpub. data). In the former study, multiple environmental factors interacted with light, and this may have lowered the light saturation value. Another difference between the studies was that Pezeshki and Hinckley (1982) measured the response of $g_{wv}$ to light, whereas S. S. N. Chan (unpub. data) and J. H. Dukes (unpub. data) both measured $A_s$. Lu (1989) found that the upper boundary of $g_{wv}$ increased steadily with PAR over a range between 100 and 1000 µmol $m^{-2}s^{-1}$. Clearly, a variety of factors influence light-gas exchange phenomena in complex ways that deserve further investigation.

DeBell et al. (1989) showed that high temperatures (above 30°C) can limit $A_s$ of red alder under saturating light conditions. This response is similar to the temperature response of most temperate trees, in which $A_s$ peaks between 15 and 25°C. The high temperature compensation point (where respiration and photosynthesis are balanced) of temperate trees usually is reached between 35 and 40°C (Larcher 1969). DeBell et al. (1989) also found that irrigation affected the response of $A_s$ to temperature. The $CO_2$ assimilation rate of unirrigated trees declined more steeply at higher temperatures with increasing temperature than did that of irrigated trees. This interaction between drought and

*Figure 5. CO₂ assimilation of red alder versus photosynthetically active radiation (PAR) measured (A) in the field on the plants presented in Figure 4 (From: S. S. N. Chan, unpub.), and (B) in the laboratory; the same individuals were stepped through the PAR gradient (From: J. H. Dukes, unpub.).*

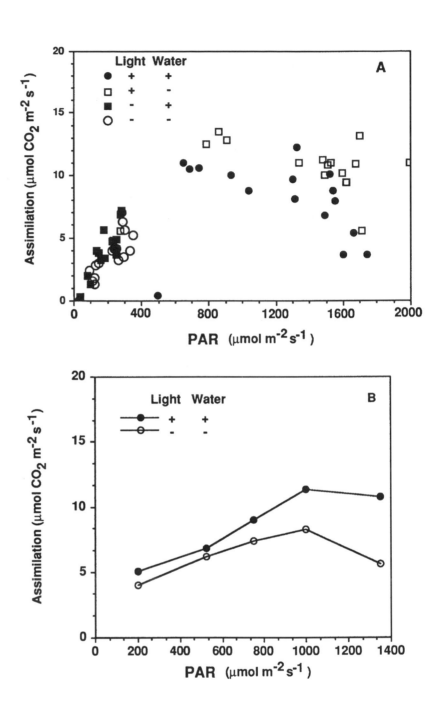

temperature is expected (Larcher 1969; Berry and Bjorkman 1980); conditions such as low soil water or high VPD lower $g_{wv}$ and cause $CO_2$ concentrations in the leaf mesophyll to decline. The low internal $CO_2$, in turn, becomes increasingly limiting to photosynthesis as temperature increases (Berry and Bjorkman 1980).

The foregoing discussion illustrates that the response of $g_{wv}$ and $A_s$ to light is highly dependent on other environmental factors. In addition, leaves grown in different light environments can acclimate to an order of magnitude of variation in PAR (Berry and Downton 1982). Comparisons between studies and between species, therefore, must be made with caution.

*Comparisons Among Species*

Some generalizations can be made regarding differences among species in responses of $A_s$ or $g_{wv}$ to radiation. Although discrepancies exist in the light saturation level reported for photosynthesis of red alder, most studies indicate a saturation level around 800 to 1000 $\mu mol\ m^{-2}s^{-1}$. Pezeshki and Hinckley (1982) observed a saturation PAR for stomatal conductance of 400 $\mu mol\ m^{-2}s^{-1}$ for both red alder and black cottonwood. Brix (1967) found that $A_s$ of Douglas-fir foliage saturated when PAR was about 400 $\mu mol\ m^{-2}s^{-1}$ (an approximate conversion from foot-candles in the original data set) at a temperature of 18°C.

Krueger and Ruth (1969) compared light saturation curves for photosynthesis of Douglas-fir, western hemlock, Sitka spruce, and red alder seedlings that were preconditioned with shade. This preconditioning limits inferences with regard to red alder, which typically grows in full sun. On a leaf-area basis, few significant differences existed in assimilation rates of the four species at any given radiation level. On a leaf-weight basis, red alder assimilation was two to three times the rates observed for the conifers. This is in part because the specific leaf area (leaf area per unit leaf dry weight) of red alder is much greater than that of the conifers.

Species differences in response to light and water regime between Douglas-fir and red alder were found by S. S. N. Chan (unpub. data; Fig. 4). The $A_s$ rates of red alder were 1.5 to 2 times greater than those of Douglas-fir, and red alder maintained

higher plant water potential. The $A_s$ rates of trees of both species growing in full sun were 2 to 2.5 times greater than those of trees grown under shaded conditions (15 percent full sun) (Fig. 4). Furthermore, species demonstrated differences in their response to the interactions between light and water. Under full sun, withholding water reduced the $A_s$ of Douglas-fir and slightly increased the $A_s$ of red alder. Under shaded conditions, $A_s$ of Douglas-fir appeared insensitive to changes in soil moisture supply, whereas that of red alder was reduced by withholding irrigation (Fig. 4).

## Carbon Dioxide

The effects of various soil water and carbon dioxide ($CO_2$) regimes on the growth and physiology of different seed sources of red alder under greenhouse conditions were explored by S. S. N. Chan et al. (unpub. data). Little variation was attributed to seed source. Unwatered trees tended to have lower $\Psi_l$ than did watered trees, although treatment separation was not more than -0.2 MPa. As expected, regardless of watering regime, $A_s$ was higher for trees growing under elevated $CO_2$ than for trees growing under ambient $CO_2$ conditions.

Chan et al. (unpub. data) also showed that enhanced $CO_2$ levels caused a small shift in the threshold $\Psi_l$ for limitation of $A_s$. Similar to the threshold noted earlier, water stress appeared to place constraints on $A_s$ as $\Psi_l$ values fell below -0.75 MPa under ambient $CO_2$ conditions. Under enhanced $CO_2$ conditions, the apparent threshold $\Psi_l$ was about -0.9 MPa. This trend suggests that red alder could achieve higher carbon gain under greater water stress conditions as $CO_2$ in the atmosphere increases.

These observations demonstrate that individual leaf- and tree-level responses to environmental conditions are complex. Interactions between environmental factors play a key role in mediating red alder physiology in ways we have not yet elucidated experimentally. Optima of photosynthesis in response to a particular environmental factor shift as other factors begin to limit physiological function. Our understanding of leaf-level responses is growing. In the next section we examine how neighboring plants may mediate individual leaf-level and tree-level photosynthesis and water relations by altering each other's functional environment.

## Physiology and Density in Monoculture and Mixed Stands

Responses of red alder to density have been intriguing. Conventional theory (e.g., Harper 1977) predicts that increasing plant density reduces the amount of space available to each individual, as well as all the resources in that space. This decline in resources can ultimately limit growth when demand outgrows supply. Most studies in which stand density is manipulated show that growth of red alder individuals declines as density increases (Smith and DeBell 1974; Smith 1986; Hibbs 1987; DeBell et al. 1989; Shainsky and Radosevich 1991, 1992). In contrast, some whole-stand spacing trials cited by DeBell et al. (1989) and most studies utilizing the Nelder design (an experimental planting that establishes a density gradient; Nelder 1962) show that growth of individuals increases as space per tree increases to some intermediate spacing, then declines at wider spacings (DeBell and Giordano, Chapter 8). Bormann and Gordon (1984) also showed parabolic responses in individual-tree and whole-stand parameters with increasing density. Similar parabolic growth responses to stand density have also been shown for black cottonwood by W. Emmingham (pers. com.). These observations suggest that, for a given site, an optimal space per tree exists across a broad gradient, and that increasing space per tree influences the growth environment, and hence physiology, in complex ways. Few studies, however, have coupled collection of data on growth with data on physiology. We would hypothesize that key physiological indicators, such as assimilation, plant water potential, and stomatal conductance, would follow patterns similar to those observed for growth.

DeBell et al. (1989) conducted several spacing studies in which red alder density was manipulated in experimental plantings. In some cases, diameter, height, and other growth variables increased as spacing increased, which suggests that competition was occurring. In other cases, height increased as spacing increased and then declined at wider spacings. Trends in assimilation rates suggested a parabolic response to density—$A_s$ was often highest at the intermediate spacing and declined at the widest spacing. Unirrigated plots displayed this trend more distinctly even though density did not appear to significantly affect $\Psi_l$. These data suggest that water is in part mediating the response to tree density, likely through both soil moisture depletion and canopy humidity mechanisms.

Giordano and Hibbs (1993) showed that as amount of space available per tree increased, predawn $\Psi$ measured late in the growing season increased from -0.42 MPa at the lowest space to -0.13 MPa at the greatest space available. The effect of density on $\Psi_l$ increased through the growing season; this suggested that late-season water resources were more limiting and could not keep pace with the evapotranspiration demands during summer dry periods. Their data indicate that a gradient in soil moisture was created by the density gradient, and that competition for water was occurring. In a related study on the same site, T. B. Harrington et al. (unpub. data) found that soil moisture depletion was significantly greater at closer spacings than at wider spacings. Giordano and Hibbs (1993) provided convincing evidence that relative growth rates become limited by plant competition for soil moisture.

In contrast, data from Lu (1989) showed declining $g_{wv}$ with increasing space per tree in the Cascade Nelder. Measurements, however, were made in such a way that time of measurement and position in the Nelder were confounded, and sampling progressed from close to wide spacings. Thus, it is likely that diurnal changes in environmental factors are also in effect. In addition, humidity may have been higher in the denser canopies in the central portion of the Nelders, which would support higher stomatal conductance at closer spacings.

Shainsky and Radosevich (1992) demonstrated clearly that red alder water stress increases as density in monoculture increases in a manner predicted by conventional theory (Harper 1977). Late-season predawn $\Psi_l$ declined as red alder density increased (Fig. 6A). The amount of soil moisture depleted per tree from the 90-cm measured profile also declined as tree density increased (Fig. 6B). This observation supports the hypothesis that resources available per plant decline as density increases. The decline in soil moisture depleted per tree was correlated with the decline in predawn $\Psi_l$ (Fig. 6C), thus indicating a

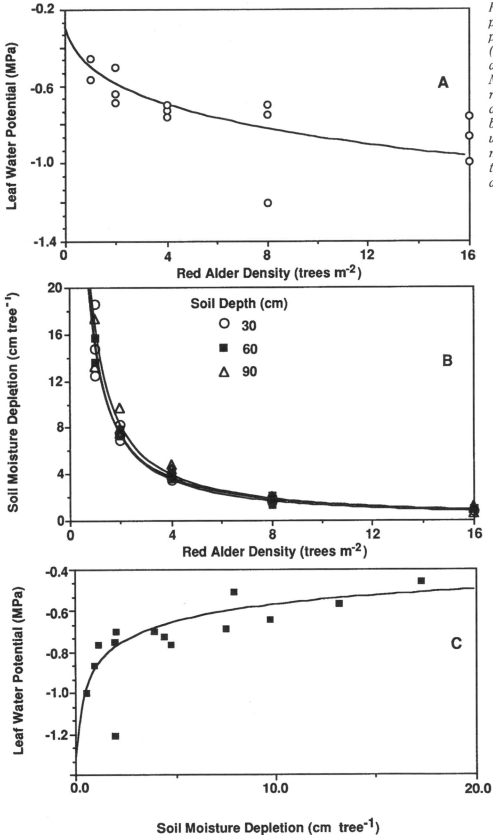

*Figure 6. Red alder (A) predawn leaf water potential in August and (B) soil moisture depleted per tree from May to August in response to red alder density; (C) relationship between predawn leaf water potential and soil moisture depletion per tree (From: Shainsky and Radosevich 1992).*

crucial link between resource availability and physiological well-being.

Links between species density, resource availability, water relations, and growth in mixed stands of red alder and Douglas-fir are illustrated in the conceptual model developed by Shainsky and Radosevich (1992; Fig. 7). They illustrated how systematic manipulation of red alder and Douglas-fir densities in mixed stands resulted in quantifiable changes in resources such as light and soil water. These changes in environment and resource conditions interactively affected leaf area, canopy structure, height, and diameter growth, as mediated by physiological factors such as plant water relations. The conceptual model was derived from analysis of the correlation matrix of all measured variables in a two-way density gradient (Shainsky and Radosevich 1992). The strength of the inter-relationships are represented by the simple correlation coefficients associated with each linkage.

As in most scenarios studied, red alder grew much faster than did Douglas-fir. The mixed-species stands thus formed a two-tiered canopy with red alder in the overstory. Whereas red alder density had the dominant effect on all parameters measured, the understory Douglas-fir also exerted significant influence. Increasing the Douglas-fir density increased Douglas-fir leaf area m$^{-2}$, and reduced the red alder leaf area m$^{-2}$. Increasing the alder density resulted in increased leaf area of red alder m$^{-2}$ and reduced Douglas-fir leaf area m$^{-2}$. The overstory red alder leaf area and canopy structure were thus modified by the opposing effects of alder and Douglas-fir densities. The balance between these effects regulated the amount of light potentially available for photosynthesis. Alder directly altered the light environment through interception, whereas Douglas-fir altered it indirectly by influencing leaf area production of the overstory alder. Douglas-fir relative growth rates were positively correlated with relative photon flux density (p < 0.001), likely a response mediated by light-induced changes in photosynthesis (S. S. N. Chan, unpub. data; J. H. Dukes, unpub. data).

Tree density also influenced the soil moisture environment by affecting transpirational leaf area, and through density effects on root biomass (Shainsky et al. 1992). Increasing the density of trees reduced the amount of soil moisture potentially depleted per individual. Reductions in soil moisture depletion per tree were highly correlated with increases in plant water stress. Relative growth rates were linearly related to plant water stress; high plant water stress exhibited by both species was associated with low relative growth rates of both species (Shainsky and Radosevich 1992). Unpublished physiological data of S. S. N. Chan gathered from trees grown at the same site also support the hypothesis that density-mediated limitations in soil water induce plant water stresses sufficient to reduce photosynthetic rates and durations, and hence growth. Species differences in access to soil moisture, moisture use efficiency, light interception, and photosynthetic responses all appear to interact to shape growth in competitive environments.

## Implications to Ecology of Red Alder

The role of red alder as an aggressive, early seral species is demonstrated by its commanding early height and biomass growth and its leaf display. How does red alder maintain such rapid growth rates? We can partially explain the rapid growth rate of red alder by differences in $A_s$. Data from S. S. N. Chan (unpub. data; Figs. 4C, 4D) demonstrated the greater $A_s$ rates of red alder compared to its associate Douglas-fir at the seedling stage. Krueger and Ruth (1969) demonstrated that on a per unit biomass basis, $A_s$ rates of red alder were higher than those of its common conifer associates, while rates were only 20 percent greater on a per unit leaf area basis. In Krueger and Ruth's experiment, red alder displayed a greater capacity to respond to increased light than did the conifers. It should be noted that the trees were measured after being pretreated with 31 or 79 percent shade. More field measurements on these species under ambient light conditions might help to elucidate physiological differences.

Most early seral species tend to be shade intolerant. The importance of radiation in red alder ecology is expressed in reported responses of $A_s$, $g_{wv}$, and $\Psi_l$ to temperature, PAR, and VPD. Data presented here indicate a PAR optimum of approximately 800 to 1000 $\mu$mol m$^{-2}$s$^{-1}$ and a temperature optimum of approximately 20°C (although both

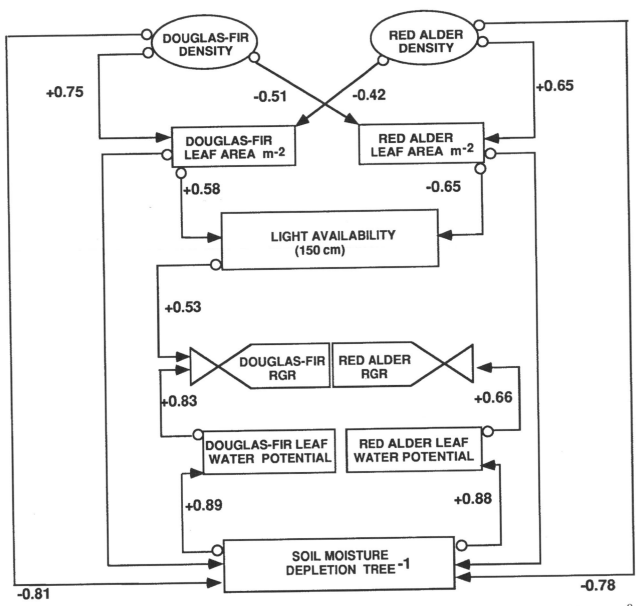

Figure 7. Conceptual model for interrelationships between stand-level factors such as density and leaf area m$^{-2}$, resource availability such as light and soil moisture depletion, and physiology and growth in mixed and monoculture stands of red alder and Douglas-fir (From: Shainsky and Radosevich 1992). Numbers linking the variables represent significant correlation coefficients ($p < 0.05$, $n = 80$ to $88$).

were variable). Gas exchange appears to be quite sensitive to environmental variability, and this is evidenced by declines in $A_s$ with PAR and temperature at high values of both (DeBell et al. 1989; S. S. N. Chan, unpub. data; J. H. Dukes, unpub. data). Chan's data show that light starvation has a greater impact on biomass accumulation of red alder than on that of Douglas-fir.

The preference of red alder for "humid or superhumid" climates (Harrington 1990) and the importance of water in regulating distribution and growth of red alder are supported by data presented here. In particular, studies clearly demonstrate the dependence of the duration of gas exchange processes on VPD and the limiting nature of $\Psi_l$. Red alder $\Psi_l$ rarely has been observed below -1.5 MPa. The data of S. S. N. Chan (unpub. data) and B. J. Yoder et al. (unpub. data) demonstrate that duration of $A_s$ is limited by a threshold $\Psi_l$ of approximately -0.8 MPa. Water supply and VPD interact to affect $\Psi_l$ and $A_s$. Reduced water supply to the roots appears to enhance the sensitivity of stomata of red alder to fluctuations in VPD (Pezeshki and Hinckley 1982, 1988; B. J. Yoder et al., unpub. data). How $\Psi_l$ drops throughout the course of a day or season depends on complex processes involving soil water inputs, relative humidity, the progression of $g_{wv}$, soil delivery (texture), and previous stress history. The data of S. S. N. Chan (unpub. data) suggest that narrow ranges in predawn $\Psi_l$ are the result of shifts in biomass allocation: red alder maintains high plant water potential by investing relatively more biomass to roots. Mortality of trees or parts of trees (e.g., leaf abscission) at lower water potentials and extensive exploration of soils by a given root biomass may also contribute to the narrow range of $\Psi_l$. During daylight, these factors plus stomatal closure at a relatively high $\Psi_l$ may keep the range of diurnal $\Psi_l$ narrow compared with that of conifers.

If soil water supply drops below the point at which red alder can no longer balance water loss with water gained through shifts in allocation and physiology, premature abscission of leaves occurs. Shedding of leaves is a classic adaptation for drought avoidance and is frequently observed in red alder under droughty conditions (Pezeshki and

Hinckley 1982, 1988; DeBell et al. 1989; Shainsky and Radosevich 1992). This is one important way in which stress history can feed back on subsequent carbon gain.

Soil disturbances can induce dense monoculture thickets of red alder seedlings. In these dense thickets, an array of different plant interactions occur that mediate individual tree growth, self-thinning, invasion of species into the understory, and other system dynamics. Presence of proximate, transpiring neighbors could increase local humidity and reduce VPD around the leaves. Interior trees would experience lower wind speeds and a greater aerodynamic boundary layer, which would further reduce VPD. These conditions could lead to reduced transpirational water loss relative to carbon gain (i.e., increased water use efficiency). These effects may be subtle and are likely offset by lower light, water, and nutrients available to each plant in dense stands. Competitive processes work to reduce carbon gain through limitations on photosynthesis and allocation as stands become more dense. The interactions between soil water, humidity, and radiation discussed here for individual trees under experimental conditions may underlie some of the intriguing density responses discussed above and elsewhere in this volume (DeBell and Giordano, Chapter 8; Newton and Cole, Chapter 7).

Data from Cole (1984), Cole and Newton (1986, 1987), Shainsky and Radosevich (1992), S. S. N. Chan (unpub. data), and T. B. Harrington et al. (unpub. data) highlight key species differences that bear upon the mechanisms of compatibility of red alder and Douglas-fir in mixtures. When the two species are established at the same time, and when red alder composes more than 10 percent of the stand, red alder clearly dominates the depletion of light by virtue of its superior height growth, and the depletion of water resources by virtue of its greater mass and extent of roots (Newton et al. 1968; Cole 1984; Shainsky et al. 1992; S. S. N. Chan, unpub. data). These consumptive effects contribute greatly to the reduction of Douglas-fir growth. Shainsky and Radosevich (1992), however, also showed that very high densities of Douglas-fir in the understory exert a limitation in water and light, thus reducing the leaf area and modifying the canopy architecture of overstory red alder

(Shainsky 1988). This effect on the overstory, in turn, influences the degree of suppression of Douglas-fir by red alder. This research underscores the importance of feedback and interactions between biotic and abiotic factors in mediating competition. As with the work of Bormann and Gordon (1984), it suggests how individual plant responses can create whole-stand features that interact to mediate changes in resources and conditions, and subsequently feed back on physiology and ultimately growth.

These observations imply that there is likely a unique optimal mix and spacing for red alder and Douglas-fir for every site and planting scenario. Effects on and responses of physiology to density and species proportion are complex and iterative, and can produce nonlinear growth responses not predicted by conventional theory. Positive effects from neighbors—raised humidity, far-red stimuli, reduced wind speed, herbivore saturation—interact with or add to negative effects of competing neighbors to influence the carbon balance of individual plants.

## Conclusions and Future Research

The dynamic nature of tree, stand, and forest processes presents us with a challenge when interpreting physiological data. We have made great progress in research of red alder physiology since the last synthesis. We must, however, continue to examine physiological behavior and to identify thresholds and limits so that we may understand red alder's potential and actual ranges, appreciate its unique character, and evaluate the suitability of sites for productive stands with red alder. Even though physiological indicators of stress provide some insight into the mechanisms of ecological processes, physiological data are of limited use to management unless direct links to productivity can be demonstrated. Although statistical and conceptual models are being developed, we are far from the elaboration of quantitative mechanistic models that predict growth of pure and mixed stands from physiological variables. Few data sets are complete enough to produce robust relationships between resources, physiology, and growth. Those that do exist are limited to young, dense plantings.

Relationships between plant $\Psi_l$ and growth have been documented in experiments designed to examine the mechanisms of competition (Giordano and Hibbs 1992; Shainsky and Radosevich 1992). Relationships between growth and $A_s$ are more problematic, because $A_s$ can be transient during any given measurement period. Unusual patterns in data from Lu (1989) and T. B. Harrington et. al. (unpub. data) compel us to recommend pursuit of measurements with more mindful protocol. Treatments must be measured randomly through time. In addition, estimates of $A_s$ and $g_{wv}$ should come from a variety of leaves on an individual tree and be coupled with measurements of both humidity of the air (not the chamber), and PAR on leaves from different parts of the canopy (not just the most fully exposed).

A truly integrated view of red alder physiology is beyond our reach at this time. Although information about leaf water potential and gas exchange is emerging, most of it represents small snapshots of parameters observed on individual leaves. More mature specimens must be examined. Analysis of dynamics in physiological parameters integrating information at the levels of whole trees and plant canopies is completely lacking. Physiological mechanisms of red alder ecology will more likely be elucidated with the development of systems that can simultaneously and continuously make multiple measurements of $A_s$ and environmental factors. More long-term mechanistic studies examining the balances between competition and benefits from nitrogen fixation and associated effects on the soil component must be conducted with a broader range of species densities and proportions, planted at more conventional spacings.

# Literature Cited

Berry, J., and O. Bjorkman. 1980. Photosynthetic response and adaptation to temperature in higher plants. Annu. Rev. Plant Physiol. 31:491-543.

Berry, J., and W. J. S. Downton. 1982. Environmental regulation of photosynthesis. *In* Photosynthesis, development, carbon metabolism, and plant productivity, vol. 2. *Edited by* R. Govindjee. Academic Press, New York. 263-343.

Bormann, B. T., and J. C. Gordon. 1984. Stand density effects in young red alder plantations: productivity, photosynthate partitioning, and nitrogen fixation. Ecology 65:394-402.

Brix, H. 1967. An analysis of dry matter production of Douglas-fir seedlings in relation to temperature and light intensity. Can. J. Bot. 45:2063-2072.

Cole, E. C. 1984. Fifth-year growth responses of Douglas-fir to crowding and other competition. M.S. thesis, Oregon State Univ., Corvallis.

Cole, E. C., and M. Newton. 1986. Nutrient, moisture, and light relations in 5-year-old Douglas-fir plantations under variable competition. Can. J. For. Res. 16:727-732.

———. 1987. Fifth-year responses of Douglas-fir to crowding and non-coniferous competition. Can. J. For. Res. 17:181-186.

DeBell, D. S., M. A. Radwan, D. L. Reukema, J. C. Zasada, W. R. Harms, R. A. Cellarius, G. W. Clendenen, and M. R. McKevlin. 1989. Increasing the productivity of biomass plantations of alder and cottonwood in the Pacific Northwest. Annual technical report submitted to the U.S. Dept. of Energy, Woody Crops Program.

Giordano, P. A., and D. E. Hibbs. 1992. Morphological response to competition in red alder: the role of water. Functional Ecology. (In press.)

Harper, J. L. 1977. Population biology of plants. Academic Press, New York.

Harrington, C. 1990. *Alnus rubra* Bong.—red alder. *In* Silvics of North America, vol. 2, Hardwoods. *Technically coordinated by* R. M. Burns and B. H. Honkala. USDA For. Serv., Agric. Handb. 654. Washington, D.C. 116-123.

Hibbs, D. E. 1987. The self-thinning rule and red alder management. For. Ecol. Manage. 18:273-281.

Hsiao, T. C. 1973. Plant responses to water stress. Annu. Rev. Plant Physiol. 24:519-570.

Krueger, K. W., and R. H. Ruth. 1969. Comparative photosynthesis of red alder, Douglas-fir, Sitka spruce, and western hemlock seedlings. Can. J. Bot. 47:519-527.

Larcher, W. 1969. The effect of environmental and physiological variables on the carbon dioxide exchange of trees. Photosynthetica 3:167-198.

Lu, S. 1989. Seasonal and diurnal trends of leaf water potential and stomatal conductance of red alder (*Alnus rubra* Bong.) growing along a density gradient in western Oregon. M.S. thesis, Oregon State Univ., Corvallis.

Nelder, J. A. 1962. New kinds of systematic designs for spacing experiments. Biometrika 18:283-307.

Newton, M., B. A. El Hassan, and J. Zavitkovski. 1968. Role of red alder in western Oregon forest succession. *In* Biology of red alder. *Edited by* J. M. Trappe, J. F. Franklin, R. F. Tarrant, and G. H. Hansen. USDA For. Serv., PNW For. Range Exp. Sta., Portland, OR. 73-84.

Nobel, P. S. 1974. Introduction to biophysical plant physiology. W. H. Freeman and Co., San Francisco.

Pezeshki, S. R., and T. M. Hinckley. 1982. The stomatal response of red alder and black cottonwood to changing water status. Can. J. For. Res. 12:761-771.

———. 1988. The water relations characteristics of *Alnus rubra* and *Populus trichocarpa*: responses to field drought. Can. J. For. Res. 18:1159-1166.

Running, S. W. 1976. Environmental control of leaf water conductance in conifers. Can. J. For. Res. 6:104-112.

Shainsky, L. J., M. Newton, and S. R. Radosevich. 1992. Effects of intra- and inter-specific competition on root and shoot biomass of young Douglas-fir and red alder. Can. J. For. Res. 22:101-110.

Shainsky, L. J. 1988. Competitive interactions between Douglas-fir and red alder seedlings: growth analysis, resource use, and physiology. Ph.D. thesis, Oregon State Univ., Corvallis.

Shainsky, L. J., and S. R. Radosevich. 1991. Analysis of yield-density relationships in experimental stands of Douglas-fir and red alder seedlings. For. Sci. 37:574-592.

———. 1992. Mechanisms of competition between Douglas-fir and red alder seedlings. Ecology 73:30-45.

Smith, J. H. G., and D. S. DeBell. 1974. Some effects of stand density on biomass of red alder. Can. J. For. Res. 4:335-340.

Smith, N. J. 1986. A model of stand allometry during the self-thinning process. Can. J. For. Res. 16:990-995.

Tibbits, T. W. 1979. Humidity and plants. BioScience 29:358-363.

Waring, R. H., J. J. Rogers, and W. T. Swank. 1981. Water relations and hydrologic cycles. *In* Dynamic properties of forest ecosystems. Int. Biol. Programme 23. *Edited by* D. E. Reichle. Cambridge Univ. Press, New York. 205-264.

Waring, R. H., and W. H. Schlesinger. 1985. Forest ecosystems, concepts and management. Academic Press, Orlando.

Webb, W. L., M. Newton, and D. Starr. 1974. Carbon dioxide exchange of *Alnus rubra*: a mathematical model. Oecologia 17:281-291.

Yoder, B. J., T. B. Harrington, and D. E. Hibbs. 1988. Allometric relationships of young red alder and Douglas-fir on four sites in western Oregon and Washington. Abstracts of the Annual Meetings of the American Association for the Advancement of Science, Pacific Division, Oregon State Univ., Corvallis.

Zobel, D. B. 1974. Local variation in intergrading *Abies grandis—Abies concolor* populations in the central Oregon Cascades. II. Stomatal reaction to moisture stress. Bot. Gaz. 135:200-210.

## Appendix: Summary of Designs of Unpublished Studies on Red Alder

Chan, S. S. N. One-year-old transplants of Douglas-fir and red alder were studied over three years at a field site located near Corvallis, Oregon. A total of 128 trees of each species were planted and grown under treatment conditions consisting of two levels of light (full light or 15 percent of full sunlight) and soil moisture (irrigated weekly or unirrigated). Growth, gas exchange, and plant water potential were measured during each growing season. Trees were harvested at the end of the second and third growing seasons to determine structural allocation patterns in response to light and water.

Chan, S. S. N., M. Castellano, D. E. Hibbs, and C. H. Niu. A controlled environment study on the effects of elevated $CO_2$ and soil moisture availability on whole-tree responses of four red alder genotypes. Four sources of red alder seeds were collected along a latitudinal gradient spanning approximately 1200 km from the central coast of Oregon to Queen Charlotte Island, British Columbia. Seeds were stratified, germinated, and grown for seven months in growth chambers under either ambient (300 µl $l^{-1}$) $CO_2$ or elevated (700 µl $l^{-1}$) $CO_2$ conditions. Two levels of soil water input were imposed on the seedlings approximately six weeks after germination. The two watering regimes consisted of (1) a well-watered condition in which seedlings were maintained near field capacity at all times, and (2) a moisture-limiting regime (approximately one-third as much water as near field capacity) with seedlings subjected to droughts before rewatering. Growth, morphological development, rhizosphere development, biomass and nutrient allocation, nitrogen fixation, plant water potential, and gas exchange responses were measured.

Dukes, J. H. A study of the effects of altering light, plant water potential, and relative humidity on the photosynthetic rates of red alder under controlled laboratory conditions. Shoot specimens used in the study were obtained from field-grown 4-year-old trees established under designated light and water treatments (S. S. N. Chan, unpub.). The shoots, which ranged in length from 15 to 20 cm, were excised while immersed in water, from the upper third of the tree canopy shortly after dawn. Plant water potential gradients ranging from -.03 to -2.20 MPa were induced by immersing the excised end of the individual shoots in various concentrations of polyethylene glycol solution and allowing the plants to equilibrate. Effects of exposing the trees to various gradients of light and water potential on photosynthesis were measured in a semi-enclosed gas exchange system with constant temperature and humidity. Plant photosynthetic response was measured over a light (PAR) gradient that ranged from 200 to 1600 µmol $m^{-2}s^{-1}$.

Harrington, T. B., and B. J. Yoder. A study of the photosynthetic rates of red alder seedlings at different densities (in a Nelder design) at Cascade Head Experimental Forest (Lincoln City, OR). Measurements were made on sunlit foliage with a LiCor 6000 portable gas exchange system on 8 and 14 July 1987, in the late morning and early afternoon. During measurements, air temperature averaged 20°C and $CO_2$ averaged 320 µl $l^{-1}$; both varied over a narrow range. Photosynthetically active radiation averaged 930 µmol $m^{-2}s^{-1}$ and varied between 500 and 2000 µmol $m^{-2}s^{-1}$.

Harrington, T. B., B. J. Yoder, and D. E. Hibbs. A field study of the predawn and midday $\Psi_l$ of 3- to 4-year-old Douglas-fir and red alder seedlings grown at three different sites: Belfair, WA; H. J. Andrews Experimental Forest (Blue River, Oregon); and Cascade Head Experimental Forest (Lincoln City, OR). Measurements of $\Psi_l$ were taken with a pressure chamber on a single day in each of four months—April, June, July, and August. Predawn measurements were taken prior to sunrise. Midday $\Psi_l$ measurements were taken between 12:00 and 17:00 PST.

Yoder, B. J., L. J. Shainsky, and D. E. Hibbs. A controlled-environment study analyzing the water relations and gas exchange response of red alder seedlings to different levels of applied water, VPD, and light. Two-year-old trees in 15-cm pots (trees had been pruned so that average height was about 60 cm) were randomly assigned to one of four watering levels: 25, 80, 250, or 500 ml $day^{-1}$. Trees were placed in a controlled environment chamber. At noon each day during the experiment, VPD was changed to one of the following settings: 0.63, 1.10, 1.11, 1.31, or 1.64 kPa. After equilibration overnight, measurements (with plants chosen in a random order) of $A_s$ and $g_s$ were made with a LiCor 6000 portable gas exchange system in the morning. Concurrent $\Psi_l$ was measured with a pressure chamber. Air temperature averaged 30°C, PAR averaged 680 µmol $m^{-2}s^{-1}$, and $CO_2$ averaged 340 µl $l^{-1}$ during measurements.

# 6

# Genetics of Red Alder and Its Implications for Future Management

ALAN A. AGER & REINHARD F. STETTLER

Red alder invites genetic study. It is abundant, grows rapidly, reproduces early and in large numbers, occurs in more or less even-aged stands permitting convenient phenotypic comparisons, and produces wood of desirable quality for a range of products from pulp to furniture. Yet, information on the genetics of red alder and such aspects as its mating system characteristics, population structure, and variation patterns and their evolutionary significance is limited. From a genetic perspective, red alder is clearly understudied, especially when compared with its sympatric conifer species. Likely reasons for this neglect are the lower market value of red alder, its aggressive establishment through natural regeneration and current abundance on logged-over former conifer sites, and, concomitantly, the low incentive to develop breeding programs for this species. Nonetheless, current interest in a more integrated approach to stream side management, increasing concerns about riparian and wetland habitats and their biodiversity, and new developments in biomass production for fuel and fiber (Lynd et al. 1991) give new prominence to red alder and call for more basic information on its genetics.

Our review draws on two previous reviews by Stettler (1978) and Hall and Maynard (1979). It adds new information in two areas in which significant advances have been made during the 1980s, namely, the description and interpretation of variation patterns in the species, and the documentation of genetic variation in symbiotic fixation of atmospheric nitrogen. In addition, this chapter addresses several basic questions regarding long-term tree improvement of red alder. Finally, suggestions are offered for future research on the genetics of this species.

## Evolution, Taxonomy, and Distribution

The genus *Alnus* evolved during the rise of the angiosperms in the Cretaceous period. Fossil evidence from that period indicates that the family Betulaceae arose from an early line of the Hamamelidaceae (Furlow 1979a, 1979b). *Alnus* and *Betula* have been shown on the basis of anatomical features to be the most primitive members of the Betulaceae, having formed a distinct line before other members, such as *Corylus, Ostrya, Ostryopsis,* and *Carpinus,* diverged and evolved to form another complex.

The geographic origin of the genus *Alnus* is uncertain, although the best available evidence points to the Asian land mass, either Southwest, East, or West Gondwanaland before Africa and South America separated. Furlow (1979a, 1979b) proposed that the three major divisions of the genus evolved in distinct regions of Asia before their migration to the New World via the Bering land bridge. Red alder presumably evolved after that migration from either white alder (*A. rhombifolia*) or thinleaf alder (*A. incanca* subsp. *tenuifolia*) (Murai 1968).

Alder taxonomy in general is somewhat confusing, owing largely to high levels of intraspecific variation, isolated relict populations, and frequent zones of introgression (Czerepanov 1955; Murai 1964; Furlow 1979a, 1979b; Hall and Maynard 1979). In the most recent revision focusing on New World species, Furlow (1979a, 1979b) distinguished three main groups to which he accorded subgenus status, namely, *Alnus, Alnobetula,* and *Clethropsis.* The basis for these subgenera are morphological and phenological traits, including leaf venation, exposure of pistillate catkins in the winter, bud and flower structure, as well as reproductive phenology. Red alder belongs to the subgenus *Alnus,* which also contains the majority of the other alder species that grow into trees.

Present-day distribution of red alder is along the west coast of North America, from Oakland, California, to Sitka, Alaska. In the southern portion of the range, the species is restricted to moist, coastal areas. Farther north, beginning in southern Oregon, red alder extends more inland to the slopes of the Cascade range, where it may grow to elevations of 1100 m. Except for several disjunct populations around Coeur d'Alene Lake, Idaho (Johnson 1968), and along the western front of the Blue Mountains of eastern Oregon (Wagner 1991), red alder does not grow east of the Cascades. Three alder species are sympatric with red alder, namely, *Alnus crispa* subsp. *sinuata*, *A. incana* subsp. *tenuifolia*, and *A. rhombifolia*.

## Breeding System

Alders are monoecious with the male and female flowers borne in separate aments. There are two reported departures from this pattern. *Alnus firma,* native to Japan and other portions of the Orient, is reportedly dioecious. A second deviation consists of hermaphroditic flowers that have been found in low frequency (0.5 to 2 percent) for individuals of *A. serrulata* (Rich 1889), and for *A. cordata, A. glutinosa,* and *A. rubra* (Millar 1977) in the University of Washington Arboretum. A low frequency of hermaphroditic flowers has also been observed in natural populations of red alder in western Washington (Ager, unpub. data).

Reproductive structures in red alder are preformed in the growing season preceding flowering. Male meiosis occurs in late summer, with anthesis preceding leafing out in the spring. Female meiosis occurs only after pollination, and the seed matures in the fall (Furlow 1979a). Flowering of male and female flowers within a tree is asynchronous, with peak receptivity preceding peak anthesis by a day or two. Presumably this is a mechanism that promotes outcrossing.

Mating patterns in alder species parallel those of other forest trees: alders are typically obligate outcrossers and, to a large extent, self-incompatible (see Hall and Maynard 1979). Most cases in which controlled self-pollinations were performed resulted in seed of low viability (1 to 12 percent) and seedlings that exhibit slower growth rate than

outcross progenies. Self-pollination of trees in natural populations of red alder yielded less than 3 percent filled seed (Ager, unpub. data). Hagman (1970, 1975) presented evidence that the mechanism that prevents selfing appears to operate in stylar tissues and manifests itself as reduced or inhibited pollen tube growth.

Observations on self crosses made with a scanning electron microscope have indicated severely reduced adhesion and germination of self pollen as compared to outcross pollen (Ager, unpub. data). The growth of putative self materials has been found to be generally good in several other alder species, in some instances equaling the growth of the outcrossed progenies (Chiba 1966; Weisgerber 1974). In selfed red alder, the data of Stettler (1978) indicate relatively poor growth.

The alder breeding system exhibits a rare capability among tree species for apomictic or agamous reproduction (the production of viable seed without gametic contribution from the pollen parent). Apomixis has been documented cytologically for only one species in the genus; Woodworth (1930) found, in what was probably a hybrid swarm of *Alnus incana* subsp. *rugosa* and *A. incana* subsp. *serrulata* (Hall and Maynard 1979), that normal seed set despite the absence of viable pollen production. Cytological examination implicated nucellar budding as one of the mechanisms responsible for viable seed production.

The occurrence of apomixis has been suspected in red alder (Stettler 1978) and *Alnus glutinosa* (Hagman 1975) when seeds were found in female strobiles that had been isolated but not pollinated. This phenomenon needs to be better documented via cytological and genetic analysis. A useful tool for this purpose would be the simply inherited *pinnatisecta* marker (Stettler, unpub. data). The capacity for apomixis in alders demands special precautions in the interpretation of breeding and hybridization results.

## Natural and Artificial Hybridization

Natural hybridization is common in alder; zones of introgression between some species can be found wherever ranges overlap. As would be expected, hybridization is more common among closely related taxa. Natural hybrids between some alder species were reported almost 150 years ago (see Chiba 1966). In fact, many of the taxa recognized as species today are probably the products of past introgression.

Putative natural hybrids of red alder have been reported with only one of the three sympatric alders, namely, thinleaf alder (*Alnus incana* subsp. *tenuifolia*) (Ager 1987; Wagner 1991). These were found in the Oregon Cascade range (Ager 1987) and in the Blue Mountains of eastern Oregon (Wagner 1991). In these observations, individuals with morphology intermediate between red and thinleaf alder were observed along with the parental species.

The apparent reproductive barriers between red alder and other alders are not fully understood. Both Hagman (1975) and Millar (1977) observed that pollen tubes of foreign pollen had retarded growth in interspecific crosses of alder species. In three interspecific crosses with red alder (i.e., *Alnus cordata, A. glutinosa, A. sinuata*), Millar found slower pollen tube growth in the wider interspecific crosses. Significant differences in pollen tube growth were also observed in reciprocal crosses and among different parents. Observations on the post-pollination events in the cross *A. crispa* subsp. *sinuata* x *A. rubra* (Ager, unpub. data) also indicated reduced pollen adhesion and germination as compared to intraspecific crosses.

## Geographic Variation in Red Alder

### Growth Traits

Geographic variation in red alder was first described by DeBell and Wilson (1978) with growth data from a common garden study established in 1968 by B. Douglas and R. Peter of the USDA Forest Service. One-year-old wildlings from 10 sources ranging from Juneau, Alaska, to Port Orford, Oregon (6 degrees in latitude), had been collected and established in tests at the Capitol Forest near Olympia, Washington, and at the Cascade Head Experimental Forest near Otis, Oregon. Extensive fall frost damage to the Capitol Forest site in 1969 led to its abandonment. The Cascade Head Trial, however, exhibited good survival and development; 8-year growth measurements were reported by DeBell and Wilson (1978) and 15-year data by Lester and DeBell (1989). Both sets of data showed the same trends. Sources from the extreme parts of the range (Juneau, AK, and Sand Point, ID) per-

formed the poorest, whereas the remaining eight provenances showed modest growth differences (about 1 to 2 m). The Concrete and Sequim, Washington, provenances exhibited the fastest growth. The intermediate provenances exhibited some rank changes between ages 8 and 15 (Lester and DeBell 1989).

While the Cascade Head trial provided an initial glimpse of geographic variation in red alder, the lack of family structure and systematic sampling limited its scope. In a subsequent study of variation in red alder (Ager 1987; Ager et al. 1993), 60 locations were sampled along four major river drainages (i.e., Santiam, Nisqually, Hoh, and Nooksack rivers) in western Oregon and Washington. Open-pollinated seeds were collected from 120 parent trees (two per source) and used to establish two common garden trials, one at Washington State University's Experimental Farm Five near Puyallup, the other at the University of Washington's Pack Demonstration Forest, near Eatonville, WA. Measurements were recorded for three- and five-year height, diameter, shoot phenology, and other traits.

This study found that for all traits measured, the bulk of the variation was among stands and drainages, with relatively little variation within stands. In general, the fastest-growing sources were from the lower Hoh, and to a lesser extent, from the lower Nooksack valley. This study also revealed the existence of elevational clines in red alder similar to those found in some sympatric conifers (Burley 1966; Campbell 1979; Kuser and Ching 1981; Rehfeldt 1991). These consisted of a decrease in three-year top-weight of between 1 and 2 g/m increase in seed source elevation (Fig. 1). The slope and shape of the clines, however, differed among the four river drainages. Sources from two of the drainages (Hoh and Santiam) displayed a pronounced elevational effect, whereas those from the other two displayed only a slight (Nooksack) or no effect (Nisqually). Thus, variation patterns in growth traits were specific to subregions or individual river drainages.

No consistent relationships were found between latitude and any of the growth traits measured. The complexity of the genetic clines with elevation and latitude made generalizations about geographic

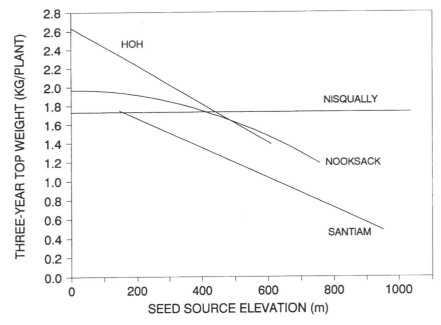

*Figure 1. Data from Puyallup red alder trial (Ager 1987) showing 3-year top weight regressed on seed-source elevation within each river drainage. Regression for the Nisqually was not significant and is shown as a horizontal line. Regression equations were calculated on family mean data, with $R^2$ values of 0.71 for the Santiam, 0.77 for the Hoh, and 0.67 for the Nooksack.*

variation patterns difficult within the region studied. This has been the case in several studies of variation in conifer species in the Pacific Northwest region (Burley 1966; Campbell and Sorensen 1978; White et al. 1981; Kuser and Ching 1981; Sorensen 1983; Sorensen et al. 1990; Rehfeldt 1991).

Evidence that geographic variation in growth traits in red alder is the product of natural selection was provided by relating variation patterns to local climate data (Ager et al. 1993). Of the climate variables analyzed, a measure of annual temperature amplitude appeared to best explain the geographic variation (Fig. 2). Other variables, such as annual precipitation and frost-free growing season, exhibited noticeable lack of fit. Faster-growing sources were from areas with long growing seasons and low temperature amplitudes. This relationship has been observed in numerous other studies of variation in forest trees (Wright 1976). The strong correlation with climatic data indicates that variation at the population level in red alder has evolved in response to spatial variation in climate. Thus the divergent elevational patterns noted above reflect the variable relationship between climate and elevation in the Pacific Northwest (Ager 1987, pp. 111-115).

Since different test conditions can lead to different conclusions about patterns of variation and their cause (Campbell and Sorensen 1978), it is important that variation patterns be studied under a variety of growing conditions. Some data of this kind are available for red alder from other test sites. One of these is the Pack Forest trial included in the study by Ager (1987) that contained 66 of the 120 families planted at the Puyallup trial. This test site differed from the Puyallup site largely in the moisture regime, since the latter site was irrigated and the Pack Forest site was not. The data from this trial largely paralleled the results discussed above, although detailed analyses of variation patterns were not performed. Family-by-site

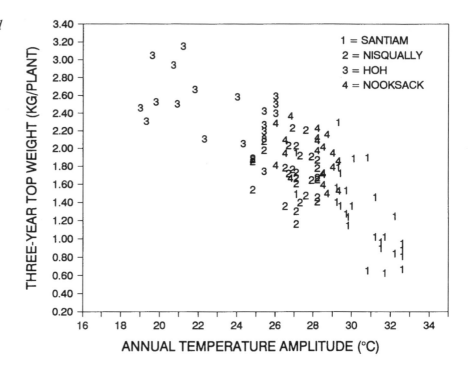

*Figure 2. Data from Puyallup red alder trial (Ager 1987) showing seed-source temperature amplitude (difference between mean January minimum and mean July maximum) versus family mean 3-year top weight.*

interactions between the Puyallup and Pack Forest sites were nonsignificant and accounted for a mere 0.6 percent of the site variance. Plots of family means (Ager 1987, p. 121), however, showed that the growth superiority of the Hoh sources was diminished at the Pack site. Other families, especially those from the lower Nooksack River, grew faster at the Pack site relative to the Hoh sources. The change in source rankings between the plantations was somewhat anticipated, since the Pack site did not receive irrigation treatments as did the Puyallup site.

Additional data on variation patterns in growth traits and family-by-site interactions have been provided by more recent red alder genetic tests in France, established in 1984 and 1985 by L. Bouvarel and E. Tessier du Cros at the Institut de la Recherche Agronomique (INRA) in Orleans. The tests included all the seedlots collected by Ager (1987) and were established on four primary test sites spanning a range of climates and soils. Data from these trials (primarily seven-year growth) were analyzed by Hibbs (in press) and indicate that, with some minor exceptions, the variation patterns observed by Ager (1987) were also observed in these trials.

With respect to source-by-site interactions, the data are also similar. In the most productive test (Rouvray), Hoh River sources grew the best, whereas in the lesser productive sites (Bardos, Facq et Jure, Luthenay) either the Nooksack or Nisqually River sources were superior. Thus, seed sources from areas with relatively good growing conditions best expressed their growth potential when tested on productive sites.

## Vegetative Phenology

Geographic variation patterns in red alder have also been described for vegetative phenology (Ager 1987; Cannell et al. 1987) and frost hardiness (Cannell et al. 1987). In the latter study, phenology and frost hardiness were measured for 15 provenances from Washington, British Columbia, and Alaska. Seedlings were grown from bulk seed collections for each provenance and outplanted in Scotland. Phenology data were recorded for spring and fall, and frost hardiness was assessed by freezing detached shoots at regular intervals while the seedlings were dormant. Strong geographic differences were noted for fall dormancy as measured with budset and frost hardening. The onset of rapid frost hardening generally occurred two days earlier for each degree of latitude northward, although considerable variation was observed among the provenances from a given latitude. In contrast, all provenances dehardened at about the same time in March. No relationships were observed between source elevation (range 15 to 521 m) and any of the measured traits, which might be expected given the confounding influence of latitude in the sampling scheme used in this study.

In the study conducted by Ager (1987; Ager et al. 1993), geographic variation was examined for date of spring budburst and fall leaf abscission among the 60 sources sampled. Fall leaf abscission was chosen to measure the onset of dormancy after it was observed that budset timing in red alder is highly influenced by environmental factors such as soil moisture. As with the growth traits examined in this study, strong genetic clines were noted. High elevation and interior sources generally showed earlier (two weeks) leaf abscission than coastal sources and low elevation sources. Leaf abscission was found to be strongly related to the average date of the first fall frost at the seed source ($R^2 = 0.99$).

The variation observed for spring bud flush was more complex than for leaf abscission. In general, coastal and northern sources leafed out earlier by about one to two weeks than the more inland and southerly sources. There was no relationship between seed source elevation and date of bud flush. In comparison to the conifers, the overall geographic pattern observed was similar to that found for western hemlock, where coastal, northern, and high-elevation sources flush first, while inland and southern sources flush later (Kuser and Ching 1981). In Douglas-fir, by contrast, southern and inland sources resume spring growth earlier than northern and coastal sources, which Campbell and Sugano (1979) have explained as a selected response to late summer drought that is characteristic of southern and inland regions. Early growers would thus capture the favorable moisture regime in the spring. Such selection pressure would not exist in mesic species like western hemlock (Campbell and Sugano 1979) and red alder.

The best climatic variable to predict bud flush was a measure of spring thermal sums calculated from long-term weather records (Ager 1987; Ager et al. 1993). Thermal sums were defined as degree-days above 4.5°C, beginning when the average monthly temperature curve exceeded 4.5°C, and ending at the mean date of the last -2.2°C fall frost (Cannell and Smith 1983). From this analysis it was shown that sources from areas with lower thermal sums, such as those from cool coastal climates with long growing seasons (e.g., Hoh River), flushed earlier ($R^2 = 0.95$).

In summary, studies on geographic variation have demonstrated that levels and patterns of variation in red alder are closely attuned to climatic variables. Variation in growth traits follows gradients in thermal stress (temperature amplitudes), whereas phenology traits are cued to spring thermal sums and fall frost dates. These climatic gradients are not always correlated with geographic features such as elevation and latitude, which leads to complex patterns of geographic variation, as observed for the sympatric conifers.

## Short-term Genetic Management

### Seed Zones and Seed Transfers

The strong gradients of genetic variation in tree species native to the Pacific Northwest led to the establishment of seed zones (Western Forest Tree Seed Council 1966) and, more recently, seed transfer guidelines (e.g., Campbell 1986; Campbell and Sugano 1987, 1989). These serve to minimize the risk of planting maladapted seedlings (e.g., Conkle 1973) in reforestation operations. The seed-transfer guidelines are based on equations developed from adaptive variation patterns in growth and phenology and are used to predict risks of seedling mortality for specific geographic transfers within a subregion. These guidelines provide a more flexible construct for matching seed source with planting site, as contrasted to the earlier seed-zone approach, which superimposed discrete boundaries on genetic gradients.

There are sufficient data for red alder to conclude that much of the variation in the species is adaptive, and therefore seed-transfer guidelines should be used to select appropriate seed sources for reforestation. Because seed-transfer risk models have not been developed, it was proposed (Ager 1987; Hibbs and Ager 1989) that seed transfers follow the seed zone map for Douglas-fir (Western Forest Tree Seed Council 1966). In addition, it was suggested that seed sources should not be transferred between north and south aspects of a given drainage, and not moved more than 215 m in elevation. These guidelines are generally conservative and should be considered as provisional. Their refinement will require additional genetic testing and the development of seed-transfer risk models.

*Tree Improvement Options*

Large genetic gains in growth rate are typically obtained in the early phases of domestication of tree species. For seed-propagated species, these first-generation gains are commonly made by field selection of parent trees, with subsequent progeny testing to further select subsets of parents for inclusion in seed orchards. Several conventional approaches for obtaining short-term genetic gains from first-generation selection among wild stands of red alder can be envisioned. Three options commonly used in tree improvement programs are presented here. The genetic gain calculations are based on data from the Puyallup field trial established by Ager (1987). Only families from sources less than 300 m in elevation (62 total) were included in this analysis since higher elevation sources would not normally be included in the base population of a tree-improvement effort. For reasons discussed below, the gain estimates are inflated and serve only to compare the relative merits of different selection strategies. The alternatives

below do not include a clonal propagation option, although the merits of such an approach are recognized, given the ease with which red alder can now be propagated vegetatively (see below).

**Field collections from select parent trees.** In this option, seed would be collected from the top 20 percent of the parent trees in the native stands, thus singling out 12 of the 62 families included in the gain calculations. This option, which is feasible since a single red alder tree can produce an enormous amount of seed, is best considered as an *"in situ* seed orchard."* This option would take advantage of family variation without establishing a seed orchard. Gain was calculated as family selection with mating taking place between unselected materials (Namkoong et al. 1966), and was estimated at 306 g per plant fresh weight at age 3, or 15.1 percent above the average.

**Clonal seed orchard.** In this option, scion material from selected parent trees would be established in a clonal orchard, either by grafting or rooting scion material. The difference in gain from option one is due to a change in parentage of the seed produced, since mating in the seed orchard is among select individuals only. The selection intensity is double that for the option above (Namkoong et al. 1966). Gain for this activity is estimated at 613 g per plant fresh weight, or about twice that for the *in situ* seed orchard (Ager 1987).

**Seedling seed orchard.** This option would involve converting the Puyallup field trial to a seedling seed orchard, as proposed by Wright and Bull (1963) for red pine. This activity combines progeny testing with seed production. Selection was arbitrarily specified to pick the top 20 percent of the families, with an additional selection of the best 10 percent of the seedlings within the best families. The combined intensity would be equal to 1 in 50. A number of other combinations of between versus within family selection intensities could be used, the optimum being that given by a selection index. Gain for this activity was estimated at 892 g per plant fresh weight, or 43.8 percent above the average at age 3 (Ager 1987).

In each of the above options, the gain estimates were significantly higher (about double) than those realized from comparable conifer programs (see

Theisen 1985). The estimates here were inflated as a result of including a wide range of geographic sources in the base population, which inflated the selection differentials and ultimately the gain estimates. Comparing the three alternatives, the seedling seed orchard approach appears to provide the greater gains and would also be the less costly of the two seed-orchard alternatives, while the first option would be an effective method for obtaining short-term genetic gains at a low cost.

## Long-term Breeding Prospects

Six features of red alder make the species particularly amenable to long-term tree breeding: early reproductive maturation (three to six years); prolific seed production ($> 1 \times 10^6$ seeds per tree); a well-diversified genus of about 30 species, including diploids and polyploid species; easy-to-perform controlled breeding via indoor potted orchards; low-cost vegetative propagation methods; and fast early growth. These characteristics open many avenues for longer-term breeding. The choice of the most appropriate strategy depends on numerous factors, since breeding programs vary widely with respect to acceptable cost, genetic gains desired, and intended conservation of genetic diversity over time. It is beyond the scope of this chapter to discuss the advantages of different advanced generation breeding schemes (see Namkoong et al. 1987). Instead, we will review research on red alder that has examined specific methods for applying genetic leverage, and traits that might be considered in a longer-term breeding program.

### Propagation of Alders by Cloning

The ability to clonally propagate selected genotypes opens many avenues for capturing genetic gains from tree breeding. Most importantly, cloning allows the capture of non-additive gene action, including heterotic effects. Techniques for cloning alder species via rooted cuttings were originally developed as a research tool to study host-by-endophyte interactions in nitrogen fixation (e.g., Huss-Danell 1981; Carpenter et al. 1984). Several methods for clonal propagation by rooted cuttings of red alder have been developed (e.g., Monaco et al. 1980; Carpenter et al. 1984; Radwan et al. 1989).

Red alder has also been cloned using *in vitro* techniques (Tremblay and LaLonde 1984; Perinet and Tremblay 1987) that had been first developed for black alder (Garton et al. 1981). These methods involve culturing nodal sections containing a lateral bud *in vitro* and subsequently rooting "microshoots" in a conventional potting medium in a warm, humid environment. Application of this technique to five alder species (13 genotypes total, of which 2 were red alder) by Perinet and Tremblay (1987) allowed them to produce 60,000 microshoots from 13 seedlings. These authors concluded that vegetative propagation was feasible on a commercial basis. No cost estimates were, however, given. As a whole, research in this area appears to have offered reliable and efficient means of cloning superior red alder genotypes.

### Interspecific Hybridization

The technology to clonally propagate alders at a commercial scale makes interspecific hybridization a promising avenue for red alder breeding. Cloning is an important, if not essential, adjunct for hybridization (Zobel and Talbert 1984, p. 365) as it helps to filter out the excessive variation commonly encountered in hybrid generations of forest trees and to focus it on the best genotypes (Stettler and Ceulemans 1992).

Interspecific hybridization has played an important role in genetics programs of many species (e.g., Stettler et al. 1988). The benefits from interspecific hybridization with red alder have yet to be fully explored and deserve further attention. As pointed out by Hall and Maynard (1979), at least three advantages to interspecific hybridization exist: useful traits from different species can be combined into hybrid offspring; a wider range of variation can be created for selection purposes; and heterosis, or hybrid vigor, can be captured.

Artificial hybridization of alders has been performed with numerous species as part of tree breeding efforts (e.g., Ljunger 1959; Chiba 1966; Hall and Maynard 1979; E. Tessier du Cros, unpub.). Ljunger (1959) in Sweden, who was probably the first to hybridize red alder by crossing it to *Alnus incana* subsp. *rugosa* and *A. glutinosa*, reported that the crosses between *A. incana* and *A. rubra* showed some heterosis and appeared to be promising from a tree-breeding perspective.

Hybridization of *Alnus rubra* and *A. glutinosa* also was attempted by L. Karki at the University of Washington, and the putative hybrids from these experiments growing in the University of Washington Arboretum (Mulligan 1976) appear to be of good form and growth. Attempts by these authors to cross *A. rubra* and *A. crispa* subsp. *sinuata* were not successful (Ager, unpub. data). In France, putative red alder hybrids have been obtained with *A. glutinosa, A. cordata, A. incana, A. hirsuta,* and *A. inokumae* (E. Tessier du Cross, unpub.). In these experiments, it was noted that several of the crosses only produced seed when red alder was used as the female parent. In some crosses, the hybrid seedlings were noted as particularly vigorous in the nursery.

Despite the extensive hybridization of red alder, a clear superiority of hybrid progeny over the parental genotypes has yet to be demonstrated. Accurate estimation of red alder hybrids for growth potential will necessitate cloning of selected $F_1$ progenies and parental genotypes in common garden tests. These experiments have not been performed to our knowledge. Given the extensive diversity in the genus *Alnus,* continued hybridization work seems warranted.

### Improvement of Wood Quality

Two studies on genetic variation in wood specific gravity in red alder have been conducted, one on the materials at the Cascade Head trial (Harrington and DeBell 1980), the other on the sources collected by Ager (1987) and outplanted by INRA at the Rouvray test near Rouen, France. In the former study, variation was assessed in wood specific gravity at age 10 for the 10 sources planted in the Cascade Head provenance trial near Otis, Oregon. Mean provenance wood density was found to range from 0.39 to 0.41 g/cm$^3$, these differences not being statistically significant. Additional work revealed no relationship between specific gravity and either radial growth rate or distance from the pith. The apparent uniformity in alder wood was viewed by these authors as an advantage for commercial utilization.

In a subsequent study (Radi 1988), specific gravity was measured on 4-year-old saplings from 105 families growing at the Rouvray plantation near Rouen, France. Variation in specific gravity was found to be significant (p < 0.01) among sources and between families within sources. Family-mean specific gravity varied from 0.35 to 0.48 g/cm$^3$. We compared the specific gravity data from this study (Radi 1988, p. 51) with nine-year growth (dry weight) for the same families (Hibbs et al., in press) and found that the faster growing families generally had the lowest specific gravity. The relationship was not tested for statistical significance for the sample as a whole.

The different results in the two studies are likely due to the expanded geographic sampling and family structure in the latter study, which enabled more precise estimation of genetic variances for specific gravity. The fact that wood density does not vary with distance to the pith (Harrington and DeBell 1980) suggests that differences between the two studies in terms of plant size and age (four vs. nine years) are not responsible for the dissimilar findings. Further studies are needed to clarify variation in wood density in red alder.

### Enhancing Symbiotic Nitrogen Fixation

A number of researchers have discussed the feasibility of enhancing symbiotic nitrogen fixation via selection of alder genotypes that are more "efficient" nitrogen fixers (e.g., Stettler 1978; Hall and Maynard 1979; Miller 1983; Tarrant et al. 1983). These schemes involved selection of alder genotypes that were more "efficient" at fixing nitrogen. Concomitant selection of the *Frankia* endophyte has also been proposed as a means of optimizing the symbiosis (e.g., Hall et al. 1979). The improvement of nitrogen fixation was seen as especially beneficial on infertile soils such as mine spoils where availability of inorganic nitrogen might severely limit growth.

Genotypic differences in symbiotic nitrogen fixation have been documented in several alder species (Gordon and Wheeler 1978; Huss-Danell 1980; Miller 1983; Palmgren et al. 1985) including red alder (Carpenter et al. 1984; Ager 1987). Acetylene reduction activity per unit nodule or stem weight, measures of "efficiency" of nitrogen fixation, also has been found to differ among alder clones in many of these experiments. For instance, Ager (1987) found that high-elevation red alder

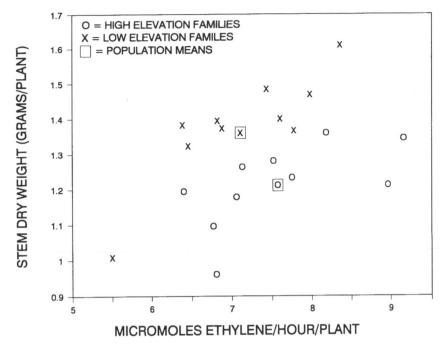

Figure 3. Data from Ager (1987) showing genetic variation in $N_2$ fixation between a low (20 m) and high elevation (1040 m) red alder population, and between families within each population. Data are family means for dry stem weight and acetylene reduction activity assayed derived from one-year-old seedlings grown from seed collected at a low and high elevation site along the Nisqually River (10 open-pollinated families per site). Seedlings were grown under greenhouse conditions and assayed in vivo (twice) for acetylene reduction. Higher elevation families exhibited higher rates of $N_2$ fixation per unit stem weight, due to increased carbohydrate partitioning to roots versus shoots. Total plant weight was not significantly different between the low and high elevation populations.

genotypes, which have higher root-to-shoot ratios, fix more atmospheric nitrogen on a top-weight basis (Fig. 3).

The vast majority of variation in nitrogen-fixation rates, however, appears to result from variation in plant growth rate, as measured by photosynthesis (Gordon and Wheeler 1978) or total dry matter production (Huss-Danell 1980). For instance, in a study of genetic variation among seven clones of *Alnus incana* (grey alder), Huss-Danell (1980) found significant variation in total plant weight, nitrogenase activity per shoot, leaves, and nodules. Several traits including percent of nitrogen, nodule weight, and nitrogenase activity per plant were not significantly different. The lack of variation in plant nitrogen content indicated that fixation of atmospheric nitrogen proceeded as needed by plant growth.

Numerous other studies with alders have also shown that nitrogen fixation is closely regulated by carbohydrate supply so that it occurs only as needed for plant growth (e.g., Gordon and Wheeler 1978). The dependency of nitrogen fixation on carbohydrate supply suggests that selective breeding for enhanced growth rate will likely improve nitrogen-fixation rates as well. Thus, on the host side of the symbiosis, it does not seem warranted to target nitrogen fixation as a specific trait for selection.

On the other hand, the opportunity to enhance fixation of atmospheric nitrogen via genetic manipulation of the *Frankia* symbiont seems more promising. This is especially true in the exotic cultivations of red alder, where native *Frankia* are absent or evolved in symbiosis with other alder species (e.g., France). Research on genetic variation on the *Frankia* symbiont has been prolific since first isolated and grown in pure culture (Callaham et al. 1978). Genetic differences among *Frankia* strains isolated from either one or more alder species have been demonstrated on the basis of infectivity, sporulation, morphology, isozymes, plasmid composition, and nitrogen fixation (Normand and LaLonde 1986; Simonet et al. 1989). Nonetheless, the selection, deployment, and testing of putatively superior strains under field conditions remain a formidable problem, and as such, the identification of clearly superior strains for commercial use has yet to be accomplished. The recent development of methods for molecular hybridization of strain-specific DNA probes to whole crushed nodules (Hahn et al. 1990) will simplify the field testing of putative superior *Frankia* strains.

## Ideotype Breeding

The concept of improving yields through improving carbon allocation to the harvested portions of the plants has been employed with much success in agriculture. The proportion of harvested-to-total plant weight has been termed the "harvest index" (HI) and is often discussed in the context of the "crop ideotype" (Donald 1968). In juvenile stands of forest trees, it is clear that genetic differences in HI exist for at least several commercial species (Cannell and Jackson 1985; Karki and Tigerstedt 1985; Dickman et al. 1990), including red alder (Hook et al. 1987). In addition, HI has a relatively high heritability and similar additive genetic variance as compared to growth traits. Genotypes with a high HI generally have fewer, more sparse branches.

The HI data for red alder were derived from the Puyallup genetic test established by Ager (1987) and consisted of 36 families measured at age 4 (Hook et al. 1990). These data indicated family heritabilities of around 0.6, with a family coefficient of variation ranging from 5 to 20 percent for various harvest indices. These authors concluded that variation in growth partitioning traits is large enough to be of practical significance in red alder breeding programs.

In an effort to further examine variation in HI in red alder, the Puyallup plantation was resampled in 1987 at age 6 (Ager, unpub. data). Thirty families were sampled, 13 of which were included in the previous study (Hook et al. 1990). The objectives were to examine how HI rankings changed over time and to determine genetic variances in HI for the fastest-growing materials in the trial. Sampling was limited to the top 30 families in the trial, all originating from sources along the Hoh and Nooksack rivers. In this way, the variance in HI would be quantified among only those families that would have been selected in a tree breeding program. In addition, the biomass contribution from leaves was not included, and HI was defined as the ratio of stem to stem plus branch weight. In the earlier sample, HI estimates included leaf weight (see Hook et al. 1990) Between the three- and six-year sample dates, the plantation had grown from 3.7 to 7.5 m in height, had undergone canopy closure, and had incurred significant mortality in the slower-growing families.

The data from the later sampling revealed that despite significant family variation in total weight, little genetic variation was present in HI (proportion stem to total weight). For instance, the family coefficient of variation (CV) for total weight was 23 percent, versus a mere 2.7 percent for HI. Family variation in HI was not statistically significant (p = 0.09). In the previous sampling, which included a wider range of sources, the family CV values were 56 and 14 percent for total weight and HI (stem over stem + branches + leaves). Although differences in the construction of the HI between the two samples do not permit exact comparisons (leaf weight was included as part of total weight in the earlier study), it appears that when the composition of the sample is restricted to the fastest-growing 25 percent of the population of genotypes, genetic variation in HI is reduced below levels that are needed to permit selective breeding for this trait.

Although additional study of growth allocation seems warranted, it will take long-term studies, including rotation-length yield trials, to determine if genetic differences in HI can be captured on a commercial scale. There are some reasons to believe that capturing genetic gains in HI in a species like red alder may be difficult. Carbon allocation is relatively plastic in species with indeterminate growth patterns like red alder, and thus is highly dependent on growing conditions at hand, especially with respect to stand density. Genetic variation in growth allocation traits such as HI can easily be masked by environmental effects.

On the other hand, red alder may well have less genetic variation in HI than in other traits such as growth rate or phenology. Since red alder almost universally grows up in dense, even-aged stands, there must have been strong recurrent selection for optimal allocation patterns under these conditions. In the process there may have been a depletion of alleles that cause significant departures from the near-optimal form in such traits as branchiness, branch angle, leaf size and distribution, leaf orientation, and so forth. The remarkable uniformity of red alder in these traits across geographic sources, and, by contrast, the greater diversity in these traits in sympatric species with greater demographic variation, certainly are consistent with that hypothesis. Thus, as with other trees, it remains unclear whether this source of genetic variation can be captured to enhance harvestable yields over a rotation.

### Molecular Approaches

Red alder, because of its reproductive features and symbiotic relationship with *Frankia*, offers several pathways to genetic improvement where molecular tools may be usefully applied, including the identification of genetic markers (chromosome mapping) and their association with important phenotypic traits, the molecular characterization ("fingerprinting") of desirable cultivars, and the genetic transformation of either host or symbiont. A good example is *Alnus viridis*, which has been transformed by both electroporation of protoplasts (Seguin and LaLonde 1988) and by the *Agrobacterium* system (Mackay et al. 1988). Regeneration of plants remains a barrier, as does the identification and cloning of useful genes that will be properly regulated in the alder genome.

## Conclusions and Future Research Needs

Red alder will receive increasing attention from the perspective of genetics as it assumes a more important role in the biological and economic health of Pacific Northwest forests. While past research has given baseline genetic data, additional studies could significantly improve our ability to wisely manage the red alder gene pool. Immediate research needs include better quantification of risks from seed transfers to prevent the planting of maladapted seedlings in reforestation projects. Careful research is needed to determine the feasibility of breeding for traits such as wood quality and harvest index. Experiments to determine the potential of interspecific hybridization also seem warranted now that cloning of alder is commercially feasible. Finally, we need a range-wide, systematic study of genetic variation in the *Frankia* endophyte to better understand the potential for enhancing symbiotic fixation of atmospheric nitrogen via manipulation of the endophyte. Hopefully, economic and other interests in the species will foster research along these lines in the foreseeable future.

## Acknowledgments

We thank Luc Bouvarel, Eric Tessier du Cros, and David Hibbs for use of their unpublished data on the red alder trials in France. Helpful comments to earlier drafts of this paper were provided by Mike Carlson, Dean DeBell, Dave Hibbs, and Frank Sorensen.

## Literature Cited

Ager, A. A. 1987. Genetic variation in red alder (*Alnus rubra* Bong.) in relation to climate and geography in the Pacific Northwest. Ph.D. thesis, Univ. of Washington, Seattle.

Ager, A. A., P. E. Heilman, and R. F. Stettler. 1993. Genetic variation in red alder (*Alnus rubra*) in relation to native climate and geography. Can. J. For. Res. 23: 1930-1939.

Burley, J. 1966. Genetic variation in seedling development of Sitka spruce (*Picea sitchensis* {Bong.} Carr.). Forestry 39:69-94.

Callaham, D., P. Del Tredici, and J. G. Torrey. 1978. Isolation and cultivation *in vitro* of the actinomycete causing root nodulation in *Comptonia*. Science 199:899-902.

Campbell, R. K. 1979. Genecology of Douglas-fir in a watershed in the Oregon Cascades. Ecology 60:1036-1050.

———. 1986. Mapped genetic variation of Douglas-fir to guide seed transfer in southwest Oregon. Silv. Genet. 35:85-96.

Campbell, R. K., and F. C. Sorensen. 1978. Effect of test environment on expression of clines in and on delimitation of seed zones in Douglas-fir. Theor. Appl. Gen. 51:233-246.

Campbell, R. K. and A. I. Sugano. 1979. Genecology of budburst phenology in Douglas-fir: response to flushing temperature and chilling. Bot. Gaz. 140:223-231.

———. 1987. Seed zones and breeding zones for sugar pine in southwest Oregon. USDA For. Serv., Res. Pap. PNW-383.

———. 1989. Seed zones and breeding zones for white pine in the Cascade Range of Washington and Oregon. USDA For. Serv., Res. Pap. PNW-407.

Cannell, M. G. R., and J. E. Jackson. 1985. Attributes of trees as crop plants. Titus Wilson & Son, Kendal, Cumbria.

Cannell, M. G. R., and R. I. Smith. 1983. Thermal time, chill days and prediction of budburst in *Picea sitchensis*. J. Appl. Ecol. 20:951-963.

Cannell, M. G. R., M. B. Murray, and L. J. Sheppard. 1987. Frost hardiness of red alder (*Alnus rubra*) provenances in Britain. Forestry 60:57-67.

Carpenter, C. V., L. R. Robertson, J. C. Gordon, and D. A. Perry. 1984. The effect of four new *Frankia* isolates on growth and nitrogenase activity in clones of *Alnus rubra* and *Alnus sinuata*. Can. J. For. Res. 14:701-706.

Chiba, S. 1966. Studies on the tree improvement by means of artificial hybridization and polyploidy in *Alnus* and *Populus* species. Bull. Oji. Inst. For. Tree Improvement, No. 1. Aji Paper Co., Japan.

Conkle, T. 1973. Growth data for 29 years from the California elevational transect study of ponderosa pine. For. Sci. 19:31-39.

Czerepanov, S. 1955. Systema generis *Alnus* Mill. S. Str. Generumque affinium. Notulae Systematicae ex Herbario Instituti Botanica Nomine V. L. Komarovi Academiae Scientiarum U R. S. S. 17:90-105.

DeBell, D. S., and B. C. Wilson. 1978. Natural variation in red alder. *In* Utilization and management of alder. *Compiled by* D. G. Briggs, D. S. DeBell, and W. A. Atkinson. USDA For. Serv., Gen. Tech. Rep. PNW-70. 193-208.

Dickman, D. I., M. A. Gold, and J. A. Flore. 1990. The ideotype concept and the genetic improvement of trees. *In* Tree crop systems. *Edited by* F. T. Last. Elsevier Press.

Donald, C. M. 1968. The breeding of crop ideotypes. Euphytica 17:385-403.

Furlow, J. J. 1979a. The systematics of the American species of *Alnus* (Betulaceae). Part I. Rhodora 81:1-21.

———. 1979b. The systematics of the American species of *Alnus* (Betulaceae). Part II. Rhodora 81:151-248.

Garton, S., M. A. Hosier, P. E. Read, and R. S. Farnham. 1981. In vitro propagation of *Alnus glutinosa* Gaertn. Hortscience 16:758-759.

Gordon, J. C., and C. T. Wheeler. 1978. Whole plant studies on the photosynthesis and acetylene reduction in *Alnus glutinosa*. New Phytol. 80:179-186.

Hagman, M. 1970. Observations on the incompatibility in *Alnus*. *In* Proc. IUFRO Sect. 22 Working Group meeting on Sexual Reproduction in Forest Trees, Vargaranta, Finland. 1(10):1-19.

———. 1975. Incompatibility in forest trees. Proc. R. Soc. Ser. B. 188:313—326.

Hahn, D., M. J. C. Starrenburg, and A. D. L. Akkermans. 1990. Oligonucleotide probes against rRNA as a tool to study *Frankia* strains in root nodules. Appl. Environ. Microbiol. 56:1324-1346.

Hall, R. B., H. S. McNabb, Jr., C. A. Maynard, and T. L. Green. 1979. Toward development of optimal *Alnus glutinosa* symbioses. Bot. Gaz. (Suppl.) 140:120-126.

Hall, R. B., and C. A. Maynard. 1979. Considerations in the genetic improvement of alder. *In* Symbiotic nitrogen fixation in the management of temperate forests. *Edited by* J. C. Gordon, C. T. Wheeler, and D. A. Perry. Oregon State Univ. Press, Corvallis.

Harrington, C. A., and D. S. DeBell. 1980. Variation in specific gravity of red alder (*Alnus rubra* Bong.). Can. J. For. Res. 10(3):293-299.

Hibbs, D. E., and A. A. Ager. 1989. Red alder: guidelines for seed collection, handling, and storage. Forest Research Laboratory, Oregon State Univ., Corvallis, Spec. Publ. 18.

Hibbs, D. E., L. Bouvard, and E. Tessier du Cros. Performance of red alder seed sources in France. Can J. For. Res. (In press.)

Hook, D. D., M. D. Murray, D. S. DeBell, and B. C. Wilson. 1987. Variation in growth of red alder families in relation to shallow water-table levels. For. Sci. 33(1):224-229.

Hook, D. D., D. S. DeBell, A. Ager, and D. Johnson. 1990. Dry weight partitioning among 36 open-pollinated red alder families. Biomass 21:11—25.

Huss-Danell, K. 1980. Nitrogen fixation and biomass production in clones of *Alnus incana*. New Phytol. 85(4):503—511.

———. 1981. Clonal differences in rooting of *Alnus incana* leafy cuttings. Plant and Soil 59:193-199.

Johnson, F. D. 1968. Disjunct populations of red alder in Idaho. *In* Biology of alder. *Edited by* J. M. Trappe, J. F. Franklin, R. F. Tarrant, and G. M. Hansen. USDA For. Serv., PNW For. Range Exp. Sta., Portland, OR. 1-8.

Karki, L., and P. M. A. Tigerstedt. 1985. Definition and exploitation of forest tree ideotypes in Finland. *In* Attributes of trees as crop plants. *Edited by* M. G. R. Cannell and J. E. Jackson. Inst. Terr. Ecol., Monks, Abbots Ripton, Hunting, U.K. 102-109.

Kuser, J. E., and K. K. Ching. 1981. Provenance variation in seed weight, cotyledon number, and growth rate of western hemlock seedlings. Can. J. For. Res. 11:664-670.

Lester, D. T., and D. S. DeBell. 1989. Geographic variation in red alder. USDA For. Serv., Res. Pap. PNW-409.

Ljunger, A. 1959. Al och alforadling. Skogen 46:115-117.

Lynd, L. R., J. H. Cushman, R. J. Nichols, and C. E. Wyman. 1991. Fuel ethanol from cellulosic biomass. Science 251:1318-1323.

Mackay, J. A., A. Seguin, and M. LaLonde. 1988 Genetic transformation of 9 in vitro clones of *Alnus* and *Betula* by *Agrobacterium tumefaciens*. Plant Cell Reports 7:229-232.

Millar, C. 1977. Pollen-pistil interactions in four alder species after selfing and crossing. Senior honors thesis, Univ. of Washington, Seattle.

Miller, G. A. 1983. Variation in growth, nitrogen fixation, and assimilate allocation among selected *Alnus glutinosa* (L.) Gaertn. clones. Ph.D. thesis, Iowa State University, Ames.

Monaco, P. A., T. M. Ching, and K. K. Ching. 1980. Rooting of *Alnus rubra* cuttings. Tree Planters' Notes 31(3):22-24.

Mulligan, B. O. 1976. Plants of the Washington Park Arboretum. Univ. of Washington Press, Seattle.

Murai, S. 1964. Phytotaxonomical and geobotanical studies on genus *Alnus* in Japan. III. Taxonomy of whole world species and distribution of each sect. Govt. For. Exp. Sta. Bull. (Japan) 171:1-107.

———. 1968. Relationship of allied species between northwestern U.S.A. and Japan on the genus *Alnus*. *In* Biology of alder. *Edited by* J. M. Trappe, J. F. Franklin, R. F. Tarrant, and G. M. Hansen. USDA For. Serv., PNW For. Range Exp. Sta., Portland, OR. 23-36.

Namkoong, G., E. B. Snyder, and R. W. Stonecypher. 1966. Heritability and gain concepts for evaluating breeding systems such as seedling seed orchards. Silv. Genet. 15:76-84

Namkoong, G., H. C. Kang, and J. S. Brouard. 1987. Tree breeding: principles and strategies. Springer Verlag, New York.

Normand, P., and M. Lalonde. 1986. The genetics of actinorhizal *Frankia*: a review. Plant and Soil 90:429-453.

Palmgren, K., A. Saarsalmi, and A. Weber. 1985. Nitrogen fixation and biomass production in some alder clones. Silva Fennica 19:407-420.

Perinet, P., and F. M. Tremblay. 1987. Commercial micropropagation of five *Alnus* species. New For. 3:225-230.

Radi, M. 1988. Variabilite genetique de proprietes papetieres chez differentes provenances du genre *Alnus*. INRA, Centre de Recherches Forestieres, Seichamps, France.

Radwan, M. A., T. A. Max, and D. W. Johnson. 1989. Softwood cuttings for propagation of red alder. New For. 3:21-30.

Rehfeldt, G. E. 1991. A model of genetic variation for *Pinus ponderosa* in the Inland Northwest (U.S.A.): applications in gene resource management. Can. J. For. Res. 21:1491-1500.

Rich, A. H. 1889. Abnormal sex expression in *Alnus serrulata*. Bull. Torrey Bot. Club 16:112.

Seguin, A., and M. LaLonde. 1988. Gene transfer by electroporation in betulaceae protoplasts: *Alnus incana*. Plant Cell Reports 7:367-370.

Simonet, P., N. T. Le, A. Moiroud, and R. Bardin. 1989. Diversity of *Frankia* strains isolated from a single alder stand. Plant and Soil 118:13-22.

Sorensen, F. C. 1983. Geographic variation in seedling Douglas-fir (*Pseudotsuga menziesii*) from the western Siskiyou Mountains of Oregon. Ecology 64:696-702.

Sorensen, F. C., R. K. Campbell, and J. F. Franklin. 1990. Geographic variation in growth and phenology of seedlings of the *Abies procera/A. magnifica* complex. For. Ecol. Manage. 36:205-232.

Stettler, R. F. 1978. Biological aspects of red alder pertinent to potential breeding programs. *In* Utilization and management of alder. *Compiled by* D. G. Briggs, D. S. DeBell, and W. A. Atkinson. USDA For. Serv., Gen. Tech. Rep. PNW-70. 209-222.

Stettler, R. F., and R. Cuelemans. 1992. Clonal material as a focus for genetic and physiological research in forest trees. *In* Clonal forestry: genetics, biotechnology and application. *Edited by* M. R. Ahiya and W. J. Libby. Springer, Berlin/New York. (In press.)

Stettler, R. F., R. C. Fenn, P. E. Heilman, and B. J. Stanton. 1988. *Populus trichocarpa* x *Populus deltoides* hybrids for short rotation culture: variation patterns and 4-year field performance. Can. J. For. Res. 18:745-753.

Tarrant, R. F., B. T. Bormann, D. S. DeBell, and W. A. Atkinson. 1983. Managing red alder in the Douglas-fir region: some possibilities. J. For. 81(12):787-792.

Theisen, P. A. 1985. Progeny Test status report. USDA For. Serv., national genetics workshop. Athens, GA.

Tremblay, F. M., and M. LaLonde. 1984. Requirements for in vitro propagation of seven nitrogen-fixing *Alnus* species. Plant Cell Tissue Organ Culture 3:189-199.

Wagner, D. 1991. Noteworthy collections in Oregon. Madrono 38:145-146

Weisgerber, H. 1974. First results of progeny tests with *Alnus glutinosa* (L.) Gaertn. after controlled pollination. Proc. IUFRO Meet. S.02.04.1, Stockholm. Session VI:423-438.

Western Forest Tree Seed Council. 1966. Tree seed zone map for Oregon and Washington. Western Forest Tree Seed Council, Portland, OR.

White, T. L., D. P. Lavender, K. K. Ching, and P. Hinz. 1981. First-year height growth of southwestern Oregon Douglas-fir in three test environments. Silv. Genet. 30:173-178.

Woodworth, R. H. 1930. Cytological studies in the Betulaceae. II. *Corylus* and *Alnus*. Bot. Gaz. 88:383-399.

Wright, J. W. 1976. Introduction to forest genetics. Academic Press, New York.

Wright, J. W., and W. I. Bull. 1963. A one parent progeny test and seed orchard for the genetic improvement of red pine. J. For. 61:747-750.

Zobel, B., and J. Talbert. 1984. Applied forest tree improvement. Wiley, New York.

# 7

# Stand Development and Successional Implications: Pure and Mixed Stands

Michael Newton & Elizabeth C. Cole

Many factors in the development of a stand of red alder, or a mixture of alder and conifers, will influence its appearance throughout its life. Forms of trees, nature of the understory, rate of growth, and ultimately succession after the stand has senesced can vary widely with conditions of establishment and early stand parameters. In this chapter, we will describe what is presently understood about the roles of site factors, stand density, associated vegetation, and mechanisms of competitive interactions in determining stand growth and structure, and succession in stands originating with a major alder component. We will also describe some anomalies of stand development resulting from both deliberate and accidental human activities.

Evidence used to develop the above picture is from two general sources. The first is a group of excellent reports in *Biology of Alder* (Trappe et al. 1968) and *Utilization and Management of Alder* (Briggs et al. 1978) which summarize many useful observations on life history from descriptive studies. The second is a mixture of intensive, sophisticated experiments with stand density and species composition and observations taken from the experience of forest managers, most in the Oregon Coast Range. Gary Carlton's (1990) work has added considerable depth and detail to these experiments and observations. Because most of these data are at present unpublished, they cannot be presented here in detail; however, concepts can be elaborated upon, using the data to flesh out areas where relevant information would be otherwise lacking.

In this chapter we will describe stand development after seedling establishment and follow it to maturity and beyond. We will also describe the

ranges of understories likely to be found in association with red alder and discuss their influences on stand development. The role of red alder in the determination of understory and associated tree species development is crucial in later succession and will be considered in some detail.

## Early Stages of Development

Alder typically occurs on better-than-average sites at elevations below 880 m. Most of today's alder stands originated with the disturbances associated with logging of conifers (Stubblefield and Oliver 1978; Carlton 1990; Harrington 1992; Hibbs and Cromack 1990). Because many of these conifer-dominated sites supported mature stands with understories of deciduous species, subsequent community development after logging carried over a legacy of sprouting shrubs and hardwoods, plus numerous herbs (Roberts 1975; Kelpsas 1978). In this setting, alder is not very conspicuous in its first two years.

The sites where red alder is found are typically cool and moist in spring, and survival of germinants is best where some mineral soil has been exposed (see Dobkowski et al., Chapter 13; Hibbs and DeBell, Chapter 14). Thus, the species is most abundant along the high-water lines of fast-moving streams, shady north slopes, and various disturbed areas where there is abundant cloud cover or fog in summer. Haeussler's (1988) studies of seed germination revealed that seedling mortality usually occurs within the first few weeks after germination on exposed mineral soils, and that the probability of failure is related to desiccation influences on the surface layer of soil.

The disturbances that facilitate alder establishment from seed also tend to favor other seral species, such as grasses and light-seeded forbs, especially composites. Most of these species persist through the disturbances that trigger establishment of alder (Kelpsas 1978). Thus, the opportunities for alder establishment are typically short-lived, leading to stands with narrow ranges in age in Oregon (Carlton 1990). The spread in ages appears to be greater in Washington (Stubblefield and Oliver 1978). The same sorts of disturbance are also advantageous for other woody species, of which the most prominent is Douglas-fir (*Pseudotsuga menziesii*). In recent decades, Douglas-fir has been planted shortly after most disturbances associated with harvest, leading to mixed species compositions where Douglas-fir is uniform and alder density is variable. Density of alder survivors influences the alder stand greatly, and also development of mixed stands, as described later in this chapter.

In the seedling stage, survival probability of alder in the first few months is low. Where alder is most abundant in the lower slopes of coastal mountains, cold-air drainage causes high frequencies of lethal spring frosts well after growth has started. Where seedlings occur within litter or established grass/forb cover, early onset of droughty soil conditions may also contribute to mortality (Haeussler 1988). The intense competition from herbs and shrubs on the highly productive lower slopes causes severe suppression, which leads to reduced growth rates and some mortality among alder seedlings, if unrelieved (Cole and Newton, unpub. data 1991, Department of Forest Science, Oregon State University).

The typical red alder stand is relatively sparse where soil has not been scarified. Where germination conditions and freedom from frost and competition are favorable, establishment is dense, however, resulting in stands that are heterogeneous in density and distribution. The densest areas are along skidroads, fill slopes, and landings. The less dense areas are likely to have a component of conifers where there has been little shrub cover. Elsewhere they may be dominated by *Rubus* species, bigleaf maple (*Acer macrophyllum*) or bitter cherry (*Prunus emarginata*) of sprout origin or perhaps of seedling origin where deer usage is light (M. Newton, unpub. data 1969 from Hamer Lake deer exclosures).

Alder's growth habit gives it a measure of protection from animals and from suppression by taller plants. Once established, red alder has very rapid height growth for over a decade. The rapidity of growth in the first several years permits it to escape many of the damage problems that accrue to conifers. Although palatable to deer and elk, alder grows so rapidly when young that it is only within the range of browsing animals for a matter of months after beginning its second year of growth.

Experience varies on probability of damage in various situations (see also Harrington et al., Chapter 1). In riparian bench areas of the Oregon Coast Range, elk and deer frequently cause damage by antler rubbing on saplings up to 5 m tall; damage is inconspicuous in very dense (i.e., > 5000 tpa) thickets, (E. C. Cole et al., unpub. data, Oregon State University College of Forestry). Thus, the dense stands tend to escape serious damage, while sparse stands tend to have more forked stems. Heavy deer and elk pressure tend to increase heterogeneity of spacing and frequency of associated species in the dominant canopy. This has a marked effect on the later stages of stand development. DeBell (unpub. data 1992) has observed that frequency of elk damage was greater with spacings of 1 x 1 m and 2 x 2 m than in wider spacings of 3 x 3 m and 6 x 6 m.

Riparian stands are highly subject to damage by beavers (*Castor canadensis*). These animals create clearings along streams and remove trees up to 50 m from stream channels, including small ephemeral tributaries. They are among the most serious of the constraints on riparian zone management because of their tendency to return to sites of damage and to damage or fell trees of large size.

Between the second and tenth years of development, red alder typically accumulates about a third or more of its mature height (Worthington et al. 1960; Newton et al. 1968). Whether this potential is reached is influenced substantially by density and type of its associates. Cole and Newton (unpub. data) observed that alder growing in its first four years with salmonberry (*Rubus spectabilis*) sprouts of the same age were a tenth the biomass of those grown without competition. Overtopping by other alder or by salmonberry in the first three years induced substantial mortality, and reduced vigor.

Shainsky and Radosevich (1992) and Shainsky et al. (Chapter 5) observed that planting dense Douglas-fir with alder of the same age tended to reduce the latter's growth rate despite the greatly inferior height and biomass of the Douglas-fir. Thus, even though faster growing than most of its associates, individual alders can be substantially delayed in development by competition from other alders, by overtopping of shrubs when small, or when associated with an associate of lower stature with some specific antagonistic attribute which, in this instance, is unknown.

The rapid height growth of young alder, combined with the high densities achieved on mechanically disturbed areas, provides a highly competitive environment for associates (Newton et al. 1968; Miller and Murray 1978; Stubblefield and Oliver 1978). Dense stands of alder 5 to 15 years old typically have relatively simple understories in structure and composition, with very low understory biomass. During this stage of development, alder mortality is heavy and the competitive conditions leading to death of intermediate and suppressed alders also leads to suppression and mortality of most other plant species. Typically, such dense stands have very low densities of sword fern (*Polystichum munitum*), red elderberry (*Sambucus racemosa*), and salmonberry, with minor occurrences of bedstraw (*Galium* spp.), sorrel (*Oxalis oregona*), miner's lettuce (*Montia* spp.), and occasional English holly (*Ilex aquifolium*). Where alder stands originate with lesser degrees of disturbance, as on mountain beaver mounds or animal trails, density is typically so low that crown closure does not occur until past the sapling stage. In this instance, there is a large variety of associates, of which trailing blackberry (*Rubus ursinus*), salal (*Gaultheria shallon*), Himalaya blackberry (*R. procerus*), thimbleberry (*R. parviflorus*), vine maple (*Acer circinatum*), sword fern, bracken fern (*Pteridium aquilinum*), and numerous herbs form dense shrub/herb communities.

The ability of the shrub and herb community to remain beneath alder is a function of overstory height and spacing. When the ratio of height to space between trees reaches 4:1 or more, intolerant understory species virtually disappear, and overall understory density decreases sharply. At about 8:1, understory is very sparse (Cole and Newton, unpub. data 1991).

Spacing is also a strong determinant of whether the eventual stand will be pure or mixed. If the space between alders is sufficient so that a conifer of the same age can persist until its height growth rate exceeds that of the taller alders, it will eventually become dominant. On a good Coast Range site, gaps of close to 6 m in the alder appear to be necessary for planted western hemlock to escape a long period of suppression (Cole and Newton, unpub. data). Even greater space is presumably required for natural conifer regeneration or intolerant planted conifers. Stubblefield and Oliver (1978) report that of Douglas-fir, western hemlock, and western redcedar, cedar is the most likely to persist if invaded by alder.

Structure of the sapling stands is strongly determined by spacing. Where stands are dense (i.e., ratio of height to spacing is above 5:1), crowns will be shallow, lower branches will be small and short-lived, and understories will be light. The absolute ratio of height to diameter at these densities often exceeds 100, and taper is gradual. Stands in this condition can be expected to have lower leaf area indices (Bormann et al. 1979, p. 473) and to develop more slowly in height than stands somewhat more widely spaced, but the dominants may temporarily increase in height more rapidly than open-grown individuals (see DeBell and Giordano, Chapter 8). Although there is some benefit to stem quality from crowding, the ratio of height to spacing changes so quickly during the first 15 years that there are few advantages to very dense stands if goals include commercial timber production.

The spatial relationships and shrub competition in widely spaced stands may lead either to poor growth or to low-quality trees in the first 10 years. These problems seem to increase where trees are more than 3.5 to 5 m apart (Kelly 1991; Hyatt 1992; Ahrens and Hibbs, unpub. data 1992). The dense stands of associated shrubs delay development of the alder stand and increase heterogeneity. If the shrubs are controlled, the open-grown alders develop basal suckers and heavy branching. Their attractiveness to antler-rubbing ungulates often causes top die-back and multiple stems. Even single-stemmed individuals have coarse branching and deep crowns until crown closure. The same factors that lead to large tree size also lead to severe taper and poor stem quality in early years for trees in which butt logs have achieved large sawlog grades. Thus, spacing and weeding are probably crucial for both growth and quality of trees, with a well-defined optimum.

All densities of alder sustain injuries that may contribute to later stem defects or mortality. Dense stands are vulnerable to damage from ice and top breakage. We have observed that stands three to five years old at a spacing of 0.8 to 1.5 m are subject to top girdling by sapsuckers, or perhaps by wasps that utilize bark fiber for paper nest construction. Girdles and breaks are often observed where stems are 5 to 8 cm in diameter, leading to branches forming new tops with resulting bayonet crooks in stems and associated entry points for decay fungi. Where stands are widely spaced (i.e., > 5 x 5 m), these trees may remain dominant but carry the stem defect into maturity. Where antler rubbing has caused multiple stems, trees may form two dominants or resolve themselves into single-stemmed individuals, and basal defects are not usually serious. Growth impacts from such injuries remain, however, but have not been quantified to maturity.

During the first 15 years, alder grows so rapidly that virtually no other tree species can be maintained in the stand unless alder density is very low. Early height increments of red alder exceed all other species in the Pacific Northwest except black cottonwood (*Populus trichocarpa*) and balsam poplar (*P. balsamifera*), which are common associates only in river bottoms, especially from the Columbia River north. Bitter cherry (*Prunus emarginata*), a vigorous sprouter most common where and when deer populations are low, is capable of remaining as a minor component of an alder stand for several decades and is a common associate in widely spaced stands. Bigleaf maple is also a common associate in many sites as sprouts or residual trees at the time of alder establishment. Bigleaf maple sprout growth is very rapid and competes well with alder seedlings, so large maples are not uncommon in alder stands on warm sites. Yet maple seedling growth rates are greatly inferior to those of red alder in the field, and seedlings are highly vulnerable

to browsing. Thus, seedling bigleaf maple seldom becomes dominant among alders. The persistence of bigleaf maple in shade, however, may cause it to outlive the alder to become dominant later.

Conifers usually occur in forest zones where alder is abundant, but they are not prominent in the first decade of stand development. Conifer associates of alder saplings seldom persist more than a decade when alder is in dense stands. Typically, where understory conifers occur, they meet the following criteria:

1. Understory conifers are the same age as the overstory alder, but grow much more slowly than the overstory. Occasionally western redcedar (*Thuja plicata*) seedlings are somewhat younger than the overstory if there has been a minor disturbance in the understory and if there is some side light.

2. Understory conifers consist of Douglas-fir (rarely grand fir) if away from the coastal fog-belt, and western hemlock (*Tsuga heterophylla*) or Sitka spruce (*Picea sitchensis*) within the fog-belt. The spruce or hemlock are almost universally found on rotten conifer logs or stumps, the Douglas-fir in gaps where alder did not establish initially.

3. Understory conifers are restricted to low-density portions of the alder stand where shrub density has also been low early in the life of the stand.

4. Understory conifers occur where heavy grazing or deer/elk have kept shrubs or hardwoods in check for several years.

Where relatively shade-tolerant conifers are observed within the fog-belts of the Coast Range or Cascades midslope, their location on elevated rotten stumps or logs provides some protection from animals and falling litter, and also some elevation toward the light.

## Pole Timber to Young Sawtimber

By the age of 15 years, red alder has reached half or more of its mature height (Worthington et al. 1960; Newton et al. 1968) and half its mature stand biomass (Zavitkovski and Stevens 1972). Height increments have begun to decline substantially. The upper branches have a rather erect habit, and the ratio of crown expansion to height growth is relatively low. Thus, the ability of alder crowns to expand to fill gaps decreases during the transition from pole timber to small sawtimber. Smith (1978) suggested that the overlap of crowns breaks up at this point and implies loss of ability to fully utilize maximum canopy density to capture light. We presume the increasing sway with height contributes to intercrown abrasion. Crown overlap virtually ceases as mortality leaves openings that cannot be filled. Understory vigor increases with onset of canopy openings, leading to increased coverage by shrubs, particularly salmonberry, vine maple, and red elderberry, plus sword fern (Newton et al. 1968; Henderson 1971; Carlton 1990). This same phenomenon is also associated with declining periodic growth, results of which appear in yields (Puettman, Chapter 16).

The tendency of red alder to occur on steep lower slopes along roads and streams and in other places (i.e., stand gaps) where light may come primarily from one side leads to stem lean during this stage of stand development. Often alder trees in the center of a strip are erect with straight stems, while trees on the edges of the strip lean away from the center. On steep slopes, trees tend to lean away from the slope, in general, and alder appears to lean away from the slope more than most species. Similarly, they tend to lean and fill in space over streams, even when the stream is 10 m or more wide. This ability can be used to advantage where stream temperature needs moderating. The disadvantage, however, is that leaning produces stresses in wood. The ability to create a bend in the bole while remaining rigid is the function of gelatinous fibers in the wood. These fibers can create very substantial internal stresses in the wood, which may cause warping during lumber seasoning. Nevertheless, leaning of alders is an important mechanism with which heterogeneous stands may occupy canopy space with remarkable uniformity. The frequency of leaning trees is probably a matter of heterogeneity of stand density and openings during the formative years of the stand.

## Young Sawtimber to Maturity

Red alder attains nearly all its mature height before the age of 40 years. After that, top damage of various sorts and reduced height growth rate lead to relatively static height development. Whereas the tree may have the capacity to grow beyond this time, it appears to not achieve a net increase in height with great regularity, especially if the stands are not protected from wind. The tallest trees are found typically where there are a few conifers for protection from wind or where they are in sheltered draws.

As upland stands mature, it is not uncommon for scattered conifers to appear above the alder canopy. Newton et al. (1968), Stubblefield and Oliver (1978), and Miller and Murray (1978) demonstrated that a small number of same-aged conifers, especially Douglas-fir, may persist in small openings in the heterogeneously established alder. At the age of 25 to 40 years, provided the conifers have not been suppressed severely, their height growth paths tend to intersect those of the hardwoods, and they appear in or above the canopy. If the conifers survive to this stage, the prognosis for their development improves radically.

After the age of 40 years when alder height becomes static, low-density Pacific coastal conifer stands still have great potential for height increases and crown spread (Newton and Cole 1987). These increases occur at the time alder vigor is decreasing and alder height has reached a maximum (Newton et al. 1968). What might have been described as a pure stand of alder at age 25, because of conifers few in numbers and shorter in stature, becomes a mixed stand of alder/conifer. Mixed stands with as few as 10 to 20 conifers/ha add structural diversity to the alder stand, and may equal the pure alder stand in net board foot volume by the age of 70 to 90 years (Williamson 1968; Newton, unpub. data). Interviews with major landowners on the subject of harvestable volumes of alder indicate that stands of more than 10 Mbf Scribner of net harvestable alder per acre are rare in natural stands. Four to eight Douglas-fir per acre (10 to 20/ha) at 90 years should yield more than 10 Mbf/a (25 m/ha). After that age, mortality in the alder causes it to decline rapidly in volume proportion, and by the age of 130 years, little alder remains. The resulting stand is a low-density conifer stand with very large individual trees having relatively smooth lower stems (up to about 30 m) and very full, rough crowns (Stubblefield and Oliver 1978; Newton and Cole 1987). Beneath these will be an understory dominated by the shrubs that have increased with the breakup of the alder; this understory is most frequently dominated by salmonberry, vine maple, and elderberry.

Stands of alder growing on south slopes or in riparian zones often include a component of bigleaf maple of sprout origin and, in hydric, warm sites, Oregon ash (*Fraxinus latifolia*) and willows (*Salix* spp.). The maple and ash, in particular, are longer-lived than alder. If stands remain undisturbed, attrition and succession will leave the associated hardwoods dominant, along with any conifers that persist. In the upland sites, common understories will include salmonberry and elderberry, as well as California hazel (*Corylus cornuta* var. *californica*), red huckleberry (*Vaccinium parvifolium*), ocean-spray (*Holodiscus discolor*), and vine maple, along with sword fern and numerous other herbs (Franklin and Pechanec 1968; Newton, unpub. obs.). In riparian areas with hydric soils, green indian plum (*Osmaronia cerasiformis*), willows, sedges (*Carex*), and salmonberry are abundant. On boggy coastal terraces, scattered Sitka spruces of immense size become established on hummocks, and eventually become the dominant features of the landscape, with understories primarily of salmonberry.

Red alder trees in mature stands on good sites have average diameters of 45 to 50 cm and are 30 to 38 m tall. After age 60, the development of annual rings may be irregular; Newton, Zavitkovski, and Krahmer (unpub. data) observed microscopic detail in annual ring formation at several heights in trees approaching 106 years, as nearly as could be determined from the ages of dominant Douglas-fir in the stand. The largest number of annual rings counted was 79 at 18 m aboveground. At breast height, 71 annual rings were counted, of which only 56 were complete rings and many existed only on one buttress. The upper crown was still producing rings with over a millimeter increment and was flowering and apparently in good health. We conclude from this that mature red alder, especially if dominated by conifers, could not be relied upon to form complete annual rings, and that many rings could be missing from mature individuals when observed only at breast height.

The above is not a universal observation. Worthington et al. (1960) reported yields based on trees up to 80 years old, presumably by annual ring count during stem analysis. Carlton (1990) observed ring counts up to 87 years. Williamson (1968) reported alder with 90 annual rings. Nonetheless, annual ring counts above 80 are rare, and alders often are found as 80-year-old (or older) stands and as relict individuals beneath even-aged stands of widely spaced Douglas-fir well over 100 years of age. Thus, when studying mature alders, an investigator should confirm the age by means other than ring count if precision is of crucial importance. Alder will seldom be more than five years younger than an even-aged stand of Douglas-fir on a good site in Oregon, though Stubblefield and Oliver (1978) reported up to 17 years spread. Larger age differences are more of a possibility on a poor site with secondary disturbances.

## Understory Variability With Site

Alder understories have remarkable similarity from cool sites in northern California to southeastern Alaska. That is, salmonberry is nearly ubiquitous, but, from south to north, other associates are common and shift with latitude.

Understories of natural alder stands vary considerably with habitat type.

Carlton (1990) described in detail the manner in which understories vary with slope, elevation, and physiographic position. In brief, steep slopes at relatively high elevations tend to be dominated by vine maple, whereas valley bottoms and terraces are dominated by salmonberry. Swordfern dominates understories beneath young stands. Franklin and Pechanec (1968) indicated that the understories of pure coastal alder stands 40 years old were much richer in species than mixed stands or pure conifer stands. Newton et al. (1968), Henderson (1971), and Carlton (1990) have observed in a variety of situations that the shrub layer may outlive the hardwoods, and perhaps even the conifers. In that event, the ultimate successional phase would be dominated by shrubs until large rotten logs from adjacent stands could provide a medium for encroachment of conifers. The brief life expectancy of alder snags and down logs apparently does not permit conifer recruitment. We have observed large-scale die-off of salmonberry attributable to *Armillaria* root rot. Shrubs, hardwoods, and conifers tolerant of the pathogen have increased markedly in dominance. Such events appear to be uncommon. Hot wildfires also have the ability to kill rhizomes of salmonberry, hence to provide some opportunity for recruitment of other species.

In the southern part of red alder's range, common understory plants include tanoak (*Lithocarpus densiflorus*) and evergreen shrubs, *Rhododendron macrophyllum,* evergreen huckleberry (*Vaccinium ovatum*), and salal. Rhododendron is less common north of Newport, Oregon, at low elevations. Salal is common on mature soil terraces and remnants, and on many cool coastal soils of dune or basaltic origin as far north as Vancouver Island and southern mainland British Columbia. Red huckleberry is also common in the same range and soils as salal. Scattered devils club (*Oplopanax horridum*) begins to occur on wet seeps beneath alder on basaltic soils near the Alsea River, Oregon, and becomes more common with northerly latitude, until it is abundant along coastal British Columbia and southeast Alaska, especially along smaller creeks and in boggy places.

## Mixed Stands

The preceding mention of conifers in maturing stands identifies the role of scattered conifers in late succession. The formation of mixed stands of very different growth habits warrants closer attention to the earlier stages of development.

Typically, we have observed that the same types of disturbances that cause alder stands to form also recruit substantial numbers of conifers if there is a source of seed. Thus, most stands of alder are actually mixtures during their earlier years, and what is observed at any given time is the result of differential mortality and growth as the alder and associated shrubs dominate the seedling conifers.

Among the conifers that may be found in the understory, Douglas-fir is at a severe disadvantage because of its intolerance, and it seldom survives except in the openings. Sitka spruce appears to be the most tolerant of alder domination of all conifers (Franklin and Pechanec 1968) although Stubblefield and Oliver (1978) indicated western redcedar may be the most advantageous in mixed stands. Sitka spruce can survive with very little growth in the coastal fog zone under conditions le-

thal to western hemlock (Newton et al. 1993) and eventually emerges erratically as alder matures and breaks up. At this time, trees reaching dominance are widely spaced and rough. Moreover, form of the spruce is likely to become highly distorted by the terminal weevil (*Pissodes strobi*), hence few trees emerge from a suppressed condition in good form. Western redcedar is occasionally found under alder, but seldom in numbers able to form a stand except in northwestern Oregon, Washington, and British Columbia. It has excellent tolerance of alder-induced shading. Hemlock, also, has substantial tolerance of alder shading. Despite this tolerance, all conifers grow very slowly at best under closed-canopy alder, and mortality is heavy. Stubblefield and Oliver (1978) reported that the form of western redcedar was less influenced by suppression than that of hemlock, hence it was in better condition at the time of emergence as a dominant to resist wind and other damage. They also observed that in mixed stands, the successful conifers are almost always older than the alders.

The formation of mixed alder/Douglas-fir stands on good sites relies on gaps in the alder in which shrubs are also absent or nearly so. On poorer alder sites other than bogs, the disadvantages of slow conifer growth are proportionally less, and conifers survive in smaller gaps (i.e., < 6 m). It is not uncommon for stands up to 15 years old on good sites to contain 10 to 50 (or more) conifers per hectare in gaps in the alder, while in general terms, the stands are considered pure alder. At age 15, the conifers are substantially shorter than the alder, although by that time those receiving overhead light are growing more rapidly in height. Basal area of conifers is at this point a small part of the stand. On poorer alder sites, including those above 600 m and those on upland sites in fringes of the Willamette Valley, alder loses vigor at approximately age 10, so that understory Douglas-fir may persist in better vigor than on Coastal sites. Valley-edge stands with severely suppressed Douglas-fir at ages 4 to 5 show enough loss of vigor in alder to approximate some degree of conifer release (Cole and Newton, unpub. data). This may be an extension of the apparent competition from below reported by Shainsky et al. (Chapter 5). By the age of 30 years on good sites, aerial views reveal the conifers as scattered individuals, still a small minority of the stand. In terms of stand volume, average conifer tree volume approaches that of alder by the age of 30 years, but growth of individual trees surpasses that of alder if the conifers have achieved dominance. On poor upland alder sites, conifer domination may be nearly complete by age 30.

By the age of 50 years on good alder sites, the conifers existing to age 30 have become the dominant feature of the stand. If there are 20 or more conifers per hectare, conifer value already surpasses that of the alder on a stumpage basis. By the age of 80 years, conifers are 45 to 60 m in height, and alders are hardly visible from a distance, except from aerial view. Tree numbers still favor alder, but volume of conifers is substantially greater. At about 130 years, the last alder succumbs, and the stand is then classed as pure Douglas-fir or other conifer. Meanwhile, the understory remains in salmonberry and red elderberry on many sites, and vine maple/salal on many others. By the time a Douglas-fir stand with 25 trees per hectare reaches 200 years (assuming Site Class II or better for Douglas-fir on a productive alder site), the stand has nearly closed canopy and trees have an average diameter of 150+ cm. If the conifer component is western hemlock or western redcedar, more conifers are needed to achieve the same picture.

Tree form reflects its competitive history. When a stand originates with conifers as a minority mixture and emerges as the dominant and finally as a pure stand, the conifers show evidence of this history. First, the initial bole is characterized by slender branches and small knots. Although crowns are surprisingly long in conifers dominated by alder, the branches are small and do not contribute later to large knots. Above the crowns of the maturing hardwoods, this changes radically. Douglas-fir, Sitka spruce, and to a lesser extent, hemlock have a great capacity for expanding crowns above a hardwood canopy. Whereas this ultimately contributes to sustained growth rates and very large tree sizes, the boles above about 35 m are characterized by very large limbs and poor wood quality. The alder stands dominated by large conifers, however, often begin as low-density, poor-quality trees, but eventually develop tall, clear boles and small crowns.

The transition of conifer crown form at the tops of the hardwood crowns is remarkable. The first three 10-m long logs may be nearly knot-free, but branches at the 40 to 45-m level may be over 15 cm in diameter by the age of 150 years, and the top logs have little lumber value despite their very large sizes. These branches contribute to very deep, broad crowns, but with a discontinuity between the tops of the dense shrub layer (10 to 13 m) and the bottom of the live conifer crowns. Despite tree sizes well above the average diameters of typical old-growth conifer stands, the low-density conifer remnant stands lack conifer understories that would provide complex vertical structure. Thus, they remain quite distinctly different from stands originating as primarily conifer stands. Moreover, until large conifer logs begin to fall (a rare occurrence in these low-density, vigorous stands until very old), conifer regeneration seldom occurs in appreciable numbers. Thus, the hardwood role in the early development of the stand projects a shrub legacy and a structural arrangement that can be expected to last for several centuries.

Nutritional status of the site may make some difference in competitive advantage of alder. Newton et al. (1968) demonstrated that there are sites where growth advantage of alder ranged from four to nine years equivalent age. We are uncertain as to the degree to which water and nutrients contribute to this range. Russell et al. (1968) demonstrated that nitrogen fixation required cobalt and other elements that must be available for nitrogen fixation but were inhibited by excesses of nitrogen, iron, and manganese. If soil mineral or acidity reduces availability of crucial elements, then the nitrogen-fixation advantage of alder would be reduced. We are not aware of the occurrence of major deficiencies or unbuffered excesses of critical elements for fixation within the commercial ranges of red alder and associated conifers.

Dryness and, perhaps because of dryness in spring, density of alder seems to affect the competitive interaction (Haeussler 1988). Low densities of alder permit conifers to reach greater growth rates before coming under domination. Dryness leads to gaps in alder establishment, hence to greater likelihood of conifer dominance. Newton et al. (1968) also identified abundant moisture as a feature of sites where the differential growth of alder is greatest. Thus, there is a moderately sharp break in the moisture regime on low-elevation sites where pure alder gives way to mixed or even pure stands of conifers. Shainsky et al. (Chapter 5) discuss alder water relations further.

There have been numerous proposals to grow alder and conifers in mixed stands of even ages. Hibbs and DeBell (Chapter 14) describe several management systems for such mixed culture.

## Conclusion

In summary, red alder is typically a fast-growing tree dependent on mineral soil and moderately moist conditions for successful germination and survival. The abundance of seed leads to dense stands if initial survival is anything but a failure on favorable sites. Thus, on physically disturbed sites, there is a high frequency of stands that appear with the following characteristics:

1. Stands are dense but patchy and heterogeneous (according to pattern of soil disturbance) in their first decade.

2. Stands are homogeneous and pure during much of their second decade, although on close examination there are spaces among the clumps of stems.

3. Crowns cover many openings but allow survival of a low density of conifers in the initial gaps in the stands if any have persisted a decade or longer.

4. The surviving conifers are taller than the alder trees, although the alder is numerically much more abundant than the conifer.

5. Conifers dominate and have greater biomass after the fifth decade if they have survived to this point.

6. Stands show increasing densities of shrub understories after the second decade, which later become the dominant cover after senescence of alder at 80 to 120 years in the absence of conifers.

7. Stands on the immediate Pacific Coast support modest numbers of Sitka spruce in the understory, especially on down logs. These spruces may form a heterogeneous but moderately dense stand as the alder senesces. Western redcedar and western hemlock are increasingly common understory plants north of the Columbia River.

In the absence of mechanical disturbance, red alder is not abundant except in riparian strips along the high-water line, and on islands. Where fire has arrested development of salmonberry and other fast-growing shrubs, alder may achieve moderate densities on north sloping or cool upland sites after wildfire. Alder is heavily influenced by abundant herb and shrub competitors in the first decade.

Red alder is highly subject to frost damage in its first three years in cold air drainages of the Coast Ranges and presumably elsewhere. Trees are also highly attractive to beavers at all stages of development if within 50 m of a stream. Deer and elk damage trees by rubbing against them with their antlers, causing multiple tops and loss of individual tree growth, which is probably inconsequential except where stand density is low.

Red alder is a moderately early successional species that will not replace itself without a major disturbance. The successional sequence it dominates in early stages, however, is likely to show the effects of initial alder domination for several hundred years.

There do not appear to be major differences of opinion about lifespans or growth habits of red alder. Miller and Murray (1978) have observed super-dominant Douglas-fir surviving some alder competition where it might have been expected to fail. It is likely that a complete history of spacing and exact timing of establishment would help explain conifer performance.

Among the important data gaps, is the question of why plantation alder appears especially vulnerable to damage by beavers and perhaps elk. Also, the relation between site quality, spacing, and ability of conifers to persist in mixtures needs attention.

# Literature Cited

Bormann, B. T., J. C. Gordon, and D. S. DeBell. 1979. The effect of density on canopy characteristics of young *Alnus rubra* stands. Abstract of Poster. *In* Symbiotic nitrogen fixation in the management of temperate forests. *Edited by* J. C. Gordon, C. T. Wheeler, and D. A. Perry. Oregon State Univ. Press, Corvallis.

Briggs, D. G., D. S. DeBell, and W. A. Atkinson, *comps.* 1978. Utilization and management of alder. USDA For. Serv., Gen. Tech. Rep. PNW-70.

Carlton, G. D. 1990. The structure and dynamics of red alder communities in the central Coast Range of western Oregon. M.S. thesis, Oregon State Univ., Corvallis.

Franklin, J. F., and A. Pechanec. 1968. Comparison of vegetation in adjacent alder, conifer and mixed alder-conifer communities. I. Understory vegetation and stand structure. *In* Biology of alder. *Edited by* J. M. Trappe, J. F. Franklin, R. F. Tarrant, and G. H. Hansen. USDA For. Serv., PNW For. Range Exp. Sta., Portland, OR. 37-44.

Haeussler, S. 1988. Germination and first-year survival of red alder seedlings in the central Coast Range of Oregon. M.S. thesis, Oregon State Univ., Corvallis.

Harrington, C. A. 1992. Ecology of red alder. *In* U.S. Forest Service silvics manual. USDA For. Serv., PNW For. Res. Sta. Olympia, WA. (In press.)

Henderson, J. A. 1970. Biomass and composition of the understory vegetation in some *Alnus rubra* stands in western Oregon. M.S. thesis, Oregon State Univ., Corvallis.

Hibbs, D. E., and G. C. Carlton. Development and succession of red alder communities. Submitted to Ecology.

Hibbs, D. E., and K. Cromack Jr. 1990. Actinorhizal plants in Pacific Northwest forests. *In* Biology of *Frankia* and actinorhizal plants. Academic Press. Ch. 17.

Kelly, L. S. 1991. Early competitive interactions between red alder and salmonberry in the Oregon Coast Range. M.S. thesis, Oregon State Univ., Corvallis.

Kelpsas, B. R. 1978. Comparative effects of chemical, fire and machine site preparation in an Oregon coastal brushfield. M.S. thesis, Oregon State Univ., Corvallis.

Miller, R. E., and M. D. Murray. 1978. The effects of red alder on the growth of Douglas-fir. *In* Utilization and management of alder. *Compiled by* D. G. Briggs, D. S. DeBell, and W. A. Atkinson. USDA For. Serv., Gen. Tech. Rep. PNW-70. 283-306.

Newton, M., and E. C. Cole. 1987. A sustained-yield strategy for old-growth Douglas-fir. West. J. Appl. For. 2(1):22-25.

Newton, M., E. C. Cole, and D. E. White. 1993. Tall planting stock for enhanced growth and domination of brush in the Douglas-fir region. New Forests 7:107-121.

Newton, M., B. A. El Hassan, and J. Zavitkovski. 1968. Role of red alder in western Oregon forest succession. *In* Biology of alder. *Edited by* J. M. Trappe, J. F. Franklin, R. F. Tarrant, and G. M. Hansen. USDA For. Serv., PNW For. Range Exp. Sta., Portland, OR. 73-83.

Roberts, C. A. 1975. Initial plant succession after brown and burn site preparation on an alder dominated brushfield in the Oregon Coast Range. M.S. thesis, Oregon State Univ., Corvallis.

Russell, S. A., H. J. Evans, and P. Mayeux. 1968. The effect of cobalt and certain other trace metals on the growth and vitamin $B_{12}$ content of *Alnus rubra*. *In* Biology of alder. *Edited by* J. M. Trappe, J. F. Franklin, R. F. Tarrant, and G. M. Hansen. USDA For. Serv., PNW For. Range Exp. Sta., Portland, OR. 259-272.

Shainsky, L. J., and S. R. Radosevich. 1992. Mechanisms of competition between Douglas-fir and red alder seedlings. Ecology 73(1):30-45.

Smith, J. H. G. 1978. Growth and yield of red alder: effects of spacing and thinning. *In* Utilization and management of alder. *Compiled by* D. G. Briggs, D. S. DeBell, and W. A. Atkinson. USDA For. Serv., Gen. Tech. Rep. PNW-70. 245-263.

Stubblefield, G., and C. D. Oliver. 1978. Silvicultural implications of the reconstruction of mixed alder/conifer stands. 307-320.

Trappe, J. M., J. F. Franklin, R. F. Tarrant, and G. M. Hansen, *eds.* 1968. Biology of alder. USDA For. Serv., PNW For. Range Exp. Sta., Portland, OR.

Williamson, R. L. 1968. Productivity of red alder in western Oregon and Washington. *In* Biology of alder. *Edited by* J. M. Trappe, J. F. Franklin, R. F. Tarrant, and G. M. Hansen. USDA For. Serv., PNW For. Range Exp. Sta., Portland, OR. 287-292.

Worthington, N. P., F. A. Johnson, G. R. Staebler, and W. J. Lloyd. 1960. Normal yield tables for red alder. USDA For. Serv., Res. Pap. PNW-36.

Zavitkovski, J., and R. D. Stevens. 1972. Primary production of red alder ecosystems. Ecology 53(2):235-242.

# 8

# Growth Patterns of Red Alder

DEAN S. DEBELL & PETER A. GIORDANO

Some understanding of growth patterns of trees and stands is prerequisite to management of red alder. The growth patterns of a species represent a significant aspect of the overall biological context for decisions on management regimes and silvicultural practices, and the economic analysis thereof. In this chapter, we will summarize current information on general growth habit, patterns of height and diameter growth, partitioning of dry matter, and stand development. Effects of environment, genetic composition, and cultural practices on many of these growth characteristics will be described. Our discussion will focus on the early stages of tree development in plantations because most of the available information was collected on planted trees less than 15 years old. Much of the discussion will concern growth of individual trees and mean trees, thereby minimizing duplication with topics discussed in chapters on natural stand development (Newton and Cole, Chapter 7), growth and yield (Puettman, Chapter 16), and stand management (Hibbs and DeBell, Chapter 14).

## General Growth Habit

The primary shoot or height growth pattern of red alder can be characterized as intermittent and indeterminate. Intermittent means that growth of alder, like that of other trees in the temperate zone, is not continuous but is interrupted by periods of rest (i.e., dormant seasons). Indeterminate growth (also known as sustained growth) refers to growth in which only a portion of the shoot and leaf primordia are pre-formed prior to budbreak; additional primordia develop during the growing season. Trees with an indeterminate growth pattern usually do not grow as rapidly in height early in the growing season as do those with pre-formed

growth (Zimmermann and Brown 1971). They continue to grow, however, as long as environmental conditions are favorable, and total height increment for the growing season may exceed that of species with pre-formed growth (Oliver and Larson 1990).

Young alder trees usually have fairly strong apical control, resulting in an excurrent branching habit in which the leader grows more rapidly than the branches beneath. Thus, young alders tend to have conical-shaped crowns. Poor planting stock and adverse environmental conditions (e.g., unseasonal frosts), however, have resulted in some stands with a high proportion of trees with multiple stems. These problems can be minimized by use of improved guidelines for seedling production and plantation establishment. Although crowns are conical in early years, they become rounded and eventually flat-topped at advanced ages.

Young alders also exhibit a sylleptic branching habit. Sylleptic branches develop from axillary buds which form and break during the current year's terminal growth. Other branches—generally the larger ones on a tree—are known as proleptic branches; they develop in spring from buds formed during the previous growing season. Both kinds of branches occur on all young trees; the larger proleptic branches generally occur near the termination of each year's growth. The smaller sylleptic branches are found in between the annual nodes and tend to die sooner than the proleptic branches. This sylleptic branching characteristic results in rapid development of leaf area at young ages and is one of the reasons that red alder offers such serious competition to Douglas-fir. Sylleptic branching and apical control tend to diminish as trees get older and larger.

Alder stems have many suppressed buds. These buds have also originated in leaf axils, but they have not developed into branches. Rather, they have continued to grow outward, remaining just beneath the bark, as additional xylem is produced. Sometimes they branch several times, resulting in clusters of buds. If trees are cut when very young (< 5 years old), suppressed buds on the stump may develop into vigorous sprouts (Harrington 1984). Sprouts may also develop from other origins—such as live lower branches and adventitious tissue. Older trees are less likely to sprout vigorously and successfully, especially if cut in mid-growing season (Harrington 1984; DeBell and Turpin 1989). Suppressed buds also may develop into epicormic branches, particularly on older, stressed trees. Epicormic branches are common on trees along recently created forest edges (e.g., clearings for harvest and regeneration, residential development, and rights-of-way). They have also been reported after pruning (Berntsen 1961b) and after thinning (Warrack 1964), particularly in older stands. Thinning a 14-year-old red alder stand did not increase the number of epicormic branches except in a chemically thinned, narrowly spaced treatment; however, existing epicormic branches became larger with thinning (Hibbs et al. 1989). Stimulation of new epicormic branches has not been mentioned, however, in other reports of thinning in stands 21 years old or younger (Berntsen 1958, 1961a, 1962; Warrack 1964).

Most of the stems in older natural alder stands have considerable lean and sweep. Such traits are exhibited at their extremes in trees growing beside streams and along roads. The lean of alder stems in irregularly spaced natural stands is probably caused primarily by heavier crown development on sunny sides of the tree rather than by phototropism per se (Wilson 1984). In managed stands and plantations, however, uniform planting stock and spacing provide more even competition. Observations in natural stands that were spaced at an early age (Bormann 1985) and in plantations (DeBell, unpub. data) indicate that stems are much straighter than in unmanaged natural stands, at least up to the time of significant competition-related mortality.

## Description of Study Plantations

Data presented in the next sections were derived primarily from three research plantations located near Apiary, Oregon; Yelm, Washington; and Cascade Head, Oregon. Most of the work has not been published to date. As background to subsequent discussion, some characteristics of these research plantings are summarized below.

**Apiary.** This study was initially established by Crown Zellerbach Corporation and is now maintained by the Olympia Forestry Sciences Laboratory. The plantation is located on International Paper Company land, about 15 km southwest of Rainier in northwest Oregon (DeBell, unpub.). The land was occupied by a natural, second-growth stand of predominantly Douglas-fir (Site Class II). The site was harvested and broadcast burned in 1974 and red alder seedlings were planted in winter 1974-75 to four specified spacing treatments (0.6 x 1.2 m, 1.2 x 1.2 m, 1.2 x 1.8 m, and 1.8 x 1.8 m). In addition, measurement plots were established in adjacent areas planted operationally at wider spacing (approximately 2.7 x 2.7 m) with the same stock. No post-planting control of competition was attempted on any plot. Height, diameter, and survival were assessed annually from age 4 to age 12.

**Yelm.** This study was established by the Olympia Forestry Sciences Laboratory with funds from the U.S. Department of Energy. The plantation is a cooperative effort with Washington State Department of Natural Resources and is located at their Meridian Tree Improvement Center, 12 km east of Olympia, Washington (DeBell, Clendenen, and Harrington, unpub.). The site was previously farmed for strawberry and hay crops, and the soil is rated as moderately productive for Douglas-fir. Precipitation averages about 1000 mm per year, and summers are periodically dry. The land was prepared for planting by plowing and disking. Red alder seedlings were planted in spring 1986 in a factorial design. Treatments included three square spacings (0.5, 1.0, and 2.0 m) and two levels of irrigation (high and low). All plots were irrigated by means of a drip system with 25 cm of water during the first year. Thereafter, the high irrigation treatment involved applications of 40 to 50 cm during each growing season; the low irrigation

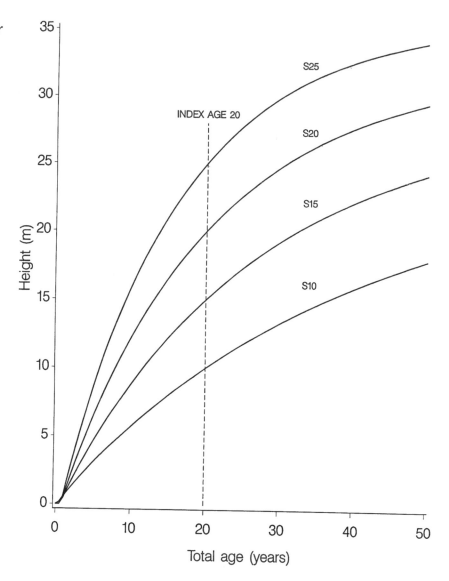

*Figure 1. Height growth curves for red alder (From: Harrington and Curtis 1986).*

treatment received no supplemental water during the second year and minimal applications (< 8 cm per growing season) in subsequent years. The intent of the minimal application was to ensure against mortality caused by drought stress alone. All plots were maintained in a weed-free condition by tilling, hoeing, and selective application of herbicides. Height, diameter, and survival were measured annually after establishment to age 5; some characteristics were assessed monthly on subsets of trees during the second through the fifth year.

**Cascade Head.** This study was established by Oregon State University. The plantation is located on the Cascade Head Experimental Forest near Lincoln City, Oregon (Hibbs and Giordano, unpub.). The site was originally occupied by an old-growth stand of Douglas-fir, western hemlock, and Sitka spruce. Site quality is estimated to be fairly high (Douglas-fir Site II) and annual precipitation averages 2500 mm. Following harvest, the area was broadcast burned in summer 1985. Alder seedlings were planted in spring 1986 in variable density plots patterned after Nelder (1962). These plots provided densities ranging from 101,000 to 238 trees per hectare. Height, diameter, and crown width were measured annually from establishment to age 6.

## Patterns of Height and Diameter Growth

**Relationship to age.** The rapid juvenile growth of red alder is well known and an important consideration in decisions to manage the species. On moderate to good sites, planted trees may be 2 to 3 m tall after 2 years, 6 to 8 m tall at 5 years, and 14 to 20 m tall at 15 years. In natural stands, such growth rates are found only on sites of better-than-average quality (cf. Fig. 1). No information is available for older plantation ages. Patterns in natural stands are reflected in site curves (Worthington et al. 1960; Harrington and Curtis 1986) and suggest that height growth rate decreases substantially beyond age 20 and becomes nearly negligible at 40 to 50 years when heights are 20 to 35 m. The newer curves (Fig. 1) prepared by Harrington and Curtis (1986) predicted an earlier and a greater decline in height growth as trees become older than did those of Worthington et al. (1960). Height, however, can exceed 40 m on the best sites.

Diameter growth increases with age as crowns expand. Subsequent peaks and declines in diameter growth are generally associated more closely with increased competition than with aging. A decline with age occurs in open-grown trees, however. Some striking accelerations in growth have been observed when slow-growing trees are released and heavy top damage occurs; this may result in formation of a new leader and "rejuvenation" of diameter growth in the lower bole. For all practical purposes though, diameter growth of red alder is unlikely to accelerate once it has begun to decline.

**Effects of site.** Growth rates differ markedly with site quality. Site index generally ranges from 10 to 25 m at a base age of 20 years (Harrington and Curtis 1986) and from 20 to 36 or 40 m at base age of 50 years (Worthington et al. 1960; Harrington 1986). Height differences among sites are apparent at age 5, continue to increase to age 20, and are maintained through age 50. Geographic and topographic position are generally the most important factors affecting height growth (Harrington 1986). Greatest height growth is found on low elevation (< 100 m above sea level), flood plain sites. More information on site quality and height growth patterns is given in the chapter on site evaluation (Harrington and Courtin, Chapter 10). Diameter growth rates are related positively to site quality, but observed patterns represent combined effects of age, stand density, and site quality.

**Effects of stand density.** Both height and diameter growth are affected by plantation density. Initially height growth is greater in close-spaced stands than in wide-spaced stands. This growth-enhancing effect of neighbor trees during the establishment phase has been observed in several species, but its cause or causes have not been assessed. Earlier canopy development, and consequent shading and mortality of competing understory vegetation, probably has some influence and may be the major factor in some circumstances. The effect seems to be most pronounced on sites where weeds were not controlled. In the Cascade Head study, 5-year-old trees planted at the equivalent of a $1.9m^2$ spacing were 41 percent taller than trees at a $6.5m^2$ spacing. A similar but less dramatic trend was noted four years after planting at Apiary; trees spaced at 0.6 x 1.2 m averaged 5.3 m tall whereas trees spaced at 2.7 x 2.7 m averaged only 4.6 m tall (Fig. 2a). Understory competition can not be the only factor, however, as similar observations have been made in plantings of several species that were maintained in a weed-free condition, including the Yelm study (Fig. 2b). The growth-enhancing effect of dense spacing was not evident in early height growth of red alder in the low irrigation treatments. It was, however, observed in the high irrigation treatments where trees in the intermediate 1 x 1 m spacing were slightly taller than those in the 2 x 2 m spacing after the second and third year.

Another phenomenon has been observed that may be related to the above matter and it may be peculiar to red alder; that is, seedlings planted at very wide spacings have grown poorly for many years. Results from two studies indicated that red alder trees planted at spacings greater than 4 x 4 m had slower growth and poorer stem form (larger

*Figure 2. Mean tree height in red alder plantations as related to (a) spacing and age, at Apiary, and (b) spacing, irrigation, and age, at Yelm.*

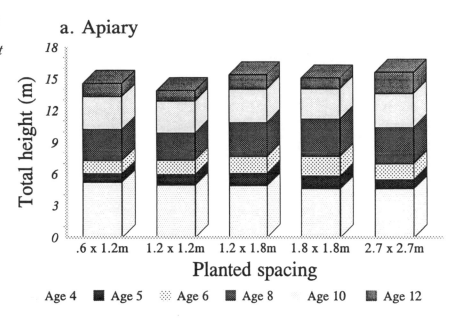

a. Apiary

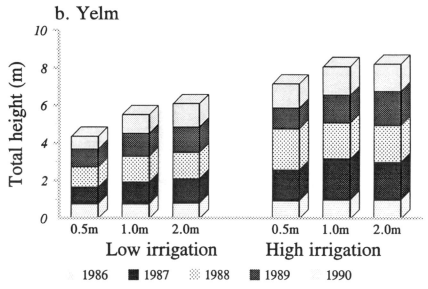

b. Yelm

branches and multiple stems) than trees planted at denser spacings. The poor performance in the wide spacings has continued through the latest measurements of both studies, age 6 at Cascade Head and age 10 in a trial established on a reclaimed strip-mine near Centralia, Washington (DeBell and Reukema, unpub. data).

As trees grow and competition increases in plantations established at spacings of 4 x 4 m or less, however, best height growth is attained at progressively wider spacings. In the Apiary study, best height growth was attained at the intermediate spacings from age 5 to age 10. At age 12, trees in the widest spacing (2.7 x 2.7 m) were tallest and were about 8 percent taller than trees in the closest spacing (0.6

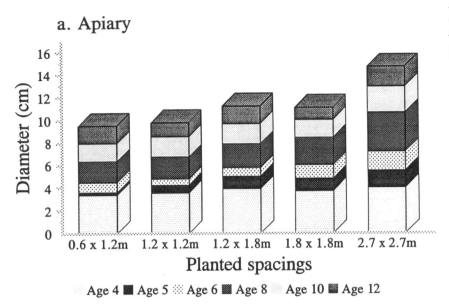

a. Apiary

Diameter (cm)

Planted spacings

Age 4 ■ Age 5 ░ Age 6 ■ Age 8 ░ Age 10 ■ Age 12

Figure 3. Mean diameter in red alder plantations as related to (a) spacing and age, at Apiary, and (b) spacing, irrigation, and age, at Yelm.

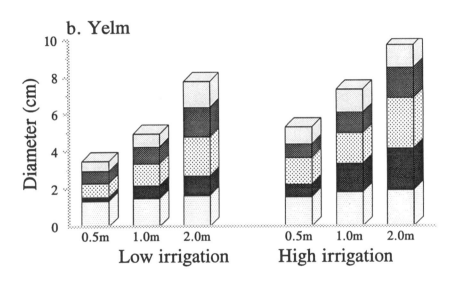

b. Yelm

Diameter (cm)

Low irrigation        High irrigation

░ 1986 ■ 1987 ░ 1988 ■ 1989 ░ 1990

x 1.2 m) (Fig. 2a). Relative differences in five-year height between the .5 x .5 m and 2 x 2 m spacing treatments at Yelm in plots receiving high irrigation were about 14 percent (Fig. 2b). Height differences associated with spacing in Yelm plots receiving low irrigation were substantially greater; trees planted at 2 x 2 m spacing were 40 percent taller than those planted at .5 x .5 m spacing (Fig. 2b).

Diameter growth is affected much more than height growth by plantation density. Effects of spacing on diameter growth may become apparent in the establishment year and increase with time (Fig. 3). Relative differences between close and wide spacings in mean diameter of 12-year-old trees in the Apiary study amounted to 60 percent

(Fig. 3a). At Yelm, 5-year-old trees spaced at 2 x 2 m were 123 percent larger in diameter than trees spaced at .5 x .5 m in plots receiving low irrigation and 80 percent larger in plots receiving high irrigation (Fig. 3b). Thus, high stand density appears to reduce both diameter and height growth to a greater degree under conditions of increased moisture stress.

**Phenology of growth.** Early work by Reukema (1965) in a 50-year-old natural red alder stand near McCleary, Washington, showed that, on average, most diameter growth occurs between mid-April and early September. There was, however, considerable year-to-year variation, much of which could be linked to weather patterns. Giordano (1990), working with 3-year-old trees planted at Cascade Head on the Oregon coast, found that budbreak occurred between late March and mid-April; measurable diameter growth generally occurred between early April and mid-September while measurable height growth tended to occur from mid-May to early September. Phenological patterns in the young red alder plantations near Yelm were similar, although diameter growth in plots receiving high irrigation continued into October. Overall, diameter growth begins earlier and continues longer in the season than does height growth.

The Yelm study has shown that spacing, irrigation, age, and weather can cause marked differences in phenology and magnitude of periodic daily increment. Some of these differences are evident in plottings based on periodic measurements of height and diameter in these plantations (Figs. 4, 5).

In the low irrigation plots, height growth for 1987 attained a maximum in late May in all three spacings (Fig. 4). Growth declined abruptly in the .5 x .5 m spacing but continued at a rapid rate through late June in the two wider spacings. In the high irrigation plots, periodic height growth for 1987 was much greater in all three spacings, and it peaked about one to two months later (in late June or July) than in the low irrigation plots. All spacings in high irrigation plots followed the same general seasonal pattern, and significant amounts of growth occurred through late August.

Effects of age and different weather conditions were evidenced in the contrasting height growth trends for low and high irrigation plots in 1987 and 1989 (Fig. 4). Height growth of trees in the low irrigation treatment peaked one month later in 1989 than in 1987. This difference in growth pattern was caused by hotter, drier weather in early 1989, leading to reduced growth before the irrigation system was turned on in June. Height growth in the high irrigation treatment, however, followed the same seasonal trend in both 1987 and 1989. The amount of growth was much less in 1989; this reduction in growth was caused in part by hot, dry weather and also by increased competition associated with larger, older trees.

Patterns of diameter growth in low irrigation treatments were similar to those for height growth, but differences among spacings were more pronounced (Fig. 5). In high irrigation treatments, growth peaked later in the wider than in the close spacings. Also, diameter growth was maintained at substantial levels much later in the season in high irrigation as compared with low irrigation treatments.

Effects of age and weather conditions were more pronounced on diameter increment than on height increment (Fig. 5). For the low irrigation treatments, growth was greater and it peaked earlier in 1987 than in 1989. For high irrigation treatments, differences in the amount of growth in the two years were even greater. Moreover, in 1989 and subsequent years, periodic diameter increment in high irrigation plots was more or less similar to that in low irrigation plots (Fig. 3b, 5). The diminished benefits of irrigation on diameter growth rates were caused in part by larger tree size and more intense competition in high irrigation than in low irrigation plots. Relative benefits of irrigation on basal area increment would also be reduced, but less so than for diameter increment. The depression in diameter growth in early 1989 was associated with hot, dry weather and was followed by a slight acceleration in growth when the irrigation system was turned on in June.

## Partitioning of Dry Matter

**General patterns.** Red alder tends to allocate the majority of its dry matter to above-ground biomass. Partitioning patterns change with age and environmental conditions. Below-ground allocation in recently established plantations may amount to about

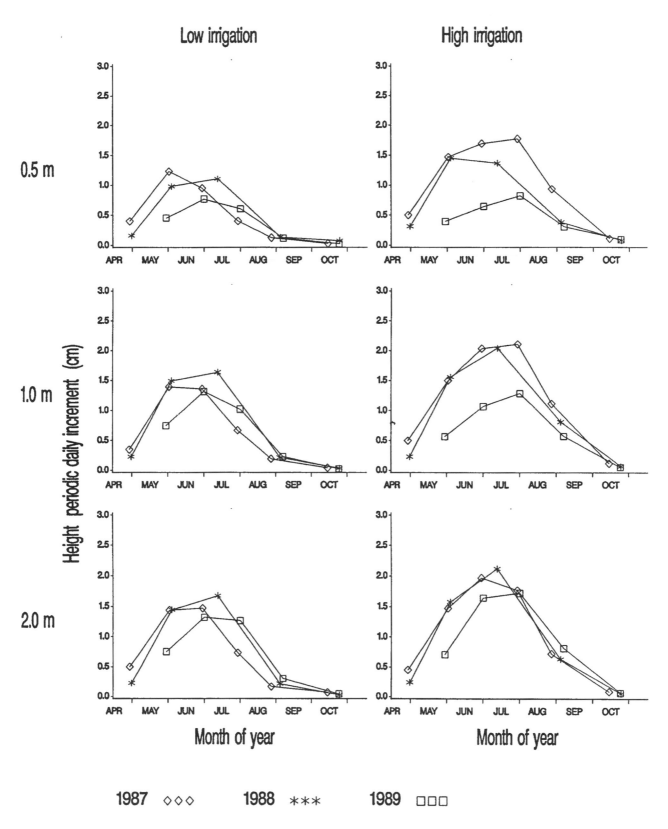

Figure 4. *Periodic daily height increment of red alder during the 1987-1989 growing seasons as related to spacing and irrigation.*

Low irrigation                    High irrigation

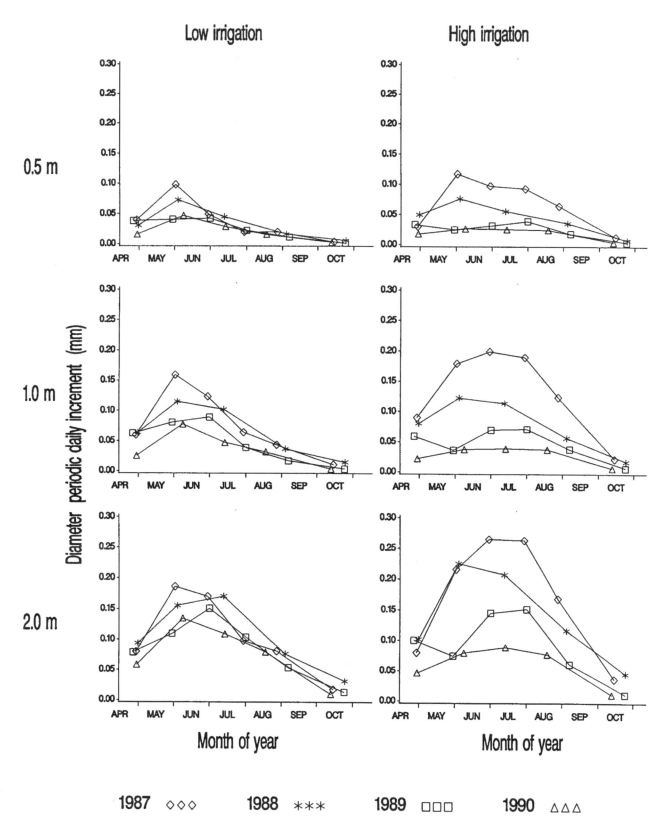

1987 ◇◇◇     1988 ✳✳✳     1989 □□□     1990 △△△

*Figure 5. Periodic daily diameter increment of red alder during the 1987-1990 growing seasons as related to spacing and irrigation.*

half of the dry matter produced (Giordano, unpub. data). Bormann and Gordon (1984) reported below-ground proportions (roots and nodules) of only 6 to 9 percent in 5-year-old plantings, but they may not have accounted for mortality occurring prior to their late-summer sample date. Work in older stands suggests that roots account for about 20 percent of total biomass (Zavitkovski and Stevens 1972; Turner et al. 1976). Of the dry matter allocated to above-ground components, branches and leaves generally account for 25 to 70 percent of the total but vary with stand age and growing conditions (Giordano 1990; DeBell et al., unpub. data). The proportion in stems increases with increased age and with increased stand density, while that in branches and leaves decreases.

Several studies have assessed the amount and distribution of dry matter in red alder trees (Snell and Little 1983; Bormann and Gordon 1984; Helgerson et al. 1988; Giordano 1990; and DeBell and Clendenen, unpub. data). Most of them developed equations for young red alder in planted stands at specific sites; the equations may not be appropriate at other sites, for older tree ages, or for estimating amounts of biomass and changes in biomass distribution due to alternative cultural practices. Snell and Little's (1983) equations were developed from trees of various sizes and ages in natural stands, which were presumably in the self-thinning stage. Although these equations may be applicable for natural stands over a wider range of sites and tree ages, they may not be suitable for use in plantations or for estimating amounts or distributions of biomass in alternative cultural treatments.

**Variation among genotypes.** Research has indicated that the genetic variation of red alder populations in Oregon and Washington shows potential for 15 to 40 percent gains in early growth rate through the selection of seed source and family (Ager 1987). In addition to genetic variation in overall growth trends, red alder exhibits considerable variation in dry matter partitioning patterns. Height/diameter ratio varied from 7.1 to 11.6 among 10 red alder provenances evaluated at age 15 (Lester and DeBell 1989). Variation between families of red alder in the ratio of leaf area or leaf biomass to woody biomass has also been identified (Hook et al. 1990). The ra-

tios for some families differed by as much as 20 percent from the overall population mean. Leaf area ratio had the highest family heritability value of any of the traits studied, suggesting a high potential for improvement of biomass yields based on this trait. See Ager and Stettler (Chapter 6) for a general review of alder genetics.

**Effects of age, stand density, and irrigation.** The most extensive data on dry matter partitioning in young plantations were collected in the Yelm study. Trees were destructively sampled in each irrigation and spacing treatment at the end of each growing season; equations were developed and harmonized over the five-year period. The equations were then used to predict components of biomass based on diameter and height measurements of individual trees. Estimates for 100 trees on each plot were summed and expanded to provide per hectare estimates. Results are displayed by irrigation and spacing treatment in Figure 6.

Among low irrigation treatments, maximum leaf weight was between 2.0 and 3.0 tonnes per hectare in the 0.5- and 1.0-m spacings and occurred at age 3 after which it remained relatively stable. Highest leaf weights in the 2.0-m spacing averaged 2.4 tonnes per hectare and were attained at age 5. Among high irrigation treatments, leaf weights tended to be highest at age 4, averaging more than 5 tonnes per hectare in the 0.5- and 1.0-m spacings and 2.7 tonnes per hectare in the 2.0-m spacing. Branch weights leveled off at about 4 tonnes in the low irrigation, 0.5-m spacings, but continued to increase in all other treatments. Branch weights in the high irrigation plots of all spacings were between 7 and 9 tonnes per hectare. Stem weights steadily increased in all treatments, attaining 32 to 42 tonnes per hectare in the 0.5- and 1.0-m spacings irrigated at the high level.

Proportions of stems, branches, and leaves are presented in Figure 7 for each irrigation and spacing treatment. As expected, the percentage of biomass in stemwood increased over the five-year period in all treatments. Proportion of leaves declined in the low irrigation treatment; in the high irrigation treatment, it declined abruptly during the first three years of growth, but remained nearly stable in the last two years in all spacings. The proportional weight in branches increased during the

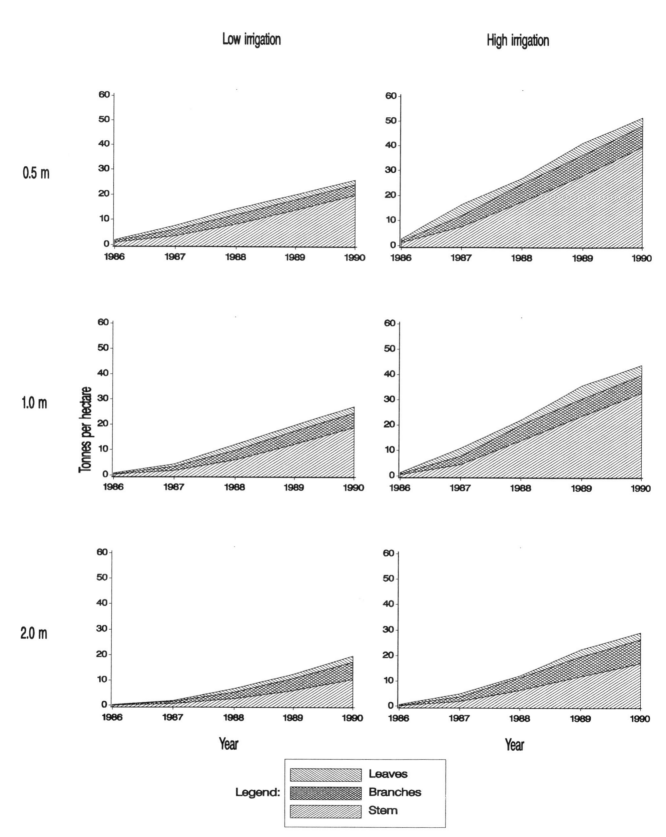

*Figure 6. Five-year cumulative biomass of stems, branches, and leaves in red alder plantations as related to spacing and irrigation.*

first two or three years, but declined steadily thereafter, especially in the two denser spacings.

## Stand Development

**Crown closure and branch mortality.** The ability of red alder to produce large amounts of aboveground biomass at a young age can lead to very quick crown closure of alder plantations. Plantations established at fairly close spacings (1000 to 1200 trees per hectare) on good alder sites can be expected to close crown within three to five years. Once crown closure occurs, leaf area index (LAI) reaches a stable level of 2.5 to 3.0 on most sites (DeBell, unpub. data; Giordano, unpub. data). In irrigated plots at Yelm, LAI peaked at much higher levels (e.g., 6 to 9), but then declined abruptly and may stabilize at about 3.0. Dense alder crowns result in low light intensities in the middle to lower portions of the canopy. As a result, leaf area and branch biomass in the lower and middle canopy begin to decrease substantially.

**Competition and differentiation.** As competition among trees for light and other resources intensifies, diameter and height growth begin to decrease, and differences among trees in leaf area and current growth become larger. These differences soon lead to differences in tree sizes and thus to stand differentiation. Dense plantings on high quality sites differentiate more rapidly than wide plantings on poor sites; and microsite and genetic variation within a stand tend to accelerate differentiation.

Managed plantations are expected to differentiate more slowly than natural stands because (1) tree spacing is more uniform; (2) stock is of uniform size and age and has been selected for rapid growth; and (3) many cultural treatments reduce microsite variation. This effect of cultural treatment on microsite variation and subsequently on stand differentiation is not widely appreciated in Pacific Northwest forestry, in part because much present knowledge and opinion is based on experience with application of cultural practices to established stands, many of which had already begun to differentiate. Fertilization at establishment, for example, resulted in greater stand uniformity in 6-year-old eucalyptus plantations in Hawaii (Whitesell et al. 1988). Moreover, relative variation in tree size at Yelm was reduced in plots irrigated at the high level as compared with those irrigated at the low level; coefficients of variation for tree diameter at age 5 in the 0.5-m spacings were 37 and 54 percent for the high and low irrigation treatments, respectively. This higher uniformity among trees in the high irrigation treatments occurred despite the fact that by most measures the plots were farther along in stand development than those receiving a low level of irrigation; that is, trees averaged 75 percent larger in dbh and suppression-related mortality was substantially greater (65 vs. 45 percent). The Yelm study also demonstrated the effect of spacing on stand differentiation; the coefficient of variation in diameter at age 5 averaged only 27 percent in the 2.0-m spacings as compared with 45 percent in the 0.5-m spacings.

Some foresters have suggested that natural red alder stands on some sites have a tendency to stagnate and may never attain merchantability (e.g., Oliver and Larson 1990). Such suggestions seem to be based on observations made mostly in older stands on low quality sites, perhaps where high soil water tables or other adverse site conditions create a uniformly poor growing environment. Any such natural tendency toward stagnation reinforced with the above-mentioned effects of plantation management could lead to undesirable reductions in stand growth rates. These potential stagnation problems can be minimized by thinning and harvesting at times and according to prescriptions set forth when plantations are designed and established, and by avoiding sites with poor growing conditions.

**Mortality and stockability.** Differentiation among trees in stands of shade-intolerant species leads eventually to death of the smaller trees. Periodic assessments in the Yelm plantings have shown that most mortality occurs during winter and soon after trees have leafed out fully in spring, presumably because energy consumed in respiration and early growth have exhausted stored energy reserves. Most managed red alder plantations, however, will be thinned or harvested before competition-related mortality (self-thinning) occurs to any significant degree.

Nevertheless, research in plantations and natural stands of red alder in advanced stages of self-thinning is of considerable interest. It has pro-

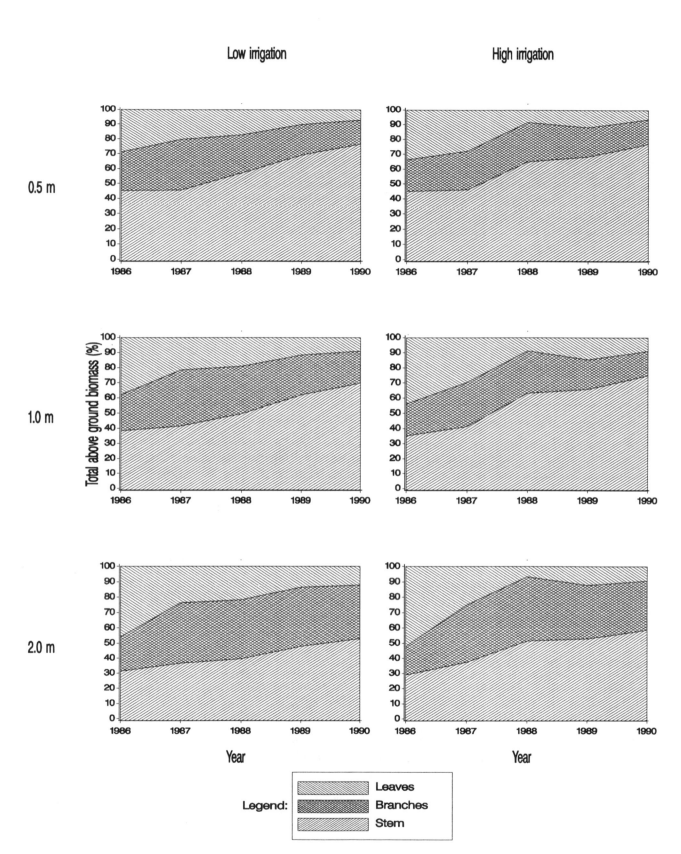

Low irrigation

High irrigation

0.5 m

1.0 m

2.0 m

Total above ground biomass (%)

Year

Year

Legend:
Leaves
Branches
Stem

Figure 7. Relative proportions of stems, branches, and leaves in red alder plantations as related to spacing and irrigation.

vided the understanding and data needed for developing guidelines for stand density management (Puettmann et al. 1993). With such information, plantation spacings can be selected so that, on average, trees will attain a specified target size before competition-related mortality occurs. Spacings of red alder plantations will be wider than those of conifers for any given target size, despite the fact that natural alder thickets commonly establish with greater stem densities than most young conifer stands. For example, recommended stand density to produce a target tree of 20-cm diameter is only about 700 stems per hectare for red alder as compared to 1000 stems per hectare for Douglas-fir.

Just as there are differences between species, differences have been observed among red alder stands and sites in the number of stems per hectare that can be grown to a given size without significant competition-related mortality; some alder stands have higher stockabilities than other alder stands (cf. DeBell et al. 1989). Why such differences occur is a matter of scientific interest and could have considerable practical significance. If factors affecting stockability were understood, some tree and site characteristics might be manipulated genetically or culturally to increase the productivity of red alder plantations.

## Implications for Management

The growth patterns of red alder offer numerous potential problems and opportunities for forest resource managers. Compared to natural stands, management of alder in plantations can improve the merchantable growth and yield, tree form and wood quality, and the economic value thereof. The probabilities of success will be increased if sites, spacings, and rotation lengths are selected to capture the benefits and minimize the problems associated with inherent growth traits of the species.

## Acknowledgments

Research conducted in the Apiary and Yelm studies was supported by the U.S. Department of Energy Biofuels Feedstock Development Program (formerly Short Rotation Wood Crops Program) via Agreement No. DE-AI05-810R20914.

## Literature Cited

Ager, A. A. 1987. Genetic variation in red alder (*Alnus rubra* Bong.) in relation to climate and geography in the Pacific Northwest. Ph.D. thesis, Univ. of Washington, Seattle.

Berntsen, C. M. 1958. A look at red alder—pure and in mixture with conifers. Proc. Soc. Am. For. 1958:157-158.

——. 1961a. Growth and development of red alder compared with conifers in 30-year-old stands. USDA For. Serv., Res. Pap. PNW-38.

——. 1961b. Pruning and epicormic branching in red alder. J. For. 59(9):675-676.

——. 1962. A 20-year growth record for three stands of red alder. USDA For. Serv., Res. Note PNW-219.

Bormann, B. T. 1985. Early wide spacing in red alder (*Alnus rubra* Bong.): effects on stem form and stem growth. USDA For. Serv., Res. Note. PNW-423.

Bormann, B. T., and J. C. Gordon. 1984. Stand density effects in young red alder plantations: productivity, photosynthate partitioning, and nitrogen fixation. Ecology 65:394-402.

DeBell, D. S., and T. C. Turpin. 1989. Control of red alder by cutting. USDA For. Serv., Res. Pap. PNW-414.

DeBell, D. S., W. R. Harms, and C. D. Whitesell. 1989. Stockability: a major factor in productivity differences between *Pinus taeda* plantations in Hawaii and the southeastern United States. For. Sci. 35:708-719.

Giordano, P. A. 1990. Growth and carbon allocation of red alder seedlings grown over a density gradient. M.S. thesis, Oregon State Univ., Corvallis.

Harrington, C. A. 1984. Factors influencing initial sprouting of red alder. Can. J. For. Res. 14(3):357-361.

——. 1986. A method of site quality evaluation for red alder. USDA For. Serv., Gen. Tech. Rep. PNW-192.

Harrington, C. A., and R. O. Curtis. 1986. Height growth and site index curves for red alder. USDA For. Serv., Res. Pap. PNW-358.

Helgerson, O. T., K. Cromack, S. Stafford, R. E. Miller, and R. Slagle. 1988. Equations for estimating aboveground components of young Douglas-fir and red alder in a coastal Oregon plantation. 1988. Can. J. For. Res. 18:1082-1085.

Hibbs, D. E., W. H. Emmingham, and M. C. Bondi. 1989. Thinning red alder: effects of method and spacing. For. Sci. 35(1):16-29.

Hook, D. D., D. S. DeBell, A. Ager, and D. Johnson. 1990. Dry weight partitioning among 36 open-pollinated red alder families. Biomass 21:11-25.

Lester, D. T., and D. S. DeBell. 1989. Geographic variation in red alder. USDA For. Serv., Res. Pap. PNW-409.

Nelder, J. A. 1962. New kinds of systematic designs for spacing experiments. Biometrics 18:287-307.

Oliver, C. D., and B. C. Larson. 1990. Forest Stand Dynamics. McGraw-Hill, New York.

Puettmann, K. J., D. S. DeBell, and D. E. Hibbs. 1993. Density management guide for red alder. Forest Research Laboratory, Oregon State Univ., Corvallis, Spec. Publ.

Reukema, D. L. 1965. Seasonal progress of radial growth of Douglas-fir, western redcedar, and red alder. USDA For. Serv., Res. Pap. PNW-26.

Snell, J. A. K., and S. N. Little. 1983. Predicting crown weight and bole volume of five western hardwoods. USDA For. Serv., Gen. Tech. Rep. PNW-151.

Turner, J., D. W. Cole, and S. P. Gessel. 1976. Mineral nutrient accumulation and cycling in a stand of red alder (*Alnus rubra*). J. Ecol. 64:965-974.

Warrack, G. C. 1964. Thinning effects in red alder. British Columbia For. Serv., Res. Div., Victoria.

Whitesell, C. D., D. S. DeBell, and T. H. Schubert. 1988. Six-year growth of *Eucalyptus saligna* plantings as affected by nitrogen and phosphorus fertilizer. USDA For. Serv., Res. Pap. PSW-188.

Wilson, B. F. 1984. The growing tree. The Univ. of Massachusetts Press, Amherst.

Worthington, N. P., F. A. Johnson, G. R. Staebler, and W. J. Floyd. 1960. Normal yield tables for red alder. USDA For. Serv., Res. Pap. PNW-36.

Zavitkovski, J., and R. D. Stevens. 1972. Primary productivity of red alder ecosystems. Ecology 53(2):235-242.

Zimmerman, M. H., and C. L. Brown. 1971. Trees: structure and function. Springer-Verlag. New York.

# 9

# Red Alder: Interactions with Wildlife

WILLIAM C. MCCOMB

Despite the abundance of red alder trees and stands in the Pacific Northwest, surprisingly few studies have focused on the relationships between red alder and wildlife species or communities. Bruce et al. (1985) summarized life history information for 412 species of vertebrates known to occur in western Oregon and Washington and they predicted that 136 species would use red alder forests for reproduction and that 178 species would use red alder forests for feeding. These numbers are underestimates of actual use of red alder stands by mammals and amphibians (McComb, Chambers, and Newton 1993). The numbers are low compared to the number of species that use mixed conifer-hardwood forests of Southwest Oregon (203 species for reproduction, 227 species for feeding), but they are higher than predicted numbers of species using lodgepole pine (*Pinus contorta*), subalpine conifer, and evergreen hardwood plant communities. Conifer-hardwood forests of the region can contain red alder as 30 to 70 percent of the crown cover, and these forests are predicted to support more species (194 for reproduction, 214 for feeding) than either temperate coniferous (182 for reproduction, 203 for feeding) or red alder forests.

This chapter describes the use of red alder trees and stands by some of these species, the role of red alder in upslope and riparian systems, and the potential contribution of red alder to influence animal diversity in conifer-hardwood stands and Pacific Northwest landscapes.

## Plant-Animal Interactions

Plants provide three dominant roles for heterotrophs in forest ecosystems. First, plants are the basis for energy input, although it is their digestible energy (not total energy) that is most important to animals. Digestibility is influenced by concentrations of complex molecules, such as lignins. Second, plants are an important source of nutrients, but astringent compounds such as phenols can bind with proteins to make these nutrients unavailable. Third, plants can provide cover for energy conservation, nesting, and escape from predators. Although little is known about the interactions between red alder and animals in any of these processes, I provide a few examples of each.

### Herbivory

The most obvious form of energy transfer from plants to animals is through consumption of the twigs and leaves (browsing). Undoubtedly some consumption of red alder catkins, fruits, and seeds occurs, but I could only find one reference to consumption of red alder seeds by pine siskins (*Spinus pinus*) (Martin et al. 1961). Red alder leaves and twigs are consumed by Columbian black-tailed deer, Roosevelt elk, and beaver (*Castor canadensis*) (Cowan 1945; Brown 1961; Taber and Hanley 1979; Leslie et al. 1984; Bruner 1989).

Red alder seems to be consumed by deer and elk most heavily in the fall (Cowan 1945; Brown 1961; Leslie et al. 1984), and consumption of abscising leaves is most common (Leslie et al. 1984). Red alder leaves and twigs are higher in crude protein in the fall and summer (12 to 15 percent) than in winter (7.5 percent), so deer and elk seem to consume this species when protein may be most abundant (Leslie et al. 1984).

Unpublished data of Edward Starkey (U.S. National Park Service Cooperative Research Unit) indicates that alder foliage is highly astringent, potentially reducing the availability of proteins to herbivores. Red alder is higher in crude protein in the fall (14 percent) than sympatric tree species such as western hemlock (*Tsuga heterophylla*) (6.2 percent) and has higher *in vitro* digestibility (33 percent) than western hemlock (21 percent) (Leslie et al. 1984). Crude protein and percent digestibility of red alder is comparable to that of trailing blackberry (*Rubus ursinus*) (16 percent crude protein, 31 percent *in vitro* digestibility) (Leslie et al. 1984), a heavily used browse plant (Friesen 1991). Because trailing blackberry also is highly astringent, its protein may be unavailable to herbivores in winter (Friesen 1991). Levels of phosphorus were lower in red alder than in nearly all other foods of deer and elk in the Hoh River Valley of coastal Washington (Leslie et al. 1984).

Beaver cut red alder stems 3 to 9 cm in diameter (43 percent were cut) disproportionate to their availability (27 percent of total stems) in two Or-

egon Coast Range riparian systems (Bruner 1989). These stems were cut not only for food but also for dam construction. The alder were cut at a time when resprouting potential for alder was highest (July through September), so alder remained a dominant plant at sites occupied by beaver (Bruner 1989; Suzuki 1992).

Based on damage to leaves, Stiles (1980) reported that red alder leaves seemed to support twice the invertebrate biomass as trailing blackberry, salmonberry (*Rubus spectabilis*), or vine maple (*Acer circinatum*), a pattern that seemed consistent regardless of the position of the leaves in the crown. Perhaps these invertebrates have evolved to cope with the astringent foliage of red alder. Foliage-dwelling invertebrates are the dominant food of many species of breeding birds in forested systems (e.g., Lack 1954).

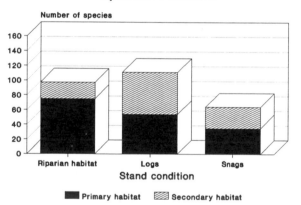

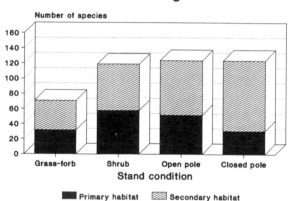

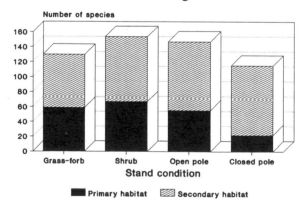

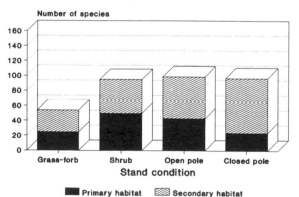

Figure 1. Predicted number of species that use specific habitat features and the four stand conditions within the red alder plant community of western Oregon and Washington (Brown 1985).

## Wildlife Cover

Little work has been done describing the importance of red alder trees or stands as cover for wildlife. Bruce et al. (1985) suggested that red alder lacked value as winter thermal cover for big game, but did not discuss the role that red alder might play as summer thermal cover. Dead conifer trees form snags or logs that provide cover for more than 100 species of vertebrates in Northwest forests (Brown 1985). Because of rapid decay, however, red alder functions as a potential cavity site or log for a short time relative to Douglas-fir (*Pseudotsuga menziesii*) snags or logs (Nietro et al. 1985). Consequently, hardwoods have been largely ignored as potential cover for cavity-dwelling species, despite a large number of species that use both snags and logs in alder stands (Fig. 1).

Gumtow-Farrior (1991) reported that bigleaf maple (*Acer macrophyllum*) and Oregon white oak (*Quercus garrayana*) are more cavity-prone (from heartrot and excavation of dead limbs by cavity-nesting birds) than Douglas-fir. The degree to which red alder may contribute to cavity abundance in Northwest forests has not been investigated.

Table 1. Species of birds (detections/700-m transect/6 visits), small mammals (captures/1000 trap nights), and amphibians (captures/1000 trap nights) that were more abundant (P < 0.10) in alder-dominated riparian stands (n = 6) than in adjacent conifer-dominated upslopes (n = 6), central Oregon Coast Range, 1988 (McGarigal and McComb 1992, McComb, Anthony, and McGarigal 1993).

| Species | Streamside | Upslope |
|---|---|---|
| Pacific water shrew (*Sorex bendirii*) | 9.4 | 0.6 |
| White-footed vole (*Phencomys albipes*) | 1.2 | 0.1 |
| Long-tailed vole (*Microtus longicaudus*) | 0.6 | 0.0 |
| Pacific jumping mouse (*Zapus trinotatus*) | 11.4 | 1.3 |
| Dunn's salamander (*Plethodon dunni*) | 1.7 | 0.4 |
| Winter wren (*Troglodytes troglodytes*) | 17.8 | 12.1 |
| Swainson's thrush (*Catharus ustulatus*) | 9.6 | 7.3 |

## Alder Stands as Wildlife Habitat

### Riparian Stands

Nearly 100 species of vertebrates were predicted to use red alder stands in association with riparian areas in western Oregon and Washington (Fig. 1). Because red alder is often associated with riparian areas in the Pacific Northwest, it is difficult to separate the effects of streamside conditions from red alder dominance on patterns of species occurrence in the region. McGarigal and McComb (1992) found that diversity of breeding bird species and number of small mammal species were lower in mixed-age alder-dominated riparian areas of the central Oregon Coast Range than in adjacent 120-year-old Douglas-fir—dominated uplands. Only two species of breeding birds were more abundant along streamsides than in upslope areas in that study (Table 1).

McComb, Anthony, and McGarigal (1993) found higher species diversity of small mammals along alder-dominated riparian areas than in upslope, unmanaged conifer forests in the central Oregon Coast Range, but species richness of small mammals and amphibians did not differ between riparian and upslope. Four species of mammals and one species of amphibian were more abundant along streams than upslope sites in this study (Table 1). Gomez (1992) also reported higher capture rates for Townsend's voles (*Microtus townsendii*), tailed frogs (*Ascaphus truei*), and rough-skinned newts (*Taricha granulosa*) in alder-dominated riparian areas than in upslope conifer stands. Alder averaged 17 to 23 percent of the canopy cover in upslope conifer stands in the above studies (vs. 37 to 40 percent along the streams), so it is difficult to know if these wildlife species were responding to streamsides, to the amount of alder cover in the stands, or to other related factors.

Jenkins and Starkey (1984) reported that Roosevelt elk in the Hoh River Valley of coastal Washington selected old red alder stands on the valley floor disproportionate to availability in the summer and late winter (especially March). Elk used these stands for feeding during the early morning and evening, and they moved to conifer-dominated sites to rest during midday. Jenkins and Starkey (1984) hypothesized that hardwood stands were selected because they provided more biomass

*Table 2. Detections/4 ha of selected bird species in four age classes (n = 2 in each age class) of alder stands, western Washington, 1971 (Stiles 1980).*

| Species | Stand age (years) | | | |
|---|---|---|---|---|
| | 4 | 10 | 35 | 60 |
| Downy woodpecker (*Picoides pubescens*) | 0 | 0 | 0.8 | 1.2 |
| Pacific slope flycatcher (*Empidonax difficilis*) | 0 | 0 | 3.0 | 3.0 |
| Brown creeper (*Certhia familiaris*) | 0 | 0 | 1.0 | 1.8 |
| Warbling vireo (*Vireo gilvus*) | 0 | 0 | 2.0 | 5.0 |
| Black-throated gray warbler (*Dendroica nigrescens*) | 0 | 0 | 0.3 | 0.5 |
| Western tanager (*Piranga ludoviciana*) | 0 | 0 | 1.3 | 1.0 |
| Dark-eyed junco (*Junco hyemalis*) | 0 | 0 | 4.0 | 3.3 |
| Rufous-sided towhee (*Pipilo erythrophthalmus*) | 5.5 | 2.0 | 0 | 0 |
| Song sparrow (*Melospiza melodia*) | 17.5 | 5.8 | 4.5 | 7.0 |
| Swainson's thrush (*Catharus ustulatus*) | 6.5 | 11.0 | 7.0 | 8.5 |
| Other species | 28.7 | 22.2 | 17.1 | 23.4 |
| Total birds | 58.2 | 41.0 | 41.0 | 54.7 |
| Number of species | 12 | 13 | 16 | 17 |
| Foliage height diversity | 0.361 | 0.722 | 1.072 | 1.069 |

of herbaceous vegetation than do conifer-dominated stands and because the nitrogen fixed by alder may improve forage quality on these sites (see Leslie 1983).

## Upslope Alder Stands

The number of wildlife species that use red alder stands for resting or reproduction was predicted to be higher in shrub, open sapling-pole, and closed sapling-pole-sawtimber stands than in grass-forb dominated stands (Fig. 1; Brown 1985). The number of species predicted to use red alder stands for feeding was higher in shrub-dominated stands than in other stand conditions (Fig. 1; Brown 1985). Stiles (1980) compared bird community composition among four age classes of red alder stands: 4, 10, 35, and 60 years of age, roughly corresponding to grass-forb, shrub, open sapling-pole, and closed sapling-pole-sawtimber stand conditions described by Brown (1985). This study found that both diversity of bird species and foliage height increased with stand age in red alder stands (Table 2). Seven species were found exclusively in 35 and 60-year-old stands (Table 2). Only rufous-sided towhees were found in 4 and 10-year-old stands but absent from older stands. Morrison (1981) found that Wilson's warblers (*Wilsonia pusilla*) and MacGillivray's warblers (*Oporornis tolmiei*) were associated with deciduous trees (including red alder) and shrubs in Coast Range clearcuts.

Gomez (1992) compared capture rates of small mammals, amphibians, and reptiles among alder stands and four seral stages of conifer stands in the central Oregon Coast Range. Two mammal species and two amphibian species had higher capture rates in the alder stands than in conifer stands (Table 3). Gomez (1992) also found that deciduous tree cover, of which red alder was the dominant deciduous species, was associated positively with the number of captures of vagrant shrews (*Sorex vagrans*) and differentiated capture sites from non-capture sites for white-footed voles, Pacific jumping mice, Pacific water shrews, Pacific shrews (*Sorex pacificus*), and shrew-moles. The white-footed vole, which is considered a candidate species for listing under the Endangered Species Act (ONHP 1991), was found to be uncommon in the Coast Range but most abundant in riparian areas dominated by hardwoods (Gomez 1992).

## Alder in Mixed-species Stands

Old-growth habitat relationships studies in western Oregon and Washington (Ruggiero, Aubry et al. 1991) provide information on the potential role of hardwoods in structuring animal communities in unmanaged conifer-dominated stands. Huff and Raley (1991) used multiple regression to identify the habitat features that best predicted the number of bird species and bird abundance in the region. The basal area of deciduous trees (including red alder) was an important predictor of the number of bird species in the region and in the Oregon Coast Range, total bird abundance in the Oregon Coast Range, and the abundance of residents, migrants, and canopy-foraging birds in the Oregon Coast Range. The specific role that alder plays in these relationships is unknown and these relationships do not represent

Table 3. Species of small mammals and amphibians (captures/1000 trap nights) with higher capture rates in red alder stands than in conifer stand conditions, central Oregon Coast Range (Gomez 1992).

| Species | Conifer shrub | Conifer pole | Conifer sawtimber | Conifer old-growth | Red alder |
|---|---|---|---|---|---|
| Shrew-mole (*Neurotrichus gibbsii*) | 8.2 | 8.6 | 14.9 | 18.5 | 19.9 |
| Trowbridge's shrew (*Sorex trowbridgii*) | 116.1 | 240.3 | 161.8 | 153.5 | 261.3 |
| Western redback salamander (*Plethodon vehiculum*) | 6.0 | 3.3 | 6.9 | 5.4 | 15.4 |
| Rough-skin newt (*Taricha granulosa*) | 0.6 | 3.9 | 5.6 | 9.9 | 35.7 |

cause-and-effect relationships, but Gilbert and Allwine (1991) found a positive correlation between red alder cover and the number of detections of several flycatcher species.

Deciduous tree cover also was found to be correlated positively with captures of Cascades frogs (*Rana cascadae*), Dunn's salamanders, and rough-skinned newts in unmanaged conifer-dominated stands in the southern Washington Cascades and the Oregon Coast Range (Aubry and Hall 1991; Corn and Bury 1991). The contribution of red alder foliage for insects (Stiles 1980) might provide added food biomass for insectivores if red alder is present in conifer stands, but this is a hypothesis to be tested.

## Landscape Considerations

No research has been conducted on the role that alder stands may play in structuring landscape and regional patterns of biodiversity. Because red alder is a relatively short-lived (compared to Douglas-fir), seral species, I assume that most of the wildlife species associated with large patches of alder would be well-adapted to colonization of these stands following disturbance. There is no information, however, describing the use of alder stands along a chronosequence spanning more than 60 years. Old stands of alder may have been important components of some landscapes prior to human disturbance (E. Starkey, pers. com.). Eight of the species listed as associated either with alder stands compared to conifer stands, or with deciduous tree cover in conifer-dominated stands were associated with old-growth Douglas-fir forests in the Oregon Coast Range, Oregon Cascades, or southern Washington Cascades (Ruggiero, Jones, and Aubry 1991): Pacific slope flycatcher, winter wren, Swainson's thrush, Pacific water shrew, shrew-mole, Dunn's salamander, rough-skin newt, and tailed frog.

Stiles's (1980) findings indicated that bird communities seemed to differ most between stands up to 10 years old and those more than 30 years old. Because of the initial rapid growth rate of red alder following disturbance, red alder patches may develop into acceptable habitat for these species faster than conifer stands in a landscape that has been subjected to stand replacement disturbance. Further, alder-dominated riparian areas may act as corridors connecting upland alder stands or old-growth conifer stands for these eight species. Alder-dominated riparian areas may not act as adequate corridors for other species in the region, however (McComb, Anthony, and McGarigal 1993; McGarigal and McComb 1992). These hypotheses should be tested.

## Research Needs

At least seven major gradients could influence habitat quality for indigenous vertebrates in Pacific Northwest forests, including red alder or mixed conifer-hardwood stands:

1. Tree and shrub species composition (especially hardwood and/or conifer presence)

2. Stand density and crown closure

3. Tree size class distribution (tree dbh and height)

4. Dead wood presence or abundance (snags and logs)

5. Edge contrast

6. Moisture (the transriparian gradient from stream to upslope)

7. Stand or patch size (especially relative to home range or territory size)

Gradients 1 to 4 represent "within-patch" gradients that also might influence within-stand vertical complexity and microclimate, gradients 5 and 6 are "between patch" gradients, and gradient 7 is an "among patch" gradient. I am aware of no work that addresses the role of red alder trees and stands in structuring wildlife communities along gradients 2, 4, 6, or 7. Only a few studies have addressed some aspects of the other three gradients (e.g., Stiles 1980; Morrison 1981; Jenkins and Starkey 1984; Gomez 1992).

There seem to be three general steps to the development of reliable knowledge in wildlife science (Murphy and Noon 1991; Nudds and Morrison 1991; Sinclair 1991): (1) observation (gaining knowledge about life-history attributes of organisms based on descriptive studies), (2) pattern recognition (correlational studies that identify patterns and offer reasonable bases for hypotheses), and (3) establishment of cause and effect (use of deductive approaches to test hypotheses, usually in manipulative experiments). Because there is inadequate information to suggest management strategies for red alder trees and stands that would benefit wildlife, there are research needs at all three levels of scientific investigation. For example, observational studies of foraging by wildlife and invertebrates on alder flowers, fruits, seeds, leaves, twigs, and roots are needed to determine the role that red alder plays in Pacific Northwest forests relative to energy transfer to heterotrophs. The influence of phenolic compounds on red alder digestibility also should be described. Cavity production by red alder compared to other hardwoods and to conifers and cavity use by vertebrates should be documented.

A number of patterns have yet to be elucidated. Except for bird communities along a sere of red alder stands (Stiles 1980), comparisons among red alder and conifer stands (Gomez 1992), and characterization of transriparian effects (Gomez 1992; McGarigal and McComb 1992; McComb, Anthony, and McGarigal 1993), patterns of wildlife abundance and fitness need to be investigated along all the gradients listed above.

Finally, no cause-and-effect studies have been conducted yet, although correlational studies now offer hypotheses for testing. For instance, based on Huff and Raley's (1991) regional summary and on Stiles's (1980) results, I would hypothesize that red alder trees support higher invertebrate biomass than do other tree species and that removal of red alders from conifer stands will reduce the number of insectivorous wildlife species and/or the number of individuals of certain insectivorous species. Testing such hypotheses would have to be replicated over areas large enough to encompass the home ranges of many individuals of each species of interest. Hence, many of the experiments that need to be conducted will have to be coordinated with silviculturists to achieve the appropriate experimental design.

## Acknowledgments

I thank D. Hibbs, K. McGarigal, and E. Starkey for their helpful comments on an early draft of this manuscript. Much of the information included in this chapter is a result of work conducted within the Coastal Oregon Productivity Enhancement (COPE) program, a cooperative research and technology transfer effort among Oregon State University, USDA Forest Service, USDI Bureau of Land Management, other state and federal agencies, industry, county governments and resource associations. This is Paper No. 2968 of the Oregon State University Forest Research Laboratory.

## Literature Cited

Aubry, K. B., and P. A. Hall. 1991. Terrestrial amphibian communities in the southern Washington Cascade Range. *In* Wildlife and vegetation of unmanaged Douglas-fir forests. *Edited by* L. F. Ruggiero, K. B. Aubry, A. B. Carey, and M. H. Huff. USDA For. Serv., Gen. Tech. Rep. PNW-285. 327-338.

Brown, E. R. 1961. The black-tailed deer in western Washington. Washington State Dept. of Game, Biol. Bull. 13. Olympia, WA.

———, ed. 1985. Management of wildlife and fish habitats in forests of western Oregon and Washington. USDA For. Serv., Publ. No. R6-F&WL-192-1985.

Bruce, C., D. Edwards, K. Mellen, A. McMillan, T. Owens, and H. Sturgis. 1985. Wildlife relationships to plant communities and stand conditions. *In* Management of wildlife and fish habitats in forests of western Oregon and Washington. *Edited by* E. R. Brown. USDA For. Serv., Publ. No. R6-F&WL-192-1985. 33-55.

Bruner, K. L. 1989. Effects of beaver on streams, streamside habitat, and coho salmon fry populations in two coastal Oregon streams. M.S. thesis, Oregon State Univ., Corvallis.

Corn, P. S., and R. B. Bury. 1991. Terrestrial amphibian communities in the Oregon Coast Range. *In* Wildlife and vegetation of unmanaged Douglas-fir forests. *Edited by* L. F. Ruggiero, K. B. Aubry, A. B. Carey, and M. H. Huff. USDA For. Serv., Gen. Tech. Rep. PNW-285. 305-317.

Cowan, I. McT. 1945. The ecological relationships of the food of the Columbian black-tailed deer, *Odocoileus hemoinus columbianus* (Richardson), in the coast forest region of southern Vancouver Island, British Columbia. Ecol. Monogr. 15:109-139.

Friesen, C. A. 1991. The effect of broadcast burning on the quality of winter forage for elk, western Oregon. M.S. thesis, Oregon State Univ., Corvallis.

Gilbert, F. F., and R. Allwine. 1991. Spring bird communities in the Oregon Coast Range. *In* Wildlife and vegetation of unmanaged Douglas-fir forests. *Edited by* L. F. Ruggiero, K. B. Aubry, A. B. Carey, and M. H. Huff. USDA For. Serv., Gen. Tech. Rep. PNW-285. 145-155.

Gomez, D. 1992. Small-mammal and herpetofauna abundance in riparian and upslope areas of five forest conditions. M.S. thesis, Oregon State Univ., Corvallis.

Gumtow-Farrior, D. L. 1991. Cavity resources in Oregon white oak and Douglas-fir stands in the mid-Willamette valley, Oregon. M.S. thesis, Oregon State Univ., Corvallis.

Huff, M. H., and C. M. Raley. 1991. Regional patterns of diurnal breeding bird communities in Oregon and Washington. *In* Wildlife and vegetation of unmanaged Douglas-fir forests. *Edited by* L. F. Ruggiero, K. B. Aubry, A. B. Carey, and M. H. Huff. USDA For. Serv., Gen. Tech. Rep. PNW-285. 177-205.

Jenkins, K. J., and E. E. Starkey. 1984. Habitat use by Roosevelt elk in unmanaged forests in the Hoh Valley, Washington. J. Wildl. Manage. 48:642-646.

Lack, D. 1954. The natural regulation of animal numbers. Clarendon Press, Oxford.

Leslie, D. M., Jr. 1983. Nutritional ecology of cervids in old-growth forests in Olympic National Park, Washington. Ph.D. thesis, Oregon State Univ., Corvallis.

Leslie, D. M., Jr., E. E. Starkey, and M. Vavra. 1984. Elk and deer diets in old-growth forests in western Washington. J. Wildl. Manage. 48(3):762-775.

Martin, A. C., H. S. Zim, and A. L. Nelson. 1961. American wildlife and plants: a guide to wildlife food habits. Dover, New York.

McComb, W. C., R. G. Anthony, and K. McGarigal. 1993. Small mammal and amphibian abundance in streamside and upslope habitats of mature Douglas-fir stands, western Oregon. Northwest Sci. 67:7-15.

McComb, W. C., C. L. Chambers, and M. Newton. 1993. Small mammal and amphibian communities and habitat associations in red alder stands, central Oregon Coast range. Northwest Sci. 67:181-188.

McGarigal, K., and W. C. McComb. 1992. Streamside versus upslope breeding bird communities in the central Oregon Coast Range. J. Wildl. Manage. 56:10-23.

Morrison, M. L. 1981. The structure of western warbler assemblages: analysis of foraging behavior and habitat selection in Oregon. Auk 98:578-588.

Murphy, D. D., and B. R. Noon. 1991. Coping with uncertainty in wildlife biology. J. Wildl. Manage. 55:773-782.

Nietro, W. A., V. W. Binkley, S. P. Cline, R. W. Mannan, B. G. Marcot, D. Taylor, and F. F. Wagner. 1985. Snags (wildlife trees). *In* Management of wildlife and fish habitats in forests of western Oregon and Washington. *Edited by* E. R. Brown. USDA For. Serv., Publ. No. R6-F&WL-192-1985. 129-169.

Nudds, T. D., and M. L. Morrison. 1991. Ten years after "reliable knowledge": are we gaining? J. Wildl. Manage. 55:757-760.

ONHP. 1991. Rare, threatened and endangered plants and animals of Oregon. Oregon Natural Heritage Program, Portland, OR.

Ruggiero, L. F., K. B. Aubry, A. B. Carey, and M. H. Huff. 1991. Wildlife and vegetation of unmanaged Douglas-fir forests. USDA For. Serv., Gen. Tech. Rep. PNW-285.

Ruggiero, L. F., L. L. C. Jones, and K. B. Aubry. 1991. Plant and animal habitat associations in Douglas-fir forests of the Pacific Northwest: an overview. *In* Wildlife and vegetation of unmanaged Douglas-fir forests. *Edited by* L. F. Ruggiero, K. B. Aubry, A. B. Carey, and M. H. Huff. USDA For. Serv., Gen. Tech. Rep. PNW-285. 447-462.

Sinclair, A. R. E. 1991. Science and the practice of wildlife management. J. Wildl. Manage. 55:767-773.

Stiles, E. W. 1980. Bird community structure in alder forests in Washington. Condor 82:20-30.

Suzuki, N. 1992. Habitat classification and characteristics of small mammal and amphibian communities in beaver-pond habitats of the Oregon Coast range. M.S. thesis, Oregon State Univ., Corvallis.

Taber, R. D., and T. A. Hanley. 1979. The black-tailed deer and forest succession in the Pacific northwest. *In* Sitka black-tailed deer: proceedings of a symposium in Juneau, Alaska. *Edited by* O. C. Wallmo and J. W. Schoen. USDA For. Serv. Series No. R10-48. 33-52.

# Section 2
# Management of Alder

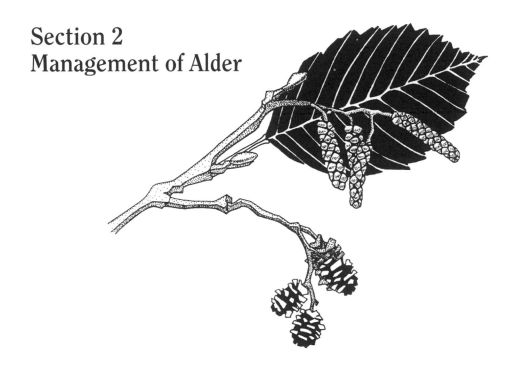

# 10

# Evaluation of Site Quality for Red Alder
CONSTANCE A. HARRINGTON & PAUL J. COURTIN

Information on the relationships between site quality and tree or stand productivity is needed by foresters to determine the sites where alder can be successfully planted, to predict growth and yield, and to evaluate the potential efficacy of silvicultural practices designed to increase productivity. Although using total yield as a measure of productivity (and thus as a measure of site quality) is intuitively appealing, determining total yield is often complicated and difficult (Carmean 1975). Site index (or height at a specified age) is the most commonly used index of site quality in the United States and Canada. This chapter (1) explains how to determine site index if suitable trees are present on the site, (2) presents information on relationships between alder site index and other measures of productivity, (3) discusses methods to estimate site index when it cannot be measured directly, (4) summarizes information on sensitivity of red alder to edaphic and physiographic factors, and (5) speculates on the potential influences of management activities on site quality.

## Determination of Site Index
For red alder, site index has been defined as the mean height of free-to-grow, dominant and codominant trees of seed origin at a specified index age. Based on this definition, site index can only be *directly* measured when stands are at the index age. Site index tables or curves allow site index to be estimated when heights and ages are available from suitable trees. The first curves for red alder (Bishop et al. 1958; Worthington et al. 1960) were anamorphic (i.e., the basic shape of the curves was the same for all sites), used an index age of 50 years, and were based on data from only western Washington; in addition, these publications did not include information on stands below 10 years of age. More recent site estimation curves are available (Harrington and Curtis 1986) that use an index age of 20 years (total age). The new curves have several advantages over previous curves: (1) they are based on a much larger sample of trees covering a greater geographic range, (2) the reference age of 20 years is more appropriate to future short rotation management of the species, and (3) the curves are polymorphic and provide a better expression of observed height growth trends. If a value of site index at 50 years is needed, the curves of Harrington and Curtis (1986) were modified by Mitchell and Polsson (1988) to provide estimates of height, given age and site index at age 50. These height growth curves (or tables) can be used to estimate site quality (given age and height) although the estimates will be somewhat less accurate.

Trees to be measured for site index determination (site trees) should have been in a free-to-grow (non-overtopped) position throughout their life; thus, trees showing evidence of past suppression in their ring pattern should not be measured. Site trees should be in dominant or codominant crown positions. Wide spacing at time of planting has been shown to depress height growth of young red alder (DeBell et al. 1989), but it is not known if early growth depressions will persist to index age or if significant effects will be detectable within the range of densities in managed stands. Site index estimates based on open-grown trees, however, may underestimate potential site index and should be used cautiously.

As a general rule, site trees should not show evidence of past major damage; however, as with any general rule, there are exceptions. Some topographic positions are very susceptible to top damage and may have several episodes of top breakage during a rotation; such positions would include areas where high winds occur (e.g., ridge tops, near mountain passes, by the ocean), or areas that receive wet snow or ice storms. On these sites it may not be possible to select trees that do not exhibit some past top damage. Site index values can be estimated for such sites by using damaged site trees; however, we would expect these estimates to be poorer predictors of (i.e., more poorly correlated with) other measures of site quality than will occur for sites without top damage.

Site index estimates should be based on a minimum of four trees per plot; more are desirable if plot size permits (Curtis 1983). Trees to be measured should represent the area to be sampled; that is, they should be well distributed throughout the plot (Curtis 1983). The greatest uniformity in site index estimates among individual trees occurs when the area is stratified based on site conditions and each stratum (e.g., upper, mid and lower slope positions) is sampled separately. The need for such stratification depends on what uses will be made of the estimate. For example, if future management activities (such as determining rotation length or thinning regimes) will vary based on differences in site index of 2 to 3 m, then areas on the ground should be stratified so that differences of that magnitude are detectable. On the other hand, if management activities will not change unless site index values differ by 10 m, then little stratification is needed. The need for stratification should also consider practical matter, such as what is the smallest area that it is feasible to manage?

Site index estimates will be most accurate when stand age is within ± 10 years of the index age. Estimates are very unreliable for stands below age 5 and become increasing less reliable at older ages. In addition, almost no information beyond age 65 was used in developing the site index curves of Worthington et al. (1960) or Harrington and Curtis (1986).

Unless total age is known (e.g., from stand records), age will commonly be determined by boring at breast height (1.3 m), then adding a correction factor. Harrington and Curtis (1986) recommended a two- or three-year correction factor for the poorest sites ($SI_{20} < 15$ m), a one-year correction factor for average sites ($SI_{20}$ 15 to 25 m), and no correction factor for the best sites ($SI_{20} > 25$ m). These recommendations should be modified if local experience indicates other values are more appropriate.

Red alder is a diffuse-porous species; thus, to obtain accurate counts of annual rings, more care is needed in preparing and examining the increment cores or cross sections from red alder than those from temperate-zone conifers or ring-porous hardwoods. Newton and Cole (Chapter 7) refer to unpublished data on alder trees approximately 106 years old that had missing or incomplete rings; however, there is no evidence that missing rings reduce the accuracy of site index estimates for stands within the age range (tree age < 70 years) used in constructing the available red alder site index curves. DeBell et al. (1978) found few abnormal rings (false, partial, or missing) in cross sections from red alder trees ages 29 to 88, especially when sections were examined from breast height rather than root collar locations. They concluded that counts made carefully on increment cores extracted at breast height should provide a reliable assessment of age. They recommended making a smooth razor cut along one or two sides of the freshly extracted core, then counting the rings when the ring boundary darkens (the darkening is caused by oxidation of phenols and occurs in less than 30 minutes).

## Relationships Among Measures of Site Productivity

It has long been recognized by mensurationists that many of the stand characteristics used to describe tree stands are interrelated. Thus, mature stands on highly productive sites have fairly high values for site index, basal area, and volume production (if the stands have not been thinned or if the basal area or volume removed in the thinning has been accounted for in some manner) and mature stands on sites of low productivity have correspondingly low values for site index, basal area, and volume production. Many land managers would like to know the current or predicted future wood volume of red alder stands without having to directly measure it. Unfortunately, volume equations for red alder are limited. Chambers (1983) developed predictions for unmanaged red alder stands; these predictions are based on percentage of normal basal area, site index and age, or site index, observed basal area, and average diameter. Earlier tables for normal (i.e., well-stocked) stands predicted basal area or cubic volume as a function of site index and age (Worthington et al. 1960).

We examined the relationships among several stand descriptors by using our unpublished data from naturally regenerated, unmanaged red alder stands in western Oregon, Washington, and British Columbia. Table 1 presents pair-wise correlation coefficients for those stands. For these stands—many of which were less than 35 years—site index was poorly correlated with total basal area or total stand stem volume but was significantly correlated ($R = 0.61$, $p = 0.0001$) with mean annual increment of stem volume (mai). A linear regression equation with site index, age, and basal area accounted for 90 percent of the variation in mai.

## Other Methods for Estimation of Site Index

When no suitable red alder site trees are present, there are several methods that can be used to estimate site index for red alder. The most commonly used methods involve known relationships between alder site index and (a) soil series, soil mapping unit, or soil classification information, (b) site index or other site quality measures for another species, (c) mathematical or biologically based site prediction models, or (d) British Columbian biogeoclimatic zones and related nutrient and moisture regimes. Each of these methods has advantages and disadvantages. The method chosen for a specific situation is probably more dependent on what information is available (and thus, which methods could be used) and which method has been adopted for a given region than on the theoretical advantages and disadvantages of a specific method. Users should be aware, however, of the relative merits of each method.

### Estimation Based on Soil Series or Taxonomic Classification

A frustration commonly expressed by foresters is that all categories of soil taxonomy—from soil order down to soil series—seem to encompass a wide

Table 1. Pair-wise correlations between stand statistics for unmanaged red alder stands in western Oregon, Washington, and British Columbia (n = 64). Age = mean age of site trees, $SI_{20}$ = site index at base-age 20, BA = basal area, Vol = total stem volume (based on Snell and Little, 1983), MAI = Vol ÷ Age. (Correlation coefficients ≥ 0.24 significant at p = 0.05).

|       | Age  | $SI_{20}$ | BA    | Vol   | MAI   |
|-------|------|-----------|-------|-------|-------|
| Age   | 1.00 | -0.45     | 0.53  | 0.60  | -0.40 |
| SI    |      | 1.00      | -0.13 | 0.09  | 0.61  |
| BA    |      |           | 1.00  | 0.91  | 0.45  |
| Vol   |      |           |       | 1.00  | 0.47  |
| MAI   |      |           |       |       | 1.00  |

range of site quality. Thus, the question has been raised, "What useful information can be gained by knowing the soil series or other level of classification information for a site?" The answer is, "A little or a lot, depending on what is known and how the information is used." No *one* piece of soil-site information (e.g., soil order, soil series, slope position, texture of the A horizon) will ever provide foresters with accurate estimates of site quality for a *wide* range of sites. Combining several pieces, though, can often be effective. We may not have enough information to "complete the puzzle," but often when we fill in some of the pieces, we can see enough of the picture to make a good guess at the result.

Information on site index for red alder on specific soils is limited. Only a fraction of the potential soil series have been examined for alder site relations and the information we have is restricted in terms of geographic spread and soil conditions. In addition, none of these soil series have been subdivided into phases based on characteristics specifically selected to reduce the variability in red alder site index within phases. Many existing or potential alder sites in western Washington and Oregon, however, have been soil surveyed and the resulting soil mapping units or soil series have been classified using the U.S. system (Soil Survey Staff 1975). We present the available soil-site index information within this taxonomic framework in the hope that it will enable managers to make inferences about soils with similar classifications. The relationship between Canadian taxonomy and red alder site productivity is also discussed below.

The following discussion assumes that the soil series or taxonomic classification available for a specific site is accurate. Users should be aware, however, that soil information presented on a map will not be correct for all sites. This occurs because it is not practical to determine or to map all existing soil units. If it is important to know that the mapped soil information is correct, users should verify that the profile description associated with the mapped soil unit is consistent with the soil profile on the site.

We summarized information from soil-site plots for which both red alder site index and soil classification were known. This 200-plot data base was collected by us (Pacific Northwest Research Station and British Columbia Ministry of Forests) or by other organizations (the former Crown Zellerbach Corporation, USDA Soil Conservation Service, and the Washington State Department of Natural Resources). The following discussion emphasizes the *maximum* site index values for each classification unit. This approach has been useful in organizing other types of soil-site data (e.g., Harrington 1991). Emphasizing the maximum values helps screen out the negative effects of factors which are not part of the classification system (and may not occur equally across all classification units).

The first level of classification in U.S. soil taxonomy is soil order; the orders are based on pedogenesis (i.e., soil-forming processes as indicated by presence or absence of major diagnostic horizons). Seven of the eleven soil orders were rep-

*Figure 1. Relationship between red alder site index (base-age 20 years) and soil order for stands in western Oregon, Washington, and British Columbia. Data collected by the authors or from the USDA Soil Conservation Service Soil-Woodland data base. Soil-site index information collected in British Columbia was classified to soil order based on the U.S. system.*

resented in the data (Fig. 1, plots from British Columbia were classified to soil order in the U.S. system for this figure). Most of the soils were Entisols or Inceptisols, but four other orders were also represented with 4 to 17 plots. Site quality information is summarized below by order. When available, information is also presented for other levels of classification (suborders, great groups, and subgroups). Site index information for red alder by soil series (or other level of classification) is presented in Table 2 (U.S. information only).

**Entisols** have little or no evidence of development of pedogenic horizons. All red alder stands sampled on Entisols were on stream terraces or in flood plains. The highest $SI_{20}$ values (> 27 m) were on Entisols but the order encompasses soils with a wide range in productivity (Fig. 1). The best great groups (maximum values for $SI_{20} \geq 24$ m) were Fluvaquents (Typic or Mollic) or Udifluvents with friable or very friable surface horizons containing little or no rock or gravel, and loamy or clayey textures. Xeropsamments (drier, sandy soils) were less productive.

**Inceptisols** are immature soils with few diagnostic features. Over half of the soils with alder soil-site information are in this order; five suborders are represented—Andepts, Aquepts, Ochrepts, Tropepts, and Umbrepts. Inceptisols are found in all slope positions and cover a wide range in site quality. The highest $SI_{20}$ values for Inceptisols represented good sites (23 to 24 m) but were about 3.5 meters less than the highest values for Entisols.

Some of the highest $SI_{20}$ values for Inceptisols are in the Andept suborder (Typic Dystrandepts). The best Andept sites ($SI_{20} > 23$ m) are at low elevations, are on flat land or gentle slopes, and have little gravel or rock in surface horizons.

The Ochrept suborder includes a wide range in $SI_{20}$. The best Ochrepts ($SI_{20} > 23$ m) are at low elevations, have good rooting volume, are not coarse-textured, and do not contain much gravel or rock. For Durochrepts and Fragiochrepts, the depth to the duripan or fragipan is an important factor influencing site index. For Dystrochrepts and Xerochrepts, the Andic or Aquic subgroups are the most favorable for alder growth. Soils classified as loamy (e.g., fine-silty or coarse-loamy) or medial (amorphous but feels loamy) are more favorable than those classified as skeletal (having 35 percent or more rock fragments).

Umbrepts also include a substantial range in alder-site quality. The poorest sites have low rooting volume (Fragiumbrepts with shallow fragipans or soils with shallow phases) or frigid temperature regimes. The best sites ($SI_{20} > 22$ m) are Typic or Andic Haplumbrepts with loamy or medial textures and mesic temperature regimes. Within a series, elevation and slope are generally important factors influencing alder site quality.

Aquepts are found on stream terraces, in depressional areas, and on gentle slopes. Shallow Placaquepts (plac = thin pan) and Haplaquepts are less productive than Humaquepts. The poorest Humaquepts have firm horizons at fairly shallow depths, sandy A horizons, and are on southerly aspects.

Tropepts are only represented in the data by Typic and Andic Humitropepts (two series). Values for $SI_{20}$ were low to moderate, but the data were too limited to form any conclusions.

**Spodosols** have a horizon in which organic matter and aluminum have accumulated. Many people assume that all Spodosols have a thick leached horizon and low productivity, but those assumptions are not necessarily true. Typic or Aqualfic Haplorthods can have $SI_{20}$ values $\geq 21$ m. Other Haplorthods can also have moderate $SI_{20}$ values if soil textures are not sandy or skeletal. Information on alder productivity on other great groups is limited; however, great groups with fragipans, placic horizons (thin pans), frigid temperature regimes, or thick leached layers are less productive than great groups without those features.

**Alfisols** have a light-colored (ochric) surface, a clayey horizon, and high base status. Information on alder site quality is limited, but measured $SI_{20}$ values ranged from low to above average. Maximum $SI_{20}$ was 23 m (Ultic Haploxeralf).

**Ultisols** are geologically old, weathered soils with low base status. Only one plot with site index information was classified as an Ultisol. The site index for that plot was low ($SI_{20} = 17.2$ m); although other Ultisols may have higher productivity than this plot, most soils in this order would have below-average productivity.

*Table 2. Site index information for red alder in western Washington and Oregon by soil classification. Within each soil order soils are listed by the maximum site index value for the series. Soils with three or more site index plots are indicated by an asterisk.*

| Soil order and other classification information | Max. SI$_{20}$(m) |
| --- | --- |
| **Entisols** | |
| Udifluvents (without additional classification information) | 27.6* |
| Rennie: fine, montmorillonitic, nonacid, mesic Mollic Fluvaquent | 27.4* |
| Nuby: fine-silty, mixed, acid, mesic Typic Fluvaquent | 26.7* |
| Coquille: fine-silty, mixed, acid, mesic Typic Fluvaquent | 25.4 |
| Hoh: coarse-loamy, mixed, acid, mesic Typic Udifluvent | 24.8 |
| Puget: fine-silty, mixed, nonacid, mesic Aeric Fluvaquent | 22.6 |
| Juno: sandy-skeletal, mixed, mesic Typic Udifluvent | 22.0 |
| Ocosta: fine, mixed, acid, mesic Typic Fluvaquent | 21.3 |
| Pilchuck: mixed, mesic Dystric Xeropsamment | 19.5 |
| Humptulips: coarse-loamy over sandy/sandy skeletal, mixed, nonacid, mesic Typic Udifluvent | 19.0 |
| **Inceptisols** | |
| Cathcart: medial, mesic Andic Xerochrept | 23.9 |
| Cloquallum: fine-silty, mixed, mesic Aquic Dystrochrept | 23.7 |
| Mues: medial over loamy-skeletal, mixed, isomesic Typic Dystrandept | 23.6 |
| Montesa: coarse-loamy, mixed, mesic Aquic Dystrochrept | 23.0* |
| Bear Prairie: medial, mesic Typic Dystrandept | 22.7* |
| Blethen: medial-skeletal, mesic Andic Xerochrept | 22.7 |
| Alderwood: loamy-skeletal, mixed, mesic Dystric Entic Durochrept | 22.7* |
| Aabab: fine-silty, mixed, mesic Aquic Dystrochrept | 22.7* |
| Bunker: medial, mesic Andic Haplumbrept | 22.5 |
| Rinearson: fine-silty, mixed, mesic Typic Haplumbrept | 22.3* |
| Ohop: loamy-skeletal, mixed, frigid Aquic Dystrochrept | 22.3 |
| Queets: medial, mesic Andic Dystrochrept | 22.2* |
| Yelm: medial, mesic Aquic Dystric Xerochrept | 22.2 |
| Templeton: medial, mesic Andic Haplumbrept | 21.7* |
| Stimson: fine-silty, mixed, acid, mesic Typic Humaquept | 21.5* |
| Everett: sandy-skeletal, mixed, mesic Dystric Xerochrept | 21.4 |
| Skamo: medial, mesic Andic Haplumbrept | 21.3* |
| Zenker: medial, mesic Andic Haplumbrept | 21.2 |
| Nestucca: fine-silty, mixed, acid, mesic Fluvaquentic Humaquept | 20.9 |
| Tealwhit: fine, mixed, acid, mesic Aeric Haplaquept | 20.4 |
| Arta: medial, mesic Andic Haplumbrept | 20.3* |
| Cathlamet: medial, mesic Andic Haplumbrept | 19.9 |
| Skipanon: fine-loamy, mixed, isomesic Andic Humitropept | 19.7 |
| Cinebar: medial, mesic Typic Dystrandept | 19.5 |
| Mt. Baker NF MU 22: coarse-silty, mixed, mesic Fragiumbrept | 19.1 |
| Siuslaw NF MU 22: fine-silty, mixed, mesic Typic Dystrochrept | 19.1 |
| Glohm: fine-silty, mixed, mesic Typic Fragiochrept | 19.1 |
| Ecola: fine-silty, mixed, mesic Andic Haplumbrept | 19.0 |
| Wishkah: fine, mixed, mesic Aquic Dystrochrept | 18.3* |
| Mt. Baker NF MU 34: fine, mixed, slowly permeable Andic Haplumbrept | 18.0 |
| Newaukum: medial, mesic Typic Dystrandept | 17.6 |
| Nemah: fine, mixed, acid, mesic Humic Haplaquept | 17.4 |
| Grindbrook: fine-silty, mixed, isomesic Typic Humitropept | 16.7 |
| Molalla: fine-loamy, mixed, mesic Typic Haplumbrept | 16.4 |
| Boistfort: medial, mesic Andic Haplumbrept | 16.2 |
| Halbert: loamy, mixed, acid, mesic, shallow Histic Placaquept | 16.1 |
| Wilhoit: medial, frigid Andic Haplumbrept | 13.0 |

*Table 2 continued.*

| Soil order and other classification information | Max. $SI_{20}$(m) |
|---|---|
| **Spodosols** | |
| Sehome: coarse-loamy, mixed, mesic Typic Haplorthod | 22.5 |
| Skipopa: fine, mixed, mesic Aqualfic Haplorthod | 21.3* |
| Yaquina: sandy, mixed, mesic Aquic Haplorthod | 19.9 |
| Solduc: medial-skeletal, mesic Humic Haplorthod | 18.5 |
| Mt. Baker NF MU 81: coarse-loamy, mixed Typic Ferrod | 18.1 |
| **Alfisols** | |
| Melbourne: fine, mixed, mesic Ultic Haploxeralf | 23.0* |
| Bow: fine, mixed, mesic Aeric Glossaqualf | 18.5 |
| **Ultisols** | |
| Mayger: clayey, mixed, mesic Aquic Haplohumult | 17.2 |
| **Histosols** | |
| Mukilteo: dysic, mesic Typic Medihemist | 19.2* |
| Fluvaquentic Borohemist | 16.5 |
| **Andisols** | |
| Hemcross: medial, mixed, mesic Alic Hapludand | 21.4 |
| Necanicum: medial-skeletal, isomesic Alic Fulvudand | 20.6 |
| Klistan: medial-skeletal, mesic Alic Hapludand | 19.2 |
| Klootchie: medial, isomesic Alic Fulvudand | 17.8 |

**Histosols** are organic soils; most are saturated or nearly saturated with water most of the year. Productivity for alder is below average on Hemists (maximum $SI_{20}$ = 19.2 m); no information is available for other suborders.

**Andisols** are derived from volcanic materials. Several series formerly classified as Andepts are now Udands. Alic Fulvudands and Alic Hapludands have average to above-average values for $SI_{20}$ (maximum $SI_{20}$ = 21.4 m). Medial textures, gentle slopes, and low elevations are more favorable than medial-skeletal textures, steep slopes, and higher elevations.

**Canadian taxonomy.** In British Columbia there has never been an attempt to correlate categories of the Canadian System of Soil Taxonomy (Agriculture Canada 1987) and forest productivity. If there had been such an attempt, it would likely have resulted in poor relationships. Like the U.S. system, the Canadian system stresses pedogenesis in its classification rather than a functional relationship between soil taxonomy and productivity. Of the nine soil orders in the Canadian system, five occur within the range of red alder: Brunisols, Gleysols, Organics, Podzols, and Regosols. Certain generalizations can be made concerning alder growth and certain chemical and morphological features that may relate to soil orders. For example, Organic soils are too wet to give anything but poor growth and often alder is replaced by willow species, Pacific crab apple, or coniferous trees. The same comment generally applies for Gleysols, although some Humic Gleysols with thick A horizons can yield excellent growth. Poor or medium growth of alder is expected on Podzols and Regosols where the degree of eluviation and organic matter accumulation in surface horizons will result in great variability. Brunisols probably have the greatest chance of providing the best growth for alder, especially the Melanic and Sombric subgroups with thick A horizons.

When more information is available, it may be helpful to use the Canadian System of Soil Taxonomy with parent material classification to develop red alder productivity relationships. Even when taxonomic units are subdivided by parent material classes, however, the resulting subunits may still include more variability in particle size and drainage than is desirable for productivity prediction. In addition, such analysis would require fairly expert knowledge or large-scale soils mapping (e.g., 1:100,000).

### Estimation Based on Other Species

Prediction of site index for a tree species has been done based on site index of another tree species, height growth of other tree species, or the presence, abundance, or size of understory plant species (Carmean 1975). The accuracy of the prediction is dependent on the ecological similarity of the two species (e.g., red alder and the species on which the prediction is based). In addition, it is important to know the type of data used in developing the prediction equations. General relationships exist between the site index values for geographically associated species; however, species differences in dependence on "growing-season" conditions, tolerance to poor drainage or flooding, need for soil-supplied nitrogen, and geographic and elevational ranges make any predictions of this type unreliable unless they are limited to sites where differences between species in ecological tolerances are not strongly expressed.

Some past predictions have used a two-step estimation process in which the site index values for the species on which the prediction is based have not been directly measured; instead they were based on a second prediction relationship (e.g., the relationship between soil series and Douglas-fir site index). This adds another source of variability to the predicted value.

Red alder site index values from 23 stands did not differ significantly (p = 0.10) among Douglas-fir site classes (Harrington and Curtis 1986); that is, in that data set, Douglas-fir site class was not a good predictor of red alder site index. Sites classified as excellent for Douglas-fir consistently had above-average $SI_{20}$ values for red alder; however, some of the highest red alder $SI_{20}$ values were on flood plains classified as unsuitable for Douglas-fir. Compared to Douglas-fir, red alder is more tolerant of occasional flooding, poor soil drainage, and low soil nitrogen (Minore 1979). Douglas-fir, on the other hand, has greater cold hardiness, especially to unseasonable frosts. In addition, because of its ability to photosynthesize during the winter (Waring and Franklin 1979) and its determinate pattern of height growth, Douglas-fir may be able to better utilize sites with deep, well-drained to somewhat excessively drained soils.

Site index predictions for one species based on a value for another species will be most accurate if the range in site conditions is low or kept within specified bounds (e.g., low elevation, well-drained soils without major nutrient deficiencies) and if it is known that both species may be present on those sites. For example, it is probable that within plant association groups in which both species occur fairly accurate prediction equations could be developed. The relationships developed for one plant association group, however, could not be transported and used for other groups without testing. Based on actual measurements of red alder and Douglas-fir trees in low elevation, mixed-species stands or pure-species, adjacent stands in western Washington, an $R^2$ value of 0.47 (n = 29) was reported (Fergerson et al. 1978). Although these sites were geographically associated, it is not known what range in ecological conditions was represented.

### Estimation Based on Mathematical or Biologically Based Soil-Site Models

Prediction of site quality from soil and site characteristics has been a goal of many forestry studies over the years (see Carmean 1975). If the area of interest can be defined so that all site characteristics are similar except for a few (e.g., soils in the same area with similar parent material but differing rooting depths), then fairly accurate prediction equations can be developed. This approach to site index prediction requires that equations be developed for each group of interest. For example, McKee (1977) grouped loblolly pine soil-site plots first by soil series, then predicted site index within soil-series group based on soil physical and chemical properties. When regression equations developed from plots that represent a wide range in site conditions are tested with data from new plots, however, the equations have not held up well (i.e., the correlation between measured and predicted site index drops substantially) (see McQuilkin 1976). For red alder, this was demonstrated by Harrington (1984); a regression equation predicting site index had an $R^2$ value of 0.95 for the 25 plots used in equation development but was worthless in predicting site index for a new set of 15 plots ($R^2 = 0.05$).

Another approach to predicting site quality is to use a biologically based model. The best-known example of this approach is for southern hardwood species (Baker and Broadfoot 1977). Each part of the model is based on relationships that are consistent with what is known about the silvics of the species rather than on a mathematical fit from a specific data set. Such models require information on many variables, but are more robust in their performance. The model for red alder (Harrington 1986), for example, required information on 14 soil and site properties. It tested well ($R^2 = 0.92$ and 0.80) with two data sets from western Washington and Oregon (Harrington 1986), but more poorly ($R^2 = 0.42$) with one from British Columbia (Courtin 1992). The model apparently needs to be modified to function outside the geographic area where it was developed.

## Estimation Based on British Columbian Biogeoclimatic Zones

In British Columbia, an approach to determining site quality and, indirectly, site index is the characterization of forest ecosystems using biogeoclimatic ecosystem classification. This system classifies the forest landscape for the purposes of site quality evaluation on the premise that the physical environment of an ecosystem can be simplified into three elements: regional climate, soil moisture regime, and soil nutrient regime (Pojar et al. 1987). Within this classification, the biogeoclimatic subzones delineate those segments of the landscape with similar regional climate and the same zonal climax vegetation development. Then, for the purposes of site classification, units called site series group ecosystems based on similar soil moisture and nutrient regimes with the same regional climate. Site associations are site series that can occur across different regional climates; they are analogous to Daubenmire's habitat type classification or to the plant associations used by the U.S. Forest Service in many western states. Field personnel can determine biogeoclimatic units by using maps and differentiating characteristics and can key out moisture and nutrient regimes from physiographic information, soil physical and morphological criteria, and vegetation (Green et al. 1984).

The link between site quality and site index evaluation for Douglas-fir has been made by developing regression equations using site associations (Green et al. 1989) and soil moisture and nutrient regimes as categorical variables (Green et al. 1989; Klinka and Carter 1990). Ultimately, the goal is to incorporate site index into the biogeoclimatic ecosystem classification system by having a range of site index values for each site association. Some red alder site index information has been collected in British Columbia along with biogeoclimatic ecosystem classifications (data on file, British Columbia Ministry of Forestry, Burnaby), but not enough data are yet available to develop prediction equations.

## Relationships Between Site Index and Selected Site Characteristics

Currently many sites exist for which no site quality information is readily available for red alder because there are no suitable trees to measure and there is no information on soil or ecological classification (or if some classification has been made, the relationship between that classification unit and alder site quality is not known or is not known precisely enough). In the following sections we discuss relationships between site index and selected site characteristics. Knowledge of these relationships may provide managers with an approximate indication of site quality or at least with an appreciation for some of the factors that influence it. One useful way to examine the effect of a specific soil or

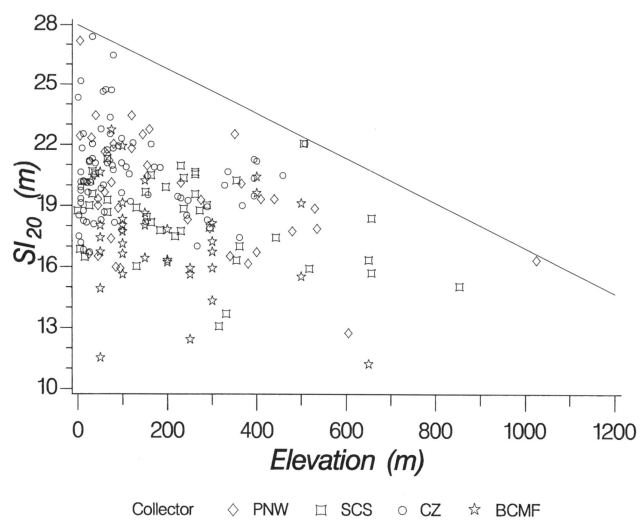

*Figure 2. Relationship between red alder site index (base-age 20 years) and stand elevation (mean above sea level) for stands in western Oregon, Washington, and British Columbia. Plotting symbols indicate organization that collected data.*

site factor (e.g., elevation) on a variable like site index that is influenced by many factors is to look at the *maximum* values of site index that are associated with each level of that factor. If sufficient data are available, this value will indicate the maximum that can be achieved when all other possible factors are at their optimum levels; this allows the effect of that factor alone to be examined. This technique, called boundary-line analysis, has been used in the analysis of plant-soil relations (see Evanylo and Sumner 1987; Harrington 1991).

*Climatic Factors*
Very little specific information is directly available on the relationships between red alder and climatic

factors; most information is based on observations of height growth or site index in natural stands or on occurrence of natural stands. The variables we would assume to have an influence on site index are temperature (mean, minimum, maximum, length of frost-free period), precipitation (amount and timing, including frequency of ice or heavy snow), wind, and radiation (light levels and photoperiod). Unfortunately, few weather stations are located in or near alder stands for which we have site index information; thus, we usually must rely on correlated variables or estimation. For example, elevation is not a climatic variable but it is correlated with many climatic variables (such as length of the frost-free growing season or minimum and maximum temperatures) and is easily obtainable. For some sites, specific information from adjacent weather stations is available and can be examined.

Elevation is strongly related to red alder site index and to species occurrence. For example, the highest values of red alder site index are at low elevations and the maximum site index value achievable declines as elevation increases (Fig. 2). Elevation alone will not accurately predict site index (R =-0.36, p < 0.01), but it appears to set the upper limit for site index values. Specific information is not available to test which climatic variables are most closely associated with the observed effects on site index; however, we assume that length of the growing season and air temperature are key factors. Elevation values must be considered in the context of latitude. For example, in northern Oregon or southern Washington, red alder trees are found as high as 1100 m, but in Alaska they generally occur close to sea level. Since alder only occurs at fairly low elevations in the most northerly portion of its range, we would assume that the effect of elevation becomes more negative as latitude increases.

Precipitation during the growing season is an important factor in influencing alder tree growth (and site index). It should be recognized, however, that precipitation does not directly affect growth but is one of the variables that influence soil moisture during the growing season, a factor that directly influences growth (see discussion below). If all other factors are equal, site index increases as growing season precipitation increases up to about 45 cm (1 April through 30 September). On well-drained sites without access to water tables, the species could undoubtedly benefit from additional growing-season precipitation (see the benefits from irrigation reported by DeBell and Giordano, Chapter 8); however, within the natural range of the species, higher precipitation is commonly associated with cooler temperatures, higher elevations, or more northerly latitudes, and does not result in higher values of site index.

Some climatic factors (e.g., high winds, the frequency of ice or heavy snow storms) influence site index for red alder by increasing the chances of top breakage rather than by directly influencing growth rates per se. These factors must be taken into ac-count when evaluating sites near mountain passes, close to the ocean (without intervening topography), at high elevations, or in areas of locally unusual climatic conditions.

## Soil Moisture

The availability of soil water during the growing season is a key factor influencing site quality for red alder. The importance of this factor can be seen in both the distribution of the species on the landscape and in its performance (Harrington 1990). Since red alder is deciduous, it is more dependent on favorable growing conditions during the summer months than its evergreen competitors (Waring and Franklin 1979). In most years, precipitation during the summer is insufficient to supply the water required for the species to achieve acceptable growth rates. Soil water available beyond that provided by current precipitation can come from storage in the soil profile, accessing soil moisture perched above temporary or permanent water tables, or from water transported by gravity from uphill positions.

Soil drainage classes used by the USDA Soil Conservation Service indicate how rapidly water applied to the soil surface will move downward through the profile. Plotting red alder site index against drainage class (Fig. 3) indicates that a wide range in site index occurs for all drainage classes. The sites with the very highest site index values were on soils classified as poorly drained. The apparent uniformity of maximum site index values for the other drainage classes, however, is somewhat misleading. Drainage classes do not take into account the availability of soil water from other sources. Thus, the maximum values for site index on well-drained or somewhat excessively drained soils are associated with plots on flood plains or stream terraces, where tree roots have access (at least for part of the growing season) to water tables, or with plots in lower slope positions, where they receive water from uphill locations.

Red alder can tolerate high water tables and flooding. Winter water tables at or close to the soil surface do not appear to reduce growth (Minore 1968; Minore and Smith 1971). The species is not hydrophytic, however, and its survival and growth will be reduced if high, unaerated water tables are

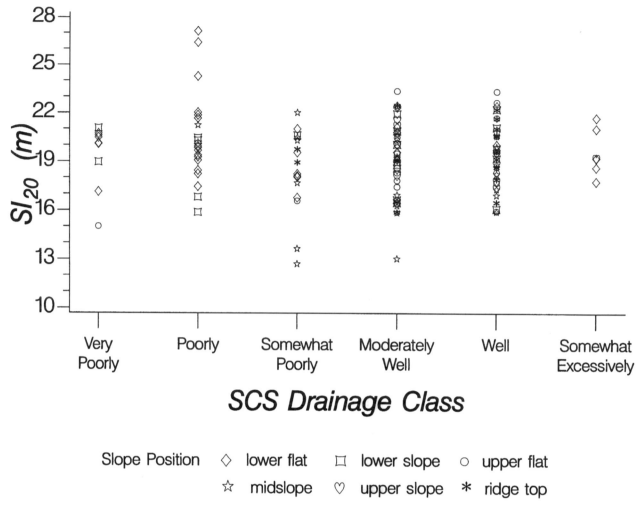

*Figure 3. Relationship between red alder site index (base-age 20 years) and soil drainage class for stands in western Oregon and Washington. Data collected by the senior author or from the USDA Soil Conservation Service (SCS) Soil-Woodland data base. Plotting symbols indicate slope position class for each stand.*

present during the growing season (for results from artificial flooding experiments, see Minore 1968, 1970; Harrington 1987). Aerated water tables (e.g., those found on sites along rapidly moving rivers and streams) are much more favorable for growth of alder than unaerated or stagnant water tables (e.g., those in bogs). One effect of high water tables (particularly unaerated ones) is to restrict the area where root growth can occur to the uppermost section of the soil profile; this concentrates roots in a low volume of soil and results in high competition for nutrients (and for water when a seasonal water table recedes). Thus the best site index values among the plots classified as very poorly drained (Fig. 3) are those that have some microrelief present. The microrelief increases the soil volume

that is suitable for growth of fine roots; this can be a critical factor influencing tree growth. For example, Tiarks and Shoulders (1982) concluded that on phosphorus-deficient soils, uptake of phosphorus and height growth of pines were controlled by the volume of soil available for root exploration. Very poorly drained soils can also be unproductive because some soil nutrients become less available or toxic substances increase under anaerobic conditions.

### Soil Physical Condition
Information on the relationship between red alder site quality and soil physical condition is limited. Minore et al. (1969) reported substantial penetration of alder seedling roots *into* as well as growth *along*

the sides of soil cores of sandy loam that had been artificially compacted to a medium density of 1.45 g/cm$^3$ (however, growth was less than at a lower density). At the highest soil density tested, seedling weight and root growth along the sides of the cores was the same as at the medium density, however, depth of root penetration into the cores was sharply reduced. This is consistent with personal observations that tree roots in high bulk density soils often grow along ped faces or at horizon boundaries. It appears that root growth can be somewhat plastic (or opportunistic) if soil structure is strongly developed. Strong soil structure provides channels for growth of roots and for movement of soil solutions and gases.

The presence of soil organic matter improves soil physical condition by promoting favorable soil structure and decreasing soil bulk density. Soil organic matter is also beneficial in increasing or retaining soil water and nutrients (see below). Red alder is a "soil builder." Because it increases soil organic matter more quickly than do many of its associated species, the negative effects of low soil organic matter may be smaller or shorter in duration than those for many other species. Young trees, however, grow more poorly in soils with very low organic matter contents.

### Nutrient Availability

Nutrient availability influences site index, but information on how specific levels of a nutrient or groups of nutrients impact site index is not available. Most information on nutrient availability is indirect in nature. For example, the low maximum site index values in some soil orders may indicate that these older weathered (and leached) soils have lower nutrient availability than the younger, less weathered soils present in other orders. Red alder is tolerant of fairly low pH conditions, but maximum site index values are not associated with the most acid soils. The reduction in site index as pH decreases probably is primarily the effect of reduced availability of the macronutrients (and some micronutrients) at low pH and not the effect of hydrogen concentration *per se*. Based on information from 129 sites, $SI_{20}$ values $\geq 21$ m are on sites with pH values between 4.5 and 5.6 (measured in water) in the 0- to 30-cm surface soil (see discussion below

and Fig. 5). This is similar to the optimum range of 4.6 to 5.5 previously reported (Harrington 1986).

One complication in the relationship between alder site quality and nutrient availability is that alder stands may cause changes in soil chemistry. Long-term presence of alder species on sites in the United States and Canada resulted in decreases in soil pH in all studies examined by Bormann et al. (Chapter 3); the decrease was greatest in basic, primary-succession soils and least in acidic, older soils. Van Miegroet and Cole (1988) attributed pH decreases to reduced base saturation resulting from nitrification. Binkley and Sollins (1990) concluded that qualitative changes in soil organic matter other than base saturation are also important factors influencing changes in soil pH. In contrast to the results in North America, Mikola (1966) reported that the presence of gray or black alder trees or the additions of alder leaf litter *increased* soil pH in acidic forest soils (pH < 5.0) or in a nursery soil (pH = 5.4). Thus, the long-term effects of alder on soil pH (and consequently on nutrient availability) may differ depending on initial soil conditions and on climatic or other site conditions. More work is needed to predict the site-specific, long-term effects of alder on soil chemistry—and on site quality.

The macronutrient that seems to be of particular importance to red alder nutrition is phosphorus (Radwan and DeBell, Chapter 15), which becomes less available at low pH. Not only are many red alder stands on soils low in pH, but they are on imperfectly drained sites where soil rooting volume and thus total available soil phosphorus is low. For example, Figure 4 demonstrates that at low levels of available soil phosphorus (defined as phosphorus concentration times rooting depth), maximum $SI_{20}$ increases linearly with increasing available soil phosphorus. Availability of phosphorus, however, can also be important on well-drained sites. In one trial on a sandy, well-drained soil, red alder trees responded to additions of phosphorus or to irrigation, but no additional increases in growth were realized when irrigation and phosphorus were combined (DeBell et al. 1990). Phosphorus is considered important in promoting root growth; under the right conditions high levels of phosphorus may promote enough additional root growth to increase the tree's capacity for water uptake.

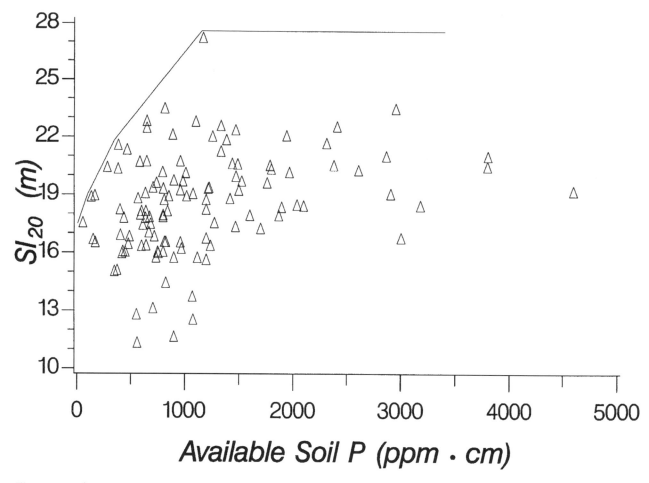

*Figure 4. Relationship between red alder site index (base-age 20 years) and available soil phosphorus [defined as (Bray #1-extractable P in 0- to 30-cm layer in ppm) x (rooting depth in centimeters)].*

Relationships between red alder site quality and micronutrients are even less well understood than those involving the macronutrients. Generally molybdenum and boron become less available at low pH, while iron, manganese, zinc, copper, and cobalt become more available. Availability of specific micronutrients is also affected by soil oxygen levels. Some poorly drained organic soils may have such low availability of micronutrients (due to both pH and anaerobic effects) that growth is significantly reduced. For example, gray alder planted on a peat bog in Sweden grew very poorly until boron was applied (Rytter et al. 1990).

Hughes et al. (1968) pointed out that red alder seedling growth responses to nutrient amendments must be interpreted carefully as fertilizer additions can alter soil pH and stimulate or depress elements other than those present in the fertilizer amend-

ment. Thus, they attributed some of the beneficial effects of calcium additions to its associated increase in soil pH and decrease in manganese uptake. Similarly, negative effects of nitrogen additions were partially attributed to the associated reduction in soil pH and the stimulation of manganese uptake. In our data, the most productive sites ($SI_{20} \geq 21$ m) have low to moderate values of soil iron or manganese. The data presented in Figure 5 indicate that the most productive sites ($SI_{20} > 21$ m) have available iron values < 30 ppm in the 0- to 30-cm soil layer. None of the most productive sites were in pH class 1 (4.00 to 4.49), and all the sites with iron values > 60 ppm were in pH class 1.

Red alder seedlings in a growth chamber grew dramatically better on forest floor material from a hemlock stand than on rotten wood (Minore 1972);

this difference was attributed to higher nutrient availability in the forest floor and to observed differences in rooting habit and nodule formation between the two media. Under natural conditions, the presence of very high soil organic matter contents (> 25 percent) is not favorable, as it indicates low organic matter turnover rates (and thus, low nutrient cycling rates) associated with cold temperatures (high elevation or northerly latitude) or low oxygen (very poorly drained soils) (Harrington 1986).

## Influences of Management Activities on Site Quality

Silvicultural decisions can influence site quality or apparent site quality (i.e., situations where measured site quality or site index has been altered but no changes have been made to the inherent productive capacity of the site) in several ways. Choice of genotype or planting stock and control of conditions at stand establishment influence early growth of red alder. Under many circumstances, these early gains will probably still be measurable at index age and thus will result in an apparent increase in site quality. For many species, separate site index and height growth curves have been developed for natural and planted stands as field observations indicated differences in growth rates.

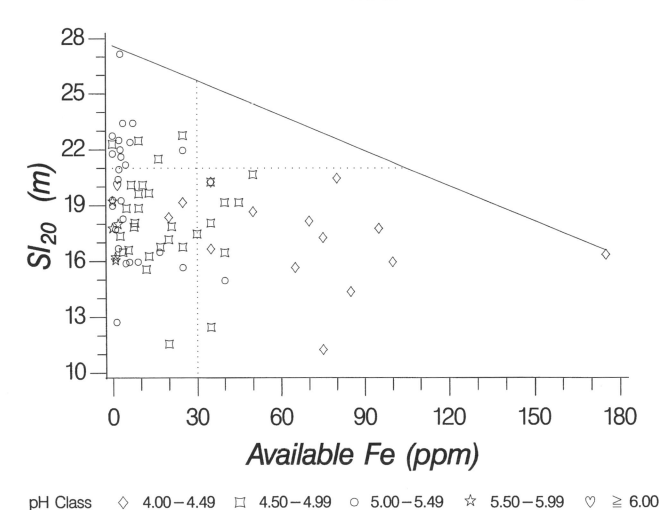

Figure 5. Relationship between red alder site index (base-age 20 years) and available iron concentration in the 0- to 30-cm soil layer. Plotting symbols indicate pH class (pH measured in water) for the same soil layer.

Spacing experiments have indicated that alder height growth is influenced by initial stand spacing with maximum height growth occurring at intermediate spacings (DeBell and Giordano, Chapter 8). It is not yet known how persistent those differences in early height growth will be or how differences in apparent site index will relate to other measures of productivity. In addition, planting specific genotypes with high growth rates or tolerances for specific site conditions may alter apparent site quality. Other management activities actually alter the inherent productive capacity of the site for short or long periods of time. Examples of such activities include drainage, irrigation, fertilization, burning, and treatments that increase or decrease soil compaction. Not enough is known to predict the specific conditions under which these activities influence site quality for red alder or the magnitude or duration of the effects. And, as discussed earlier, we do not know which sites may be degraded (decreased soil pH and increased cation leaching) or under what conditions changes in site quality may occur by growing repeated crops of pure alder (Bormann et al., Chapter 3).

## Future Research Needs
Our knowledge base from which to evaluate site quality for red alder is quite limited. Polymorphic site curves for the species (Harrington and Curtis 1986) exist, and this chapter includes recommendations on their use. A research priority should be to determine if planted or managed stands require new curves or adjustments to the existing ones. Alternative methods to evaluate site quality for red alder are in their infancy. Expanded work on biologically based soil-site or biogeoclimatic-site models could help pinpoint areas where our understanding of the underlying biological relationships is weak. This could be coordinated with more basic plant nutrition research to maximize the returns. For example, additional research is needed on the relationships among soil organic matter, pH, aeration, nutrient availability, and growth of alder. Improved understanding of these relationships could pinpoint which soils warrant testing to improve accuracy of site quality estimation. In addition, the general topic of how management activities affect short-term or long-term site quality

for red alder stands should also be a high priority for future research.

## Acknowledgments
Some of the soil-site data collected in Washington and Oregon was supported by funding from the U.S. Department of Energy, Woody Crops Program, under interagency agreement DE-AI05-810R20914. Data collection for the sites in British Columbia was supported by funding from the Canada-British Columbia Forest Resources Development Agreement, 1985-90 Backlog Reforestation Program, Research and Development Subprogram Project No. 2.54 (red alder component).

## Literature Cited
Agriculture Canada Expert Committee on Soil Survey. 1987. The Canadian system of soil classification. 2nd ed. Agric. Canada Pub. 1646.

Baker, J. B., and W. M. Broadfoot. 1977. A practical field method of site evaluation for eight important southern hardwoods. USDA For. Serv., Southern For. Exp. Sta., Gen. Tech. Rep. SO-14.

Binkley, D., and P. Sollins. 1990. Factors determining differences in soil pH in adjacent conifer and alder-conifer stands. Soil Sci. Soc. Am. J. 54:1427-1433.

Bishop, D. M., F. A. Johnson, and G. R. Staebler. 1958. Site curves for red alder. USDA For. Serv., Res. Note PNW-162.

Carmean, W. H. 1975. Forest site quality evaluation in the United States. Advances in Agronomy 27:209-269.

Chambers, C. J. 1983. Empirical yield tables for predominately alder stands in western Washington. 4th print. Washington State Dept. of Natural Resources, DNR Rep. 31. Olympia, WA.

Courtin, P. J. 1992. The relationships between the ecological site quality and the site index and stem form of red alder in southwestern British Columbia. M.S. thesis, Univ. of British Columbia, Vancouver.

Curtis, R. O. 1983. Procedures for establishing and maintaining permanent plots for silvicultural and yield research. USDA For. Serv., Gen. Tech. Rep. PNW-155.

DeBell, D. S., B. C. Wilson, and B. T. Bormann. 1978. The reliability of determining age of red alder by ring counts. USDA For. Serv., Res. Note PNW-318.

DeBell, D. S., M. A. Radwan, D. L. Reukema, J. C. Zasada, W. R. Harms, R. A. Cellarius, G. W. Clendenen, and M. R. McKevlin. 1989. Increasing the productivity of biomass plantations of alder and cottonwood in the Pacific Northwest. Annual technical report submitted to the U.S. Dept. of Energy, Woody Crops Program.

DeBell, D. S., M. A. Radwan, C. A. Harrington, G. W. Clendenen, J. C. Zasada, W. R. Harms, and M. R. McKevlin. 1990. Increasing the productivity of biomass plantations of alder and cottonwood in the Pacific Northwest. Annual technical report submitted to the U.S. Dept. of Energy, Woody Crops Program.

Evanylo, G. K., and M. E. Sumner. 1987. Utilization of the boundary line approach in the development of soil nutrient norms for soybean production. Comm. Soil Sci. Plant Anal. 18:1379-1401.

Fergerson, W., G. Hoyer, M. Newton, and D. R. M. Scott. 1978. A comparison of red alder, Douglas-fir, and western hemlock productivities as related to site: a panel discussion. In Utilization and management of alder. Compiled by D. G. Briggs, D. S. DeBell, and W. A. Atkinson. USDA For. Serv., Gen. Tech. Rep. PNW-70. 175-182.

Green, R. N., P. J. Courtin, K. Klinka, R. J. Slaco, and C. A. Ray. 1984. Site diagnosis, tree species selection and slashburning guidelines for the Vancouver Forest Region. Land Manage. Handb. 8., British Columbia Ministry of Forests, Victoria.

Green, R. N., P. L. Marshall, and K. Klinka, K. 1989. Estimating site index of Douglas-fir (Pseudotsuga menziesii {Mirb.} Franco) from ecological variables in southwestern British Columbia. For. Sci. 35:50-63.

Harrington, C. A. 1984. Site quality prediction for red alder. Agronomy Abstracts 76th annual meeting. 260-261.

———. 1986. A method of site quality evaluation for red alder. USDA For. Serv., Gen. Tech. Rep. PNW-192.

———. 1987. Responses of red alder and black cottonwood seedlings to flooding. Physiol. Plant. 69:35-48.

———. 1990. Alnus rubra Bong.—red alder. In Silvics of North America, vol. 2, Hardwoods. Technically coordinated by R. M. Burns and B. H. Honkala. USDA For. Serv., Agric. Handb. 654. Washington, D.C. 116-123.

———. 1991. PTSITE—a new method of site evaluation for loblolly pine: model development and user's guide. USDA For. Serv., Southern For. Exp. Sta., Gen. Tech. Rep. SO-81.

Harrington, C. A., and R. O. Curtis. 1986. Height growth and site index curves for red alder. USDA For. Serv., Res. Pap. PNW-358.

Hughes, D. R., S. P. Gessel, and R. B. Walker. 1968. Red alder deficiency symptoms and fertilizer trials. In Biology of alder. Edited by J. M. Trappe, J. F. Franklin, R. F. Tarrant, and G. M. Hansen. USDA For. Serv., PNW For. Range Exp. Sta., Portland, OR. 225-237.

Klinka, K. and R. E. Carter. 1990. Relationships between site index and synoptic environmental factors in immature Douglas-fir stands. For. Sci. 36:815-830.

McKee, W. H., Jr. 1977. Soil-site relationships for loblolly pine on selected soils. In Proceedings 6th Southern Forest Soils Workshop, 19-21 October 1976, Charleston, SC. USDA For. Serv., State and Private Forestry, Southeastern Area, Atlanta, GA. 115-120.

McQuilkin, R. A. 1976. The necessity of independent testing of soil-site equations. Soil Sci. Soc. Amer. J. 40:783-785.

Mikola, P. 1966. The value of alder in adding nitrogen in forest soils. Final Report for Project E8-FS-46, grant EG-Fi-131 under P.L. 480. Dept. of Silviculture, Univ. of Helsinki, Finland.

Minore, D. 1968. Effects of artificial flooding on seedling survival and growth of six northwestern tree species. USDA For. Serv., Res. Note PNW-92.

———. 1970. Seedling growth of eight northwestern tree species over three water tables. USDA For. Serv., Res. Note PNW-115.

———. 1972. Germination and early growth of coastal tree species on organic seed beds. USDA For. Serv., Res. Pap. PNW-135.

———. 1979. Comparative autecological characteristics of northwestern tree species—a literature review. USDA For. Serv., Gen. Tech. Rep. PNW-87.

Minore, D., and C. E. Smith. 1971. Occurrence and growth of four northwestern tree species over shallow water tables. USDA For. Serv., Res. Note PNW-160.

Minore, D., C. E. Smith, and R. F. Woollard. 1969. Effects of high soil density on seedling root growth of seven northwestern tree species. USDA For. Serv., Res. Note PNW-112.

Mitchell, K. J., and K. R. Polsson. 1988. Site index curves and tables for British Columbia: coastal species. Canadian Forestry Service and British Columbia Ministry of Forests, FRDA Rep. 37.

Pojar, J., K. Klinka, and D. V. Meidinger. 1987. Biogeoclimatic ecosystem classification in British Columbia. For. Ecol. Manage. 22:119-154.

Rytter, L., U. Granhall, and A. S. Arveby. 1990. Experiences of grey alder plantations on a peat bog. In Fast growing and nitrogen fixing trees. 8-12 October 1989, Marburg. Edited by D. Werner and P. Muller. Gustav Fischer Verlag, Stuttgart. 112-114.

Snell, J. A. K., and S. N. Little. 1983. Predicting crown weight and bole volume of five western hardwoods. USDA For. Serv., Gen. Tech. Rep. PNW-151.

Soil Survey Staff. 1975. Soil taxonomy. USDA Soil Conservation Serv., Agric. Handb. 436. Washington, D.C.

Tiarks, A. E., and E. Shoulders. 1982. Effects of shallow water tables on height growth and phosphorus uptake by loblolly and slash pines. USDA For. Serv., Southern For. Exp. Sta., Res. Note SO-285.

Van Miegroet, H., and D. W. Cole. 1988. Influence of nitrogen-fixing alder on acidification and cation leaching in a forest soil. *In* Forest site evaluation and long-term productivity. *Edited by* D. W. Cole and S. P. Gessel. Univ. of Washington Press, Seattle.

Waring, R. H., and J. F. Franklin. 1979. Evergreen coniferous forests of the Pacific Northwest. Science 204:1380-1386.

Worthington, N. P. 1965. Red alder. *In* Silvics of forest trees of the United States. *Compiled by* H. A. Fowells. USDA For. Serv., Agric. Handb. 271. Washington, D.C. 83-88.

Worthington, N. P., F. A. Johnson, G. R. Staebler, and W. J. Lloyd. 1960. Normal yield tables for red alder. USDA For. Serv., Res. Pap. PNW-36.

## 11

# Biology, Ecology, and Utilization of Red Alder Seed

ALAN AGER, YASUOMI TANAKA, & JIM MCGRATH

Since the last two alder symposia (Trappe et al. 1968; Briggs et al. 1978), information on the biology of red alder seed has advanced significantly. For example, two in-depth masters' theses on the basic biology of seed reproduction in red alder (Brown 1986; Lewis 1985) and recent research on the utilization of red alder seed (e.g., Berry and Torrey 1985; Tanaka et al. 1991) have provided new and valuable information on seed propagation of this species.

We have organized this chapter into two sections. The first discusses the biology and ecology of red alder seed. Basic features of the reproductive system in red alder, especially patterns of seed production in natural stands, are described. The second section reviews current knowledge of the collection, storage, and utilization of red alder seed. We conclude the chapter with a discussion of immediate research needs.

## Seed Biology and Ecology

### Morphology and Development of Reproductive Structures

The development and morphology of reproductive structures in red alder has been described in detail by Furlow (1979a) and Brown (1986). Red alder is monoecious, with the male and female aments borne on separate unisexual structures. The male catkins and female strobiles appear in groups of three to six. The flowers are arranged adaxially in twos for catkins and in threes for strobiles on shelflike bracts (Brown 1986). Four ovules are initiated for each flower, although only one ovule fully matures.

Red alder aments are preformed, differentiating in late June or early July of the preceding year. Thus, the reproductive structures are apparent throughout the winter. Meiosis in the staminate catkins occurs in late summer or early fall before flowering the following spring (Wetzel 1929, as cited in Brown 1986). When mature, pollen grains in the genus weigh on average $6.5 \times 10^{-9}$ g. Receptive stigmata are red, bilobed, and about 2 mm long.

Red alder flowers in late February or early March in the Pacific Northwest. At a given latitude, flowering is delayed with increasing elevation. This delay can amount to over a month between stands growing at sea level versus those growing at the elevational limit (ca. 1000 m) of the species. Within a stand, flowering is generally synchronous, although variation in peak receptivity and anthesis can vary among trees by several days (Brown 1986). Variation in reproductive phenology is less among trees in areas with short growing seasons, such as at higher elevations (Brown 1986). Flowering within a tree generally progresses from base to apex.

Among the male and female flowers on a given tree, flowering is generally asynchronous, with pollen dispersal preceding peak receptivity by several days. Deviations from this trend have been noted by Brown (1986). The period of time during which the stigma is receptive has not been determined but probably is in the range of a few days to a week. Once in contact with a receptive stigma, pollen germination is rapid (1 hour), and pollen tubes grow through the stigma to the ovule in less than 24 hours (Millar 1977).

### Development and Maturation of Seed

As in many other trees, the cone and seed structures of red alder develop whether or not ovules are fertilized. If ovules have not been fertilized, the seed will be empty and slightly shrunken. Some alders are also capable of producing viable seed without pollination (apomixis), although this phenomenon is only documented cytologically for one species (*Alnus rugosa*). In red alder, data from crossing studies in which aments were isolated indicated apomictic seed production at a rate of 0.75 to 1.6 percent (Stettler 1978).

Seed and strobiles of red alder develop throughout the summer. In the fall, strobiles lignify and change color from green to yellow, and finally to brown. The progression of strobile color during maturation is generally not uniform. The cones appear mottled yellow and yellow-brown in the late stages of ripening. Mature alder cones are 1 to 3 cm long and contain 50 to 100 wingless, nutlike seeds (Worthington et al. 1962). Once fully mature, strobile bracts separate and reflex to allow seed dis-

persal. Strobiles often remain on the tree for a year after maturation and are sometimes mistaken for the current year's crop.

In the Pacific Northwest, strobiles mature between early September and mid-October, depending on the location. Geographic variation in maturation appears to be closely correlated with fall frosts; strobiles mature earlier in areas with earlier fall frosts. Since climate variables are not always correlated with physiography in the Pacific Northwest (Ager 1987), geographic gradients in the date of cone maturation are inconsistent. For example, data taken in 1984 on cone maturity over an 800-m elevational gradient along the Nooksack River (east of Bellingham, WA) showed no pattern between elevation and cone ripeness (Ager, unpub. data). In contrast, data taken the same year for a similar gradient along the Nisqually River (east of

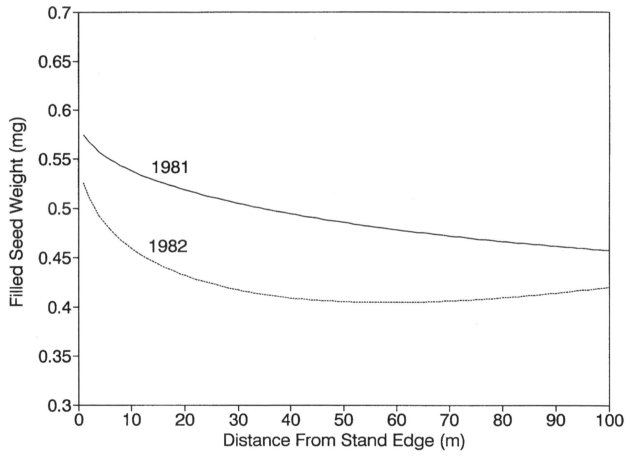

Figure 1. Plots of equations developed by Lewis to quantify the relationship between filled seed weight and dispersal distance for the 1981 and 1982 red alder seed crops. Seedfall data were collected from seed traps placed at four distances (0, 14, 60, and 100 m) from the edge of a mature alder stand near Eatonville, Washington (from Lewis 1985, pp. 63, 66).

Tacoma, WA) showed a pronounced gradient in ripeness, with strobiles at the high elevation stands maturing about three weeks before those at the low elevations. Climatic data also showed a strong gradient in fall frost dates along the Nisqually River but not along the Nooksack River.

*Seed Dispersal*

Although red alder seed is light and well suited for wind dispersal, available data indicate the bulk of a given seed crop is not dispersed very far. Lewis (1985) measured seedfall in 1981 and 1982 seed crop at varying distance from the stand edge (0, 14, 60, and 100 m) and estimated seedfall at around 60 million seeds/ha/year within the stand versus 1 to 2 million seeds/ha/year at 100 m from the stand edge. Lewis (1985) also noted that seed collected from the more distant collection sites had a lower seed weight, lower filled seed percentage, and reduced viability.

The relationship between filled seed weight and dispersal distance for the 1981 and 1982 seed crop was modeled by Lewis (1985) and found to vary somewhat between the two years of collection (Fig. 1), although in both years seed weight fell sharply with increasing distance from the stand edge. The inverse relationship between seed weight and dispersal distance is usually interpreted as an ecological tradeoff. Larger seeds have more carbohydrate reserves and germination vigor, but are less likely to be dispersed onto a suitable seedbed (i.e., mineral soil, full sunlight).

Although peak seed dispersal occurs in the late fall, significant quantities of red alder seed are disseminated throughout most of the year. Lewis (1985) found significant seed dispersal for 9.5 months near the forest edge (Fig. 2), and for 6 months at 100 m from the forest edge. Even during the spring (April), long after peak dispersal, seedfall was estimated at over 100,000 seeds/ha within the study stand (Fig. 2), and over 30,000 seeds/ha in the plots 14 m from the stand edge.

The practical implications of the extended seed dispersal in red alder are many with respect to operational seed procurement (discussed below). It is not known whether the prolonged seedfall is the result of variation within a tree or among trees within a stand in the timing of seed dispersal. Our personal observations suggest that dispersal dates can vary substantially among trees within a stand, perhaps as a result of variation in cone morphology (e.g., thicker cone bracts) that retards the dissemination of seed.

As in other trees, the timing of seed dispersal in red alder is also related to fall weather patterns. In general, drier conditions result in larger seed output, but Lewis (1985) noted two deviations from this pattern. First, heavy seed dispersal can occur during periods of significant precipitation. Second, if temperatures are below freezing, dry weather conditions may not necessarily induce seedfall. Lewis (1985) hypothesized that ice crystals can essentially freeze the seed in the cone.

*Temporal and Spatial Patterns of Seed Production*

Seed crops in red alder vary among years, stands, and trees within a stand. It is generally thought that seed production does not follow a periodic pattern in red alder as it does in many conifers, but rather red alder produces regular seed crops interspersed with more abundant crops every two to five years (Schopmeyer 1974). Seed production is also highly variable among stands and among trees within a stand. Brown (1985) recorded data on seed production among 45 mature trees of roughly equal size in natural stands and found a coefficient of variation among trees of 70 to 130 percent over three years. Seed production varied from 0 to 5.4 million seeds/tree. Variation among trees within a stand for most seed output traits was quite large in comparison to variation among stands.

Stand level variation was apparent, however, for seed and cone size, and correlated with elevation. Smaller strobiles with fewer, smaller seeds were observed at higher elevations, presumably a consequence of the shorter growing season. Total seed output, however, was found to be relatively constant, due to the production of more strobiles at higher elevations.

Genetic and environmental factors cause variability in seed output. In red alder, our collective observations suggest that environmental influences on seed output can be particularly strong, as evidenced by the fact that seed output along the edge of a stand is often significantly

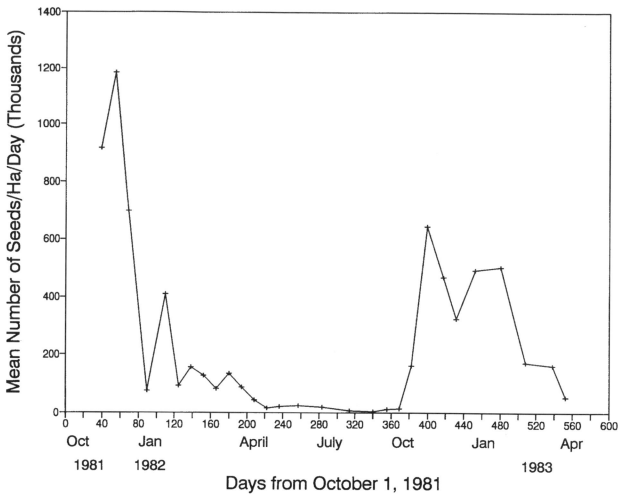

*Figure 2. Mean daily seedfall of red alder seeds in 1981 and 1982. Data were collected from seed traps placed in a mature alder stand near Eatonville, Washington (Lewis 1985, p. 30).*

greater than that within a stand. This enhanced seed output is often accompanied by only modest phenotypic differences among the edge and interior trees. Apparently, light intensity has a marked stimulatory effect on seed production in red alder. Thus nongenetic factors such as stand density and position of a tree within a stand can be major determinants of seed output.

### Temporal and Spatial Patterns of Seed Viability

The spatial and temporal patterns of seed viability largely mirror those for seed production. On a broad geographic scale, seed viability is generally lower in small, isolated stands on marginal red alder sites. These stands can be found on either low or high elevation areas, and thus correlations be-

tween viability and elevation are weak (Ager, unpub. data).

As in seed output, variability among trees within a stand has been observed to be more striking than variability among stands (Kenady 1978; Ager, unpub. data). For example, an analysis of variance on viability data for the 120 trees sampled by Ager (1987) showed that 76 percent of the total variation could be attributed to trees within stands, 20 percent among stands within a drainage, and 4 percent among drainages.

Some data suggest that year-to-year variation in viability is less than variation in seed production and other seed traits. Brown (1985) noticed that some trees consistently produced either low or high viability seeds despite large changes in total seed output. By contrast, data from bulk collections of red alder seed have shown viability to be generally higher in years of abundant seed crops (McGrath, unpub. data). Long-term studies on seed production and seed viability are needed to clarify trends over time.

Seed viability also varies within the dispersal period of a single seed crop. Seeds dispersed during the peak dispersal period for a given seed crop have a higher viability than those dispersed later in the year (Lewis 1985). In addition, seed viability varies within a tree. Seed from lower branches are of lower viability and smaller size compared to those from branches in the mid to upper portions of the crown (Brown 1985).

## Seed Germination

The ecological requirements for germination and survival of red alder seed are consistent with its colonizing life history. Although alder seed germinates in a variety of seedbed conditions (Ruth 1968; Schalin 1968; Minore 1972; Lewis 1985; Haeussler 1987), germination and early survival are enhanced on mineral soils in partial or full sunlight (Worthington et al. 1962; Ruth 1968; Schalin 1968). The stimulatory effects of light on seed germination have been demonstrated under lab conditions (Kenady 1978; Bormann 1983) and probably result from the regulation of germination by phytochrome (Bormann 1983). The affinity of alder seed to mineral soils can also be attributed to the limited carbohydrate supply available for growth of the radicle after germination, making establishment on duff layers difficult.

## Seed Procurement and Utilization

### Genetic Considerations

Red alder is locally adapted to the climatic gradients within the Pacific Northwest (Ager 1987; Ager et al. 1993). As a consequence, selecting appropriate seed sources is important to minimize risks associated with geographic transfer of seed. Provisional seed transfer guidelines and seed zones for red alder have been proposed by Ager (1987) and Hibbs and Ager (1989). These guidelines are largely based on those proposed for Douglas-fir in the Pacific Northwest (Western Forest Tree Seed Council 1966). See Ager and Stettler (Chapter 6) for a review of alder genetics as well as an elaboration on seed-transfer guidelines.

### Selection of Seed Trees

Ideal seed trees generally are found on good alder sites along the edges of large stands where a road or clearing has provided more light and stimulated cone production. Trees growing under these conditions tend to produce large, regular seed crops of relatively high viability. In addition, Hibbs and Ager (1989) suggested selecting seed trees that are evenly distributed throughout a stand within the seed collection area, and that a minimum of 15 parent trees contribute to a seed lot. We also suggest avoiding isolated trees, as they may be poorly pollinated and have low seed yields.

### Assessing Crop Quality

Assessing the quality of the strobiles and the seed they contain is important to the economics of seed collection. Three factors must be assessed: ripeness, seed quantity, and seed quality. Seed crops can be assessed in July or August by obtaining a count of mature strobiles and filled seed. Because cone production is highly variable among trees within a stand, several trees should be observed. A "good" cone crop for a single tree occurs when several hundred strobiles can be observed (with binoculars for large trees) from a single vantage point.

Filled seed count, an important part of the assessment, should be determined by sampling strobiles from the upper third of a crown, where the viability is highest (Brown 1985). A small caliber rifle is useful for this purpose. At least a few healthy appearing strobiles should be sampled from each tree. Seed quality can be assessed by cutting the cone longitudinally and counting the filled seed on one of the cut faces. Filled seed has a milky-white interior (cotyledons). Low seed vigor can be indicated by poorly developed cotyledons (shrunken and discolored). Although filled seed on a cut face can vary from 0 to 20 or more; less than 3 or 4 seeds per cut face indicate a marginal crop.

### Assessing Crop Maturity

Methods for determining ripeness as outlined by Hibbs and Ager (1989) were developed by Ager (1987) while collecting red alder seed from a variety of locations in the Pacific Northwest. The technique judges ripeness by two factors: cone color and morphology. The color of red alder strobiles progresses from green to yellow-green, to yellow-tan or yellow-gray, then to gray-brown and brown. The color is rarely uniform on a strobile, and strobiles often have a mottled appearance in the fall. (Pathogens can cause strobiles to change to a silver-gray color which is not part of the normal ripening process). Alder strobiles are apparently sufficiently ripe to harvest when the color has changed to at least 50 percent yellow.

A second test for maturity involves twisting the cone along the long axis. If the cone twists easily and the cone bracts separate when twisted, the seed is sufficiently mature to be harvested.

Once mature, strobiles can be collected as late as early January with usable quantities of seed. The later the collection date, however, the greater the likelihood that seed will have dispersed from the strobiles. As discussed previously, seed dispersal is dependent on weather. Warm, dry and windy weather induces seedfall, while wet and cold conditions delay the process. Seed collections attempted after the peak dispersal must accommodate the generally lower seed quality and fewer seeds per strobile. Late harvests should not be conducted during dry weather as the strobile bracts will be open and much of the seed may be lost during collection.

The loss of seed from late collections is illustrated by data from Weyerhaeuser seed harvests (six locations in southwestern Washington) in 1988 (Tanaka, unpub. data). In this collection, five lots collected in mid-November yielded 0.6 to 1.1 lbs of seeds per bushel of cone while one lot collected in late December gave only 0.23 lbs of seeds per bushel of cone. Germination rate averaged approximately 80 percent and did not vary significantly between the two collection dates.

### Methods for Collecting Seeds from Trees

Most of the Weyerhaeuser and U.S. Forest Service collections have been made via a destructive method, by either cutting down the selected young (approximately 20 to 25 years old), vigorous trees or by gathering strobiles from harvested trees. The quality of seeds has generally been excellent. Although this method of cone collection is quick and easy, it requires new suitable trees each year.

The U.S. Forest Service has also performed some cone collections on standing trees by climbers using spurs or a ladder and/or a pole pruner to trim strobile-bearing branches. Collected branches were then placed in the bed of a truck and raked with a custom "cone rake." The cone rake can also be used on felled trees to strip strobiles from branches.

Once the strobiles have been stripped from the branches, great effort should be made to clean all the debris from the collection before the cones start to dry and open. Leaves are especially troublesome in the cleaning process; once they are dry, they can break up into small pieces that have similar physical properties as alder seed, making seed removal difficult. Leaves and light debris can be removed by placing the unopened strobiles in front of a large fan. Heavy debris such as sticks and next year's flowers can be picked out by hand at the same time. Cleaning a collection immediately after harvest saves considerable time and expense later.

### Air Drying, Storage, and Handling

Field-collected cones must be placed in bags and air-dried to allow for after-ripening. For small lot collections, large grocery bags are sufficient. For large-scale collections, tight weave cloth bags (e.g., flour sacks) work well. Plastic bags are not suitable because they do not allow adequate ventilation. The U.S. Forest Service extractory at Wind River (Carson, WA) prefers nylon laundry sacks used by the military. The nylon mesh is fine enough to retain the seed, while allowing good air circulation for drying. The bags are expensive, but are very durable and easily hold a bushel of strobiles.

After seeds are bagged and appropriately tagging to identify the seed lots within a collection, the bags should be arranged on racks or other storage devices for air drying. At the Wind River Nursery, temperatures of less than 27°C are maintained and small fans provide good air circulation. During the first week after collection, bags are turned and checked daily to prevent strobiles from molding and to provide uniform drying. The bags can be safely held in this environment for several weeks until seed extraction. At Weyerhaeuser, the cones are placed in plastic mesh bags and laid on top of the kiln until they are sufficiently dry for extraction.

## Seed Processing: Extraction and Cleaning

Seed processing of the 1988 collections by Weyerhaeuser involved placing dry cones in a tumbler, followed by thrashing. Then, the cones were wet down, placed in a cooler for 24 hours, and redried to improve overall seed yields. The seeds were then screened to remove all large debris and finally placed in a pneumatic separator to remove empty seed.

The extractory at Wind River processes air-dried strobiles by placing them on open screen-bottomed trays and drying them in the kiln for 24 hours at 27°C. The cones then are removed and examined to see if the bracts are opened and the seed is loose. If the strobile bracts are not flared and open, they are sprayed with a fine water mist and redried in the kiln. Once the seed is loose, it is extracted using a small tumbler. If, after tumbling, significant seed remain in the strobile (15 to 25 per strobile), the lot is wet again, dried, and retumbled.

Extracted seeds are first screened to remove large debris. An air column machine such as an aspirator or pneumatic separator is then used to remove small trash and some empty seed. The extractory at Wind River has had good success in using a gravity table to separate empty seeds from the lot. An x-ray machine is necessary to monitor the cleaning processes to make sure only empty seeds are being thrown away. As mentioned earlier, the early removal of leaves from the strobiles is important.

## Seed Drying and Moisture Content

The cleaning process described earlier dries seed adequately for short-term storage of a year or less. Longer-term storage requires a seed moisture content of less than 10 percent. Further drying is accomplished at Wind River Nursery by placing seed in a room at 27°C with less than 25 percent relative humidity.

## Seed Testing

All seed should be tested prior to sowing to determine seed quality. In addition, seed lots in long-term storage should be retested at some predetermined interval. The frequency of testing depends upon the quality of the seed and the importance of maintaining that quality. For red alder seed, testing every five years or less is recommended. Storage over long periods of time can increase seed moisture content, causing deterioration of seed germination capacity. A seed lot also can show a decrease of germination percent over a period of years for no apparent reason. Thus, it is useful to record changes in germination over time so replacements can be obtained.

**Sampling.** The first step in seed testing is to draw a sample that represents the entire seedlot. A seedlot is defined as a unit of seed of reasonably uniform quality from a particular location or elevation (Bonner 1974). Seedlot size varies with testing rules and among laboratories. The International Seed Testing Association (ISTA 1985), for example, recommends that seedlots for tree species with relatively small seeds, like red alder, should be less than 1000 kg. The Western Forest Tree Council (Stein 1966) recommends that lots in excess of 227 kg be evenly divided into equal smaller lots for sampling. The sample should be composed of equal portions taken from evenly distributed volumes of the lots, each sample proportional to the size of the container. The sample should be subdivided in the testing laboratory with a mechanical divider until a subsample of the desired weight is obtained.

**Purity.** Purity tests measure the percent by weight of pure seeds of the test species. Red alder seeds are small and light (considerably lighter than the most of the conifer species in the Pacific Northwest), and it is often difficult to separate pure seeds from debris of various sizes. For this reason, it is rather challenging to consistently achieve a purity as high as some of the western coniferous species (Tanaka 1984).

**Moisture content.** Seed moisture content is most often determined with the forced-air oven-dry method. Seed samples are heated in ovens and the weight loss that occurs during drying is equated with moisture in the seed. ISTA rules prescribe oven drying at 105°C for 16 hours. As with other tree species, moisture content is critical for long-term storage of alder seeds. Data indicate that seeds should have a moisture content of less than 10 percent on a fresh weight basis. Red alder seeds can be safely kiln-dried to 4 to 8 percent on a fresh weight basis without adversely affecting seed viability (Tanaka, unpub. data).

**Seed weight.** Seed weight, required for calculating sowing rates, is a function of seed size, moisture content, and proportion of full seed in a given lot. The commonly used unit is the weight of 1000 pure seeds. It is also often expressed as the number of cleaned seeds per pound. For red alder, 383,000 to 1,087,000 clean seeds per pound have been reported (Schopmeyer 1974). The wide range is probably due to source variation and also to variation in the proportion of empty seeds in different lots. The separation of empty from filled seeds is difficult in red alder.

**Germination potential.** Germination potential, perhaps the most important quality measurement in seed testing, is used to determine sowing rates as well as to determine whether seed must be sown immediately or can be stored. This potential can be evaluated directly by germinating seeds under predetermined conditions. Excellent germination of red alder seed has been obtained under a variety of conditions, including presoak treatments of either 16 hours (Berry and Torrey 1985) or 24 hours (Elliot and Taylor 1981; Radwan and DeBell 1981; Tanaka et al. 1991) and a variety of temperature regimes: 25°C (16 hours)/ 20°C (8 hours) (Berry and Torrey 1985), 25°C (12 hours)/ 18°C (12 hours) (Elliot and Taylor 1981), 30°C (10 hours)/ 20°C (14 hours) (Radwan and DeBell 1981), and 30°C (8 hours)/ 20°C (16 hours) (Tanaka et al. 1991).

The current ISTA procedure recommends testing germination potential on wet paper for 21 days at alternating temperatures of 30°C in the light for 8 hours and then 20°C in darkness for 16 hours (ISTA 1985). Stratification of seed is not required for germination testing. The use of light was recommended earlier (ISTA 1976), but is not recommended in the current test (ISTA 1985). Nonetheless, based on the beneficial effect of light on germination observed for red alder by Berry and Torrey (1985), it seems advantageous to conduct germination testings using light.

Very little work has been done to indirectly estimate seed viability with biochemical staining, cutting tests, x-ray radiography, or hydrogen peroxide tests. This is probably due to the fact that red alder seeds germinate promptly under the ISTA conditions and so the direct method can be easily applied. In addition, the small size of red alder seed make indirect testing difficult. A test conducted in one laboratory showed that percentage of filled seeds determined by x-ray radiography was highly correlated ($r^2$ = 91 percent) with the actual germination percentage (Tanaka, unpub. data).

**Seed vigor.** Nursery-bed germination is usually slower and less complete than laboratory germination. Therefore, various laboratories have attempted to define and determine seed vigor to improve the prediction of nursery germination of mostly coniferous species (Tanaka 1984). Three major groups of expressions have been proposed: mathematical values based on standard laboratory test results, germination under stressful conditions, and biochemical testing. To our knowledge, research to correlate laboratory and nursery germinations of red alder has not been done. In one study, however, where Czabator's (1962) germination value was used as an indicator, Elliot and Taylor (1981) found significant differences in seed vigor among several geographic sources.

### Seed Storage

If red alder seed is adequately dry and sealed in containers to protect it from moisture and mold, refrigerator temperatures are adequate for short-term storage (< 1 year). Successful long-term storage requires more stringent conditions, especially with respect to storage temperature and seed moisture content. Although there is little information on long-term storage of red alder seed, available data indicate that such storage can be accomplished with little loss in seed viability. High-quality red alder seed with less than 10 percent moisture content and good storage conditions (< 12°C) should maintain germination capacity for 10 years without significant loss in viability (McGrath, unpub. data). At the Wind River Nursery, red alder seed lots have been stored up to 20 years with less than 20 percent decline in germination. In an ongoing storage experiment being conducted by Weyerhaeuser, no decline in germination was observed after two years of freezer storage (-18°C, 4 to 8 percent moisture content) for a sample of seed lots collected in 1988. As mentioned above, some data indicate that higher quality seed from good seed years retains higher viability in either long-term or short-term storage (McGrath, unpub. data).

A variety of containers are used for storing seeds, though all of them are moisture proof and can be opened easily and resealed. Bags should be able to withstand low temperature for many years without deteriorating in strength. The Wind River Nursery uses 0.007-mm plastic bags specially treated for freezer storage. The bag is heat sealed and stored within a cloth bag to protect it from punctures. Other U.S. Forest Service seed storage facilities use a cardboard box to protect the inner plastic bag. The U.S. Forest Service Dorena Tree Improvement Center uses a three-layer laminated bag with a middle layer of aluminum foil that can be heat-sealed.

All storage systems should employ a labeling method that identifies each seed lot clearly and completely with the source information, the collec-tion year, and the certification class. A duplicate la-bel should also be on the inside of the container with the seed.

*Presowing Treatments*

Tree seeds, unlike agricultural seeds, are in many cases characterized by deep dormancy. Seeds of dif-ferent species and different geographical origins often require different pretreatments and condi-tions for optimum germination. The most commonly used pretreatment to break dormancy is stratification, which usually consists of a moist cold treatment for up to several months. Stratifi-cation is generally known to bring about changes in anatomy, physiology, and metabolism. Success-ful stratification requires proper moisture content, low temperature, adequate aeration, and proper length of time.

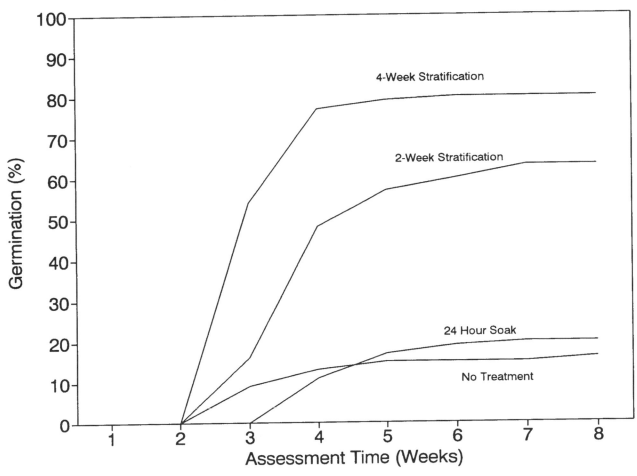

*Figure 3. Germination of red alder seeds under cool temperatures (15°C/5°C) and four pregermination treatments (data from Tanaka et al. 1991).*

At warm seed germination temperatures, stratification at low temperature (0°C to -5°C) for up to two months has no significant effect on the rate or completeness of germination of red alder seeds (Elliot and Taylor 1981; Radwan and DeBell 1981; Tanaka et al. 1991). In contrast, under cool laboratory germination temperatures, a two- to four-week stratification significantly increased the speed of germination and total germination (Tanaka et al. 1991; Fig. 3). This suggests that up to four weeks of stratification could be advantageous before sowing of red alder seeds outdoors in early spring, when temperatures are often suboptimal for quick and uniform germination. It also indicates that stratification would widen the range of temperatures for optimum germination.

Although stratification is an effective method to break dormancy, it is often time consuming. Three quick pregermination treatments have been successfully tested on red alder seed. They are giberellin (GA$_3$) (Berry and Torrey 1985), 1-percent captan (Berry and Torrey 1985), and 30-percent H$_2$O$_2$ (Neal et al. 1967). All the quick treatments, however, were tested under optimum germination conditions and may not be effective under the suboptimal temperature conditions frequently encountered in the field in early spring. Captan and H$_2$O$_2$ presumably have a sterilizing effect and, under optimum germination conditions, should be effective on seeds infected by disease.

## Research Needs

We have identified two problem areas that deserve immediate research. The first concerns the large-scale production of high quality seed. Since good seed crops of red alder only occur periodically, it be would advantageous to consider converting selected natural stands to seed production areas to help maintain a consistent seed supply. These areas might receive cultural treatments, such as thinning, fertilization, and girdling, to stimulate cone production in other tree species. Research is needed, however, to determine the effectiveness of these and other treatments to stimulate seed production in red alder. Seed production could also be enhanced by establishing seed orchards. Genetic gains in wood yield and quality, however, would be required to justify the financial investment.

The second area in need of research is sowing technology. Inoculation with the *Frankia* endophyte significantly improves seedling growth and yield of red alder seedlings in the greenhouse (Stowers and Smith 1985) and in the open-bed bare-root nursery (Hilger et al. 1991; Martin et al. 1991). The current method of inoculating container media or nursery beds, however, is quite laborious. It may be advantageous to pelletize red alder seeds and introduce *Frankia* into pelletizing materials. This approach may also improve precision and efficiency of sowing in greenhouse containerized seedling production.

## Acknowledgments

We thank Dean DeBell, Vicky Erickson, David Hibbs, Steve Omi, and Reinhard Stettler for reviewing earlier drafts of this chapter.

## Literature Cited

Ager, A. A. 1987. Genetic variation in red alder (*Alnus rubra* Bong.) in relation to climate and geography in the Pacific Northwest. Ph.D. thesis, Univ. of Washington, Seattle.

Berry, A. M., and J. G. Torrey. 1985. Seed germination, seedling inoculation and establishment of *Alnus* spp. in containers in greenhouse trials. Plant and Soil 87(1):161-173.

Bonner, F. T. 1974. Seed testing. *In* Seeds of woody plants in the United States. *Technically coordinated by* C. S. Schopmeyer. USDA For. Serv., Agric. Handb. 450. 136-152.

Bormann, B. T. 1983. Ecological implications of phytochrome mediated seed germination in red alder. For. Sci. 29(4):734-738.

Briggs, D. G., D. S. DeBell, and W. A. Atkinson, *comps.* 1978. Utilization and management of alder. USDA For. Serv., Gen. Tech. Rep. PNW-70.

Brown, S. M. 1985. A study of reproductive biology of *Alnus rubra* along three elevational transects in Washington and Oregon. Report on file, USDA For. Serv., PNW Res. Sta., Olympia, WA.

———. 1986. Sexual allocation patterns in red alder (*Alnus rubra* Bong.) along three elevational transects. M.S. thesis, Univ. of Washington, Seattle.

Czabator, F. J. 1962. Germination value: an index for combining speed and completeness of pine seed germination. For. Sci. 8:386-396.

Elliot, D. M., and I. E. P. Taylor. 1981. Germination of red alder (*Alnus rubra*) seed from several locations in its natural range. Can. J. For. Res. 11:517-521.

Furlow, J. J. 1979a. The systematics of the American species of *Alnus* (Betulaceae). Part I. Rhodora 81:1-21.

———. 1979b. The systematics of the American species of *Alnus* (Betulaceae). Part II. Rhodora 81:151-248.

Haeussler, S. 1987. Germination and first-year survival of red alder seedlings in the central Coast Range of Oregon. M.Sc. thesis, Oregon State Univ., Corvallis.

Hibbs, D. E., and A. A. Ager. 1989. Red alder: guidelines for seed collection, handling, and storage. Forest Research Laboratory, Oregon State Univ., Corvallis, Spec. Publ. 18.

Hilger, A. B., Y. Tanaka, and D. D. Myrold. 1991. Inoculation of fumigated nursery soil increases nodulation and yield of bare-root red alder (*Alnus rubra* Bong.). New For. 5:35–42.

International Seed Testing Association. 1976. International rules for seed testing. Seed Sci. Tech. 4:4-180.

———. 1985. International rules for seed testing. Seed Sci. Tech. 13.

Kenady, R. M. 1978. Regeneration of red alder. *In* Utilization and management of alder. *Compiled by* D. G. Briggs, D. S. DeBell, and W. A. Atkinson. USDA For. Serv., Gen. Tech. Rep. PNW-70. 183–192.

Lewis, S. J. 1985. Seedfall, germination, and early survival of red alder. M.S. thesis, Univ. of Washington, Seattle.

Martin, K. J., Y. Tanaka, and D. D. Myrold. 1991. Peat carrier increases inoculation success with *Frankia* on red alder (*Alnus rubra* Bong.) in fumigated nursery beds. New For. 5:43-50.

Millar, C. 1977. Pollen-pistil interactions in four alder species after selfing and crossing. Senior honors thesis, Univ. of Washington, Seattle.

Minore, D. 1972. Germination and early growth of coastal tree species on organic seed beds. USDA For. Serv., Res. Pap. PNW-135.

Neal, J. L., Jr., J. M. Trappe, K. C. Lu, and W. B. Bollen. 1967. Sterilization of red alder seedcoats with hydrogen peroxide. For. Sci. 13:104-105.

Radwan, M. A., and D. S. DeBell. 1981. Germination of red alder seed. USDA For. Serv., Res. Note PNW-370.

Ruth, R. H. 1968. First season growth of red alder seedlings under gradients in solar radiation. *In* Biology of alder. *Edited by* J. M. Trappe, J. F. Franklin, R. F. Tarrant, and G. M. Hansen. USDA For. Serv., PNW For. Range Exp. Sta., Portland, OR. 99-105.

Schalin, I. 1968. Germination analysis of grey alder and black alder seeds. *In* Biology of Alder. *Edited by* J. M. Trappe, J. F. Franklin, R. F. Tarrant, and G. M. Hansen. USDA For. Serv., PNW For. Range Exp. Sta., Portland, OR.

Schopmeyer, C. S. 1974. *Alnus* B. Ehrh. *In* Seeds of woody plants in the United States. *Technically coordinated by* C. S. Schopmeyer. USDA For. Serv., Agric. Handb. 450. 206-211.

Stein, W. I., *ed.* 1966. Sampling and service testing western conifer seeds. Western Forest Tree Seed Council, Western Forestry and Conservation Assoc., Portland, OR.

Stettler, R. F. 1978. Biological aspects of red alder pertinent to potential breeding programs. *In* Utilization and management of alder. *Compiled by* D. G. Briggs, D. S. DeBell, and W. A. Atkinson. USDA For. Serv., Gen. Tech. Rep. PNW-70. 209-222.

Stowers, M. D., and J. E. Smith. 1985. Inoculation and production of container-grown red alder seedlings. Plant and Soil 87:153-160.

Tanaka, Y. 1984. Assuring seed quality for seedling production: Cone collection and seed processing, testing, storage and stratification. *In* Forest nursery manual: production of bareroot seedlings. *Edited by* M. L. Duryea and T. D. Landis. Martinus Nijhoff/Dr. W. Junk Publishers, The Hague/Boston/Lancaster, for Forest Research Laboratory, Oregon State Univ., Corvallis. 27-39.

Tanaka, Y., P. J. Brotherton, A. Dobkowski, and P. C. Cameron. 1991. Germination of stratified and non-stratified seeds of red alder at two germination temperatures. New For. 5:67-75.

Trappe, J. M., J. F. Franklin, R. F. Tarrant, and G. M. Hansen, *eds.* 1968. Biology of Alder. USDA For. Serv., PNW For. Range Exp. Sta., Portland, OR.

Western Forest Tree Seed Council. 1966. Tree seed zone map for Oregon and Washington. Western Forest Tree Seed Council, Portland, OR.

Worthington, N. P., R. H. Ruth, and E. E. Matson. 1962. Red alder: its management and utilization. USDA For. Serv., Misc. Publ. 881.

# 12

# Seedling Quality and Nursery Practices for Red Alder

GLENN R. AHRENS

Regeneration of red alder (*Alnus rubra* Bong.) has become a primary objective on some forest lands in the Pacific Northwest. Although natural regeneration may be practical in some situations, planting is currently recommended for consistent and uniform regeneration over large (> 10 ha) management units (Ahrens et al. 1992; Hibbs and DeBell, Chapter 14). Nurseries have cultivated seedlings of several *Alnus* species for a variety of uses in the past, but large-scale production of red alder for reforestation in the Pacific Northwest is a recent objective. Even with the current expanding interest in reforestation, alder remains a minor species for Western forest nurseries.

The main steps and general principles of forest tree nursery production are well established, and have been refined and covered in detail in other publications (Duryea and Landis 1984; Rose et al. 1990a). These general principles have been applied successfully to alder in some cases; however, more specific information concerning both biological and operational aspects of alder seedling culture is needed.

The objective of this chapter is to review and synthesize available information on seedling quality and nursery practices for red alder. In addition to discussing typical cultural systems, I provide recommendations for increasing the reliability and success of nursery practices for alder.

## Seedling Quality

Similar to initial findings for other species, a substantial amount of the variation in performance among early red alder plantings can be attributed to variability in seedling quality. Several organizations have established red alder plantations since the early 1980s (e.g., the USDA Forest Service,

USDI Bureau of Land Management, Oregon State University, and industrial owners). In many cases, the production of seedlings for these plantations has involved independent, first-time efforts. An indication of the highly variable quality of seedlings is provided by persons involved in these efforts, who often observed a wide range in seedling size at planting. Thus, before proceeding with a discussion of specific nursery practices, criteria for evaluating seedling quality need to be established.

The ultimate measure of seedling quality is actual field performance. Many excellent reviews of seedling quality and assessment have been made (Cleary et al. 1978; Sutton 1979; Bunting 1980; Ritchie 1984; Duryea 1985; see also the special issue of *New Zealand Journal of Forestry Science*, vol. 10, no. 1). Ideally, standards for rating seedling quality should be based on various key attributes of seedlings, and modified with respect to the expected range of outplanting site conditions.

Recent development of the target seedling concept has provided a useful framework for defining seedling quality goals and applying appropriate nursery practices to meet them (Rose et al. 1990a). A key element in the target seedling concept is that many seedling traits operate together (and interact) to produce a desirable field response (Rose et al. 1990b). Important traits may include various measures of morphology, bud dormancy, water status, nutritional status, root growth potential, and stress tolerance. Target seedling criteria are expressed as a range of acceptable values for each attribute. Within a single nursery's crop of seedlings, many of the above attributes will be highly correlated, which allows effective grading to be based on few criteria; however, good general criteria for assessing seedling quality for any combination of nursery practices, production years, and outplanting conditions must be based on multiple attributes.

## Morphological Characteristics

Criteria for evaluating alder stock currently are based on morphological characteristics. These criteria have been developed recently and are the result of field tests with various stocktypes by Weyerhaeuser Company and observations of seedling performance in numerous other research plantings (Ahrens et al. 1992; Radwan et al. 1992; A. Dobkowski and Y. Tanaka, unpub. data). Morphological criteria for superior planting stock include the following:

1. A minimum basal diameter of 4 mm;

2. Seedling height of 30 to 100 cm;

3. Healthy branches or buds along the entire length of stem;

4. Full and fibrous root system; and

5. A plant free of disease and damage.

The importance of applying multiple criteria must be stressed. Currently, alder stock often is graded by size alone. Because of alder's potentially rapid growth rate, which depends on the specific nursery climate, length of the growing season, and nursery practices, a wide range of average crop sizes can be produced in one year. Seedlings that meet minimum size standards often do not have adequate bud development or full root systems when average crop size is large.

The occurrence of healthy buds and branches may be the best indicator of quality for seedlings of any size. Higher bud densities, especially on the lower portion of the stem, are desirable for alder, because they ensure dense foliage and self-shading of photosensitive stems. This trait has been subjectively associated with stout or "stocky" seedlings. A quantitative measure (sturdiness quotient) may provide a more useful, objective criterion related to stockiness, buds, and branches.

Sturdiness quotients, such as height/diameter ratio (H/D), have been useful for other species as an indicator of resilience to physical damage or exposure to sun, frost, and wind (Roller 1977; Ritchie 1984; Thompson 1985; Mexal and Landis 1990). Alder seedlings are particularly sensitive to these damaging agents. Relationships between H/D and field performance have not been established for alder; however, correlations between H/D, bud density, and shoot/root (S/R) ratios suggest that this sturdiness quotient is a sensitive indicator of attributes of known importance.

Samples from alder grown at the U.S. Forest Service Humboldt Nursery (McKinleyville, CA) demonstrated a high correlation between H/D and the number of buds per cm of shoot ($R^2 = 0.9$, Fig. 1A); stockier seedlings had higher bud densities. Stockier seedlings also had lower S/R ratios (Fig. 1B), which are particularly desirable (Ritchie 1984; Thompson 1985) in the case of such drought-sensitive species as red alder.

The data in Figure 1A represent combined samples from several populations of alder grown under different conditions (thinned, unthinned, fertilized, and unfertilized). The close correspondence between H/D and bud density among and within these different treatments suggests that general H/D criteria can be developed for application to alder crops raised under different nursery regimes. The essential linkage between H/D (or related attributes, such as S/R) and field performance must be made before specific criteria can be formulated for a given outplanting environment.

To illustrate the application of multiple criteria to highly variable alder, samples from five nursery crops are plotted in Figure 2 along with reference lines for height, diameter, and sturdiness criteria. For this discussion, I will assume a maximum acceptable H/D ratio of 100 (mm/mm). Yield of acceptable seedlings based only on diameter and height ranges from 38 to 82 percent. If H/D criteria are applied, the range in yield is only 38–50 percent. Although sturdiness criteria have not been established for alder, a substantial portion of the alder crops shown here appears to be excessively spindly. This is particularly apparent for the seedling sample shown in Figure 3, in which 80 percent of the trees were excessively tall and spindly. In light of the relationship between sturdiness and bud density (Fig. 1A), seedlings with excessive H/D probably do not have adequate numbers of buds to meet the third criterion listed at the beginning of this section.

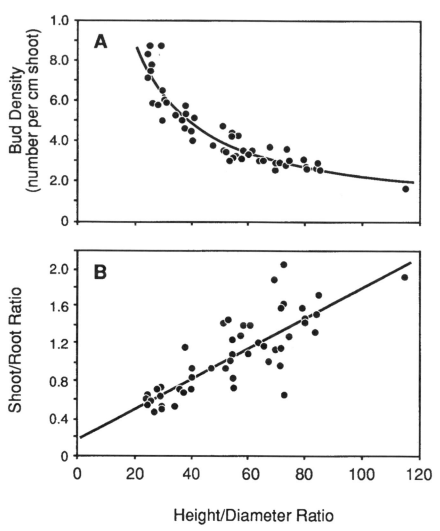

*Figure 1. (A) Bud density and (B) shoot/root (S/R) ratio of 1+0 red alder seedlings as a function of stem height/diameter (H/D)ratio. Bud density is the number of intact buds greater than 1 mm in diameter per cm stem. Equations for regression lines: ln (bud density) = 2.4225 - 0.855 \* ln (H/D), $R^2$ = 0.90, and S/R = 0.18 + 0.016 \* (H/D), $R^2$ = 0.67.*

Figures 2A and 2B represent the total population of seedlings from adjacent beds grown under similar conditions (1991 crop), except that the bed in Figure 2B was thinned. Thinning increased the uniformity of spacing and reduced density from 206 to 140 seedlings m$^{-2}$ (from 19 to 13 seedlings ft$^{-2}$). As a result, all seedlings sampled from the thinned bed had an H/D < 100. This illustrates the potential to produce stocky seedlings by managing for lower densities and uniform spacing. Management of bed density is discussed further in a later section.

Given the desirability of stocky seedlings, height may be the least important single criterion. Initial height has been a good predictor of height growth in some species (Ritchie 1984). Results from several alder plantations on good sites, however, show little correlation between initial height and height growth; a substantial portion of seedlings 20–30 cm in height grew as well as larger seedlings (Fig. 4). In stocktype comparisons, the tallest stocktype (bed-house) has shown the lowest vigor after outplanting (Radwan et al. 1992; Dobkowski et al., Chapter 13). The greater height of bed-house stock is associated

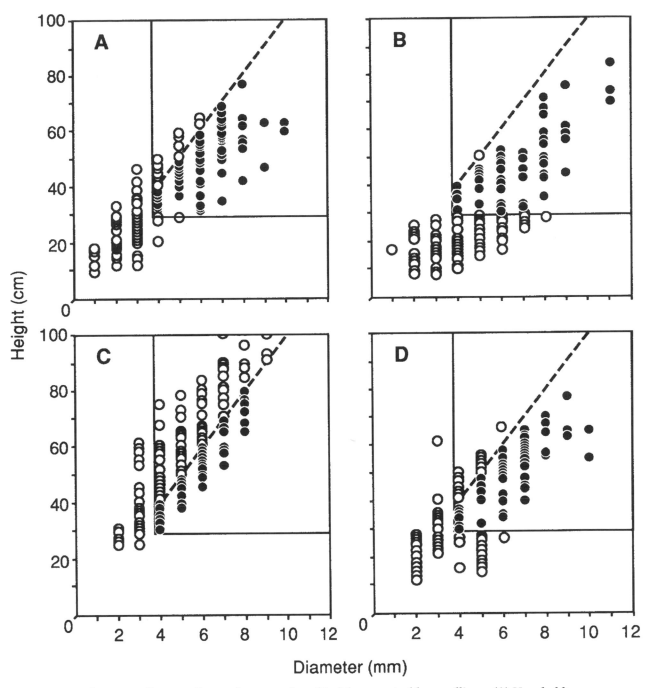

Figure 2. Height versus diameter for random samples of 1+0 bare-root alder seedlings. (A) Humboldt nursery, 1991 crop; (B) Humboldt nursery, 1991 crop thinned; (C) Wind River nursery, 1990 crop; and (D) Elkton state nursery, 1991 crop. Reference lines are shown for height, diameter, and H/D grading standards. Seedlings that pass grading standards are indicated by closed symbols.

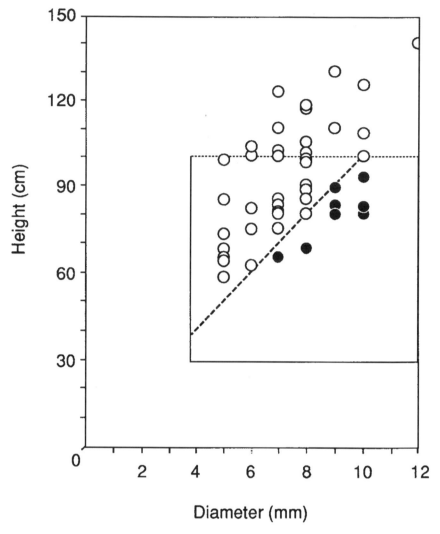

*Figure 3. Height versus diameter for a sample of 1+0 alder seedlings taken at the planting site (after grading by the nursery). Reference lines are shown for height = 30 cm, diameter = 4 mm, and H/D = 100. Acceptable seedlings are indicated by closed symbols.*

with an excessive H/D and a lack of basal buds. Greater emphasis on diameter and sturdiness, rather than height, seems warranted.

With the substantial percentage of seedlings that do not meet morphological criteria (Fig. 2), is there any basis for accepting lower grade seedlings under specific conditions? For example, reducing the minimum height standard from 30 to 20 cm would increase yield by 80 percent for the thinned seedlings represented in Figure 2B. Given that these plants met diameter and sturdiness criteria and had full root systems, it seems reasonable to expect good performance. Perhaps nonmorphological criteria (i.e., tests of vigor and physiological quality) could be applied to general morphological criteria for a specific crop.

### Nonmorphological Characteristics

Nonmorphological indicators of alder seedling quality are not well known. For other species, attributes such as bud dormancy, water status, root growth potential, and stress tolerance can vary greatly, independent of morphological traits (Duryea 1985). Reviews strongly

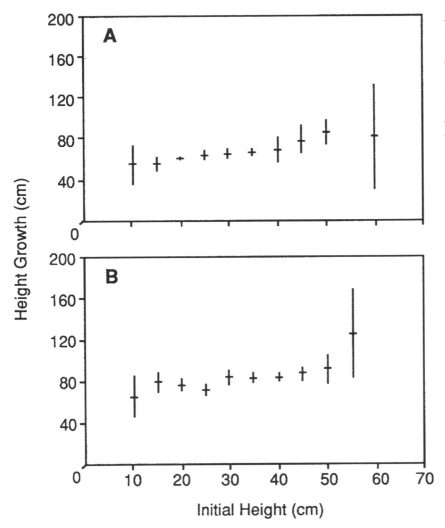

*Figure 4. Height growth versus initial height for alder seedlings at the end of the first growing season after outplanting. (A) 1+0 bare-root seedlings, 43 cm average height, 7.5 mm average diameter, and 62 average H/D. (B) 1+0 plug seedlings, 37 cm average height, 3.8 mm average diameter, and 96 average H/D.*

emphasize the importance of nonmorphological or physiological attributes for monitoring and testing seedling quality (Lavender 1989; Puttonen 1989; Rose et al. 1990a).

There is good evidence suggesting that bud dormancy and cold tolerance of alder seedlings can be improved. Rapid dehardening of alder planted in early spring can increase susceptibility to frost relative to wild alder on some sites (Ahrens et al. 1992; Dobkowski et al., Chapter 13). First-year plantings have also been damaged by cold during the following fall and winter (more so than wild seedlings or older planted seedlings), and this is presumed to be the result of inadequate frost-hardening. Dobkowski et al. (Chapter 13) found that bare-root seedlings deharden and break bud more slowly than do seedlings grown in greenhouses. Thus, research is needed to identify, monitor, and control factors affecting the onset of dormancy and the extent of chilling requirements.

Measures of root growth potential (RGP) and bud dormancy have been successfully used to test effects of lifting and storage treatments on vigor and physiological quality for other deciduous hardwoods

(Farmer 1975; Rietveld and Williams 1978; Webb and von Althen 1980). Webb and von Althen (1980) found a high correlation between xylem water potential and RGP for six different hardwood species after cold storage, thus indicating that water potential may be a quick and useful measure of physiological quality. Such measures could be applied to alder to establish relationships among dormancy, vigor, and climatic variables such as chilling hours.

## Nursery Practices

### Seedling Stocktypes

The earliest documented plantings of red alder in the Pacific Northwest were bare-root seedlings grown during the 1930s (U.S. Forest Service, Wind River Nursery. Carson, WA) for use as fire breaks (Tarrant 1961). Some bare-root nursery production of red alder continued but was discontinued after the 1940s because of lack of demand (Kenady 1978).

With increasing interest in red alder management, a variety of research plots and trial plantations were established in the 1970s. The most common stocktypes used in these trials were container-grown (plug) seedlings (DeBell and Radwan 1979; Gordon et al. 1979). Plug stock have also been used for extensive plantings of various alder species for reclamation of soils impacted by mining or dam-construction projects (Perinet et al. 1985). More recently, 1+0 bare-root seedlings have been used for many reforestation, reclamation, and research plantings (Heilman 1990; Ahrens et al. 1992) in the Northwest.

Nurseries in the Northwest have experimented with a variety of stocktypes and cultural methods for red alder. Most methods can produce seedlings with desirable characteristics, although specific conditions and limitations inherent in each method tend to produce seedlings with typical characteristics. The common red alder stocktypes for practical applications are grown in 1 year as follows:

**1+0 Open-bed bare-root.** Seedlings are grown from direct sowing into open nursery beds. This method is currently considered to provide the best combination of cost and quality for most situations (see the following section). The most difficult challenge is to achieve target densities and uniform spacing with direct sowing in nursery beds.

**1+0 Bedhouse bare-root.** Seedlings are grown from direct sowing into nursery beds under partial shelter provided by shade cloth and/or polyethylene plastic. Shelter may be removed at any time during the summer. As with direct sowing in open beds, achieving target densities is a difficult and important step. Depending on the amount and duration of shelter, seedlings may have low resistance to environmental extremes after outplanting. This method has not been widely tested. Typical bed-house stock have been tall and spindly.

**1+0 Plug.** Seedlings are grown in any of a variety of containers in the greenhouse or shelterhouse. Choice of container size and density in the greenhouse allows some control over final size and spacing. Except for the largest container sizes and with relatively wide spacings, containerized alder tend to become overcrowded and to develop excessive H/D ratios. Along with sheltered conditions in the greenhouse, this may produce seedlings that are less resilient to environmental extremes.

**Plug+0.5 or plug-transplant.** Seedlings are initially grown in small containers during the spring and transplanted to open nursery beds later during the growing season. Germination and establishment of seedlings in the greenhouse allow greater control over establishment conditions; transplanting eliminates unpredictability of final spacing in the nursery bed. This method consistently produces extremely robust, stocky, and well-branched seedlings, but does so with considerably higher costs.

Dobkowski et al. (Chapter 13) and Radwan et al. (1992) provide objective comparisons of field performance among alder stocktypes. Both papers conclude that bare-root alder seedlings grown in open nursery beds (transplanted or sown directly) are generally superior to greenhouse or bed-house seedlings in terms of survival, growth, and resilience to damage. Open-bed conditions more consistently produce stocky seedlings with buds along the entire stem. Large plug-transplant stock may be preferable for harsh sites.

### Seed Extraction, Storage, and Handling

Complete guidelines for cone collection, seed extraction, and seed storage are given by Hibbs and Ager (1989) and Ager et al. (Chapter 11).

### Germination and Seedling Establishment Rates

Pregermination treatments and environmental conditions required for germination of alder seed are covered in detail by Ager et al. (Chapter 11). Red alder has shown high variability between seedlots in rates of seed viability, germination, and seedling establishment (Kenady 1978; Elliot and Taylor 1981; Berry and Torrey 1985). Good estimates of these rates are needed for efficient use of seed (seed collection costs as estimated by Silvaseed Company, Roy, WA; Diamond Wood Products, Eugene, OR; and USDA Siuslaw National Forest, Corvallis, Oregon, are $200 to $500 per pound) and for control over spacing with direct sowing in nursery beds. Control over spacing is crucial; alder seedlings can grow rapidly and overcrowding seriously reduces seedling quality (see the following section).

Many nurseries have reported field germination rates that are substantially different (often higher, occasionally lower) than those determined in standard tests. After germination, seedling establishment and final yield rates are quite dependent on specific nursery conditions.

The best way to obtain good estimates of establishment rates is to determine the actual performance of a given seedlot for a specific nursery practice. This requires repeated use of a large seedlot and consistent cultural practices to determine actual rates of establishment and survival. With small seedlots or infrequent sowing requests for alder, the next best strategy may be to (1) estimate initial rates of germination for each seedlot with test temperatures that represent probable field conditions, or (2) to estimate average rates of establishment, survival, and yield for alder specific to the nursery and cultural regime.

More work is needed to develop germination tests that better predict field performance. One method suggested for realistic estimates of germination/establishment is to sow a sample of seeds into sand flats and count only those germinants that survive past the cotyledon stage (Berry and Torrey 1985). Temperature regimes maintained during germination tests should represent probable conditions for a given cultural method (i.e., higher temperatures for greenhouse or late spring sowing,

Table 1. Seedling yields and sowing rates for 1+0 open-bed, bare-root red alder seedlings grown at the U.S. Forest Service Humboldt (McKinleyville, California) and Wind River (Carson, Washington) nurseries.

| | Seed sown (number x 1000) | Germination (estimated % seed sown) | Bed density (gross number seedlings m$^{-2}$) | Seedling yield[1] (number x 1000) Gross | Net | Seedling yield[1] (% seed sown) Gross | Net |
|---|---|---|---|---|---|---|---|
| **Humboldt nursery** | | | | | | | |
| 1986 | 2205 | — | 678 | 492 | 146 | 22 | 6 |
| 1987 | 992 | — | — | — | 132 | — | 13 |
| 1988 | 472 | 45 | 667 | 117 | 70 | 25 | 15 |
| 1989 | 630 | 59 | 323 | 117 | 109 | 19 | 17 |
| 1989 | 454 | 67 | 667 | 183 | 80 | 40 | 18 |
| 1990 | 504 | — | — | — | 87 | — | 17 |
| | | | | | | | |
| **Wind River nursery** | | | | | | | |
| 1988 | 52 | 35 | 172 | — | 5 | 10 | — |
| 1989 | 277 | 15 | — | 4 | 3 | 2 | 1 |
| 1990 | 73 | 56 | 226 | 50 | 16 | 69 | 22 |
| 1990 | 213 | 34 | 269 | 67 | — | 31 | — |
| 1990 | 139 | 40 | — | 18 | 10 | 13 | 7 |
| 1990 | 113 | 45 | 151 | 27 | 13 | 23 | 13 |
| 1990 | 227 | 50 | 312 | 136 | — | 60 | — |

[1]Yield: gross = total number of seedlings lifted, net = number of seedlings passing grade standards.

lower temperatures for early field sowing). Seedling yields obtained from two U.S. Forest Service nurseries in the Pacific Northwest are summarized in Table 1.

*Sowing Methods*

For direct sowing in open nursery beds, small-seed drill machines, such as the Oyjord (J. E. Love Co., Garfield, WA) seed drill, have been successfully applied at several nurseries. Particularly for small orders, broadcast sowing by hand is also commonly used. Research is needed to develop sowing methods and to adapt equipment to increase accuracy in sowing rate and uniformity of spacing within drill rows. Pelletized alder seed has been sown successfully by machine in the field and in greenhouse containers.

In containerized systems, alder seed is often hand sown directly into containers. Because of the difficulty in predicting germination/establishment rates, sowing in germination flats and transferring germinants to containers has also been a common practice to ensure efficient use of seed and greenhouse space.

Seed is sown on the surface and held in place by a light covering of crushed rock or peat (not soil). When seed is sown directly into outdoor beds, it should be further secured by reemay cloth (available from Ken-Bar Inc., 24 Gould St., Reading, MA 01867), burlap, or plastic netting. The covering holds the light seeds and delicate germinants in place and protects them from desiccation and disturbance by wind and rain. This covering should not block the light completely and should be removed before germinants are large enough to be obstructed by the covering.

*Inoculation with Frankia*

The symbiosis between alder roots and the nitrogen-fixing actinomycete *Frankia* is discussed in detail by Molina et al. (Chapter 2). Adequate nodulation by *Frankia* may be necessary for good growth and vigor after outplanting (Dommergues 1982; McNeill et al. 1989; Molina et al., Chapter 2). Artificial inoculation with *Frankia* may also be necessary to obtain consistent establishment and growth of alder seedlings in the nursery. In con-

tainerized production systems, inoculation of soil media with *Frankia* can be essential (Berry and Torrey 1985). In nursery beds, the need for inoculation depends on the abundance of natural inoculum from native alder and on the impact of soil fumigation on inoculum viability (Hilger et al. 1991). If seedlings do not become nodulated from either natural or artificial sources, seedling performance may be poor (Stowers and Smith 1985; Koo 1989; Martin et al. 1991).

Hilger et al. (1991) demonstrated a 42-percent increase in the yield of acceptable seedlings following artificial inoculation with *Frankia* in beds that had been previously fumigated with methyl bromide-chlorpicrin. A second study in the same nursery (Weyerhaeuser Co. Nursery, Mima, WA) demonstrated an increase in effectiveness of inoculum when the *Frankia* culture was mixed with a peat carrier (Martin et al. 1991). Natural inoculation eventually occurred on uninoculated treatments in these studies; the major benefit of artificial inoculation in field beds was the result of early and more consistent nodulation.

No effect on seedling size or yield was produced with application of pure culture *Frankia* in fumigated beds at the Wind River Nursery (Table 2) with methods and rates similar to the intermediate inoculation level used by Martin et al. (1991). In the Wind River study, seedlings in all treatments appeared to be nodulated by midseason. Although natural inoculum levels were not assessed, background levels of inoculum probably provided adequate nodulation, in spite of the fumigation treatment.

Many nurseries have not employed artificial inoculation with *Frankia* (J. McGrath, U.S. Forest Service, Wind River Nursery, pers. com.; J. Nelson, U.S. Forest Service, Humboldt Nursery, pers. com.). Based on observations made at the end of the growing season, these nurseries have not had obvious problems as a result of inadequate nodulation; natural inoculation has taken place in most situations. Nonetheless, the results of Martin et al. (1991) suggest that there is a risk of poor performance in fumigated beds if artificial inoculation is not employed. Some of the substantial variation in seedling quality commonly observed in a given crop may result from inconsistent nodulation early in

Table 2. Size and yield of 1-year-old red alder seedlings grown in fumigated nursery beds inoculated with pure culture Frankia or fresh alder root nodule extract, and uninoculated control. Standard error of the mean in parentheses*

| Treatment | Basal diameter (mm) | Height (cm) | Density (seedlings m$^{-2}$) |
|---|---|---|---|
| Inoculated with pure culture | 4.64 ($\pm$0.07) | 46.8 ($\pm$0.5) | 231.1 ($\pm$16.2) |
| Inoculated with nodule extract | 4.71 ($\pm$0.10) | 49.4 ($\pm$1.3) | 230.0 ($\pm$13.5) |
| Control | 4.66 ($\pm$0.13) | 47.1 ($\pm$1.4) | 211.7 ($\pm$1.4) |

*G. R. Ahrens and J. McGrath, unpublished data. Methods were as follows: The nursery soil was fumigated on 7 Sept. 1989 with 66% methyl bromide and 33% chloropicrin at a rate of 325 kg ha$^{-1}$. Treatments were applied (9 May 1990) to 1.22 x 1.83-m plots randomly selected from one 36-m section of nursery bed. A 0.5-cm layer of peat was spread over the soil surface of each plot. Treatments were applied as 2.7 l liquid sprayed over each plot as follows: control—pure distilled water; pure culture inoculation—102.5 ml packed cell volume (pcv) in distilled water, which produces a rate of 46 ml pcv per m$^2$ of soil (approximating the intermediate rate in Martin et al. 1991).

After spraying, the peat was raked into the surface 10 cm of the soil and irrigated lightly. Each treatment was replicated three times. Seed was sown the next day. Measurements were made (November 1990) on seedlings in four 20 x 50-cm quadrats randomly located in each treatment plot.

the season. Frequency and intensity of fumigation, along with the proximity of natural inoculum in native alder stands, are factors that should be assessed to evaluate the risks when growing alder for the first time.

Artificial inoculation with *Frankia* can be achieved with crushed nodule homogenates or pure culture inoculum. Methods for preparation and application of inoculum are described in detail by Molina et al. (Chapter 2) and Martin et al. (1991). Inoculum should be applied at the time seed is sown.

### Seedling Spacing and Seedbed Density

Selecting and achieving target nursery bed densities are extremely important and provide a challenge to alder nursery practice. Seedbed density and uniformity of spacing may have the greatest influence of any practice on alder seedling quality. Williams and Hanks (1976) recommended seedbed densities of 110–160 seedlings m$^{-2}$ for *Alnus glutinosa* and other fast-growing hardwood species. Radwan et al. (1992) recommended densities of about 65 m$^{-2}$ for 1+0 alder sown in open beds. Typical seedbed densities recorded for red alder have often been much higher (Table 1), because of underestimates of field germination and a lack of familiarity with appropriate densities for fast-growing alder.

In general, uniform spacing and lower seedbed densities produce seedlings with greater average diameter, root mass, and shoot mass (Duryea 1984). For red alder, lower densities produce sturdier seed-

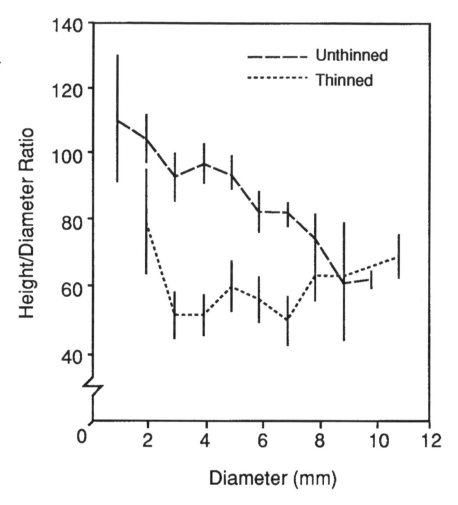

*Figure 5. Height/diameter ratio versus diameter for 1+0 alder seedlings in thinned (140 seedlings m⁻²) and unthinned (206 seedlings m⁻²) nursery beds (same samples as shown in Figs. 2A and 2B). Thinning also resulted in more uniform spacing.*

lings (lower H/D) and lower S/R (Figs. 1; Radwan et al. 1992). Lower density and increased uniformity of spacing within drill rows substantially increases stockiness (reduces H/D), particularly for smaller diameter seedlings (Fig. 5). Uniformity of spacing, particularly within drill rows, is an important aspect of seedbed spacing. Clumped distributions commonly result from sowing small-seeded alder with common seed-drill machines. Lower densities can also reduce damage from insects and disease, particularly fungal pathogens that thrive under poorly ventilated, high-density canopies.

The superior quality of plug-transplant seedlings is in large part the result of the low density (32 to 64 m⁻²) and uniform spacing that can be consistently achieved with this method. Maintenance of low densities and uniform spacing with direct sowing in nursery beds produces good quality seedlings at a much lower cost than that achieved by transplanting. Radwan et al. (1992) showed that alder seedlings grown at low density (108 m⁻²) had larger average diameter (4.1 mm) and lower H/D (76) compared to stock grown at higher density (269 m⁻², which had a diameter of 3.2 mm and an H/D = 94). No significant differences, however, were found in field performance of high-grade seedlings from the two different densities.

Because higher densities allow more efficient use of nursery space, optimal densities need to be determined by some estimate of the relative trade-off between seed and seedling production costs and stock quality. Particularly with alder, the importance of the specific outplanting site in determining true quality (performance) should be stressed. For example, small plug seedlings performed exceptionally well (height growth year 1 = 83 cm; year 2 = 170 cm) on a good site in northwest Washington. Alternatively, based on studies in southwest Washington (Dobkowski et al., Chapter 13), these seedlings would have a high risk of damage or mortality on harsher sites. Whenever possible, to balance seedling costs with minimum quality needs, minimum grading standards should be set with specific outplanting conditions in mind.

Optimal densities depend on final crop size, which can vary with climate and cultural regime. Given the target height of 30 to 100 cm, primary control over plant size should be attained by determining the necessary length of growing season (sow date) for a typical year at a given nursery site. This can be difficult because alder can grow very rapidly (1 to 2 cm per day) when climatic conditions are favorable. Some adjustments for seasonal variation can be made with irrigation and fertilization, or with mechanical methods such as wrenching or undercutting. Top pruning is not recommended for alder because it may promote multiple tops and increase susceptibility to disease.

## Irrigation

Because alder has been a minor species for most nurseries, the standard irrigation and fertilization regimes for major crop species are often applied to it, with some supplemental irrigation during germination and establishment. Specific operational irrigation and fertilization regimes are not well developed for alder. General guidelines may be derived from information on alder's response to water stress and nutrient deficiencies under experimental conditions (Radwan and DeBell, Chapter 15; Shainsky et al., Chapter 5).

Desiccation and heat are major factors limiting establishment of alder from seed under natural conditions (Haeussler 1988). Frequent irrigation may be particularly important to prevent desiccation and heat damage of small, surface-sown seed or germinants during germination and establishment.

Alder is a relatively drought-sensitive hardwood (Pezeshki and Hinkley 1988; Shainsky et al., Chapter 5), as indicated by its preference for coastal and riparian environments. Alder avoids water stress via rapid, extensive root development. Reduced leaf water potential induces stomatal closure and limits photosynthesis and transpiration in alder below a threshold level of about -1.1 MPa, although this effect is less apparent with high humidity (Pezeshki and Hinckley 1982; Shainsky et al., Chapter 5). Seedlings under field conditions, however, do not show midday plant water potential (PWP) below -1.4 MPa (minimum predawn PWP -0.8 MPa); leaf senescence or seedling mortality occurs under conditions that would induce lower PWP. This leaves a relatively narrow range between stomatal limitation and leaf senescence (from -1.1 to -1.4 MPa). Thus, whereas maintenance of moisture deficits can be used to control seedling size (particularly in drier climatic zones), deficits may be difficult to maintain in such a narrow range without risk of inducing severe stress or leaf senescence.

A strategy for irrigation of alder, then, is to employ more frequent (relative to Douglas-fir regimes) irrigation during germination and establishment. After establishment, maintenance of moderate drought (soil water potential from -0.1 to -0.5 MPa) should promote extensive fibrous root development without inducing severe stress or growth limitation. Frequent but light irrigation to increase humidity should increase stomatal conductance and photosynthesis even under moderate drought. Seedling size may be limited by withholding water, but care must be taken to avoid severe stress (PWP below -1.4 MPa at any time).

## Fertilization

Evidence suggests that nitrogen-fertilization regimes should be moderate to low for alder in order to promote development of nitrogen-fixing nodules. Periodic nitrogen amendments are still necessary for maintenance of adequate nitrogen in intensively cropped nursery soils. High rates of nitrogen fertilization, however, can reduce nodulation and growth of alder seedlings (Koo 1989; Martin et al.

1991). Although fertilization with nitrogen can compensate for poor nodulation to some extent, well-nodulated seedlings are most desirable for outplanting (Dommergues 1982; Koo 1989; McNeill et al. 1989; Molina et al., Chapter 2).

Alder is sensitive to phosphorus deficiencies; maintaining adequate supplies of available phosphorus is essential for establishment, growth, and nodulation (Hughes et al. 1968; Radwan 1987; Koo 1989; Radwan and DeBell, Chapter 15). Radwan et al. (1992) state that fertilization with phosphorus and lime is essential for production of quality alder stock. Examples of successful fertilizer regimes for both container and field nursery culture are given by Radwan and DeBell (Chapter 15).

Substantial reductions in alder seedling growth have been associated with low soil pH (4.7), although the factor limiting growth was not conclusively identified (Cole et al. 1989). Problems associated with acid soils should not be significant in the more moderate pH range typical of most nursery soils.

## Insects

A wide variety of insects feed on alder foliage and twigs; as with many deciduous hardwoods, evidence of such feeding is common towards the end of the growing season. Documentation of problems with insect pests in alder nursery crops is sparse. Destructive outbreaks of several species have been noted in natural stands and plantations (Worthington et al. 1962; Borden and Dean 1971; Gara and Jaeck 1978). Whereas mortality has been rare in older stands, the potential for mortality or serious damage may be great if a local outbreak spreads to nursery crops. Important pests are listed in Harrington et al. (Chapter 1).

The western tent caterpillar (*Malacosoma californicum* Packard) and the alder flea beetle (*Altica ambiens* Lec.) have caused significant mortality in young plantations and dense young thickets of alder, particularly where drought stress also develops (G. Ahrens, pers. obs. on young research plantations). Aphid infestations were shown to reduce basal area growth by as much as 34 percent in one test plantation (Schroeder 1984). Biological control agents (*Bacillus thuringiensis* Berliner, numerous parasites, viruses, fungi, and

bacteria) are often effective at preventing serious damage from the western tent caterpillar (Morris 1969, Gara and Jaeck 1978). Little is known about effective control methods for the flea beetle, whose damage is more severe where plantation density is high and tree crowns overlap, as is the case in nurseries. Aphid outbreaks were effectively controlled with Diazinon® on small plots in one test plantation (Schroeder 1984).

## Diseases

Important pathogens infecting alder in the nursery include *Septoria alnifolia*, *Botrytis* spp., and various damping-off fungi. *Septoria* leaf spot and stem canker has caused serious losses. *Septoria* in a bare-root crop decreased the yield of plantable seedlings by 38 percent (Y. Tanaka and W. Littke, Weyerhaeuser Company, unpub. data). This disease is generally present in native alder, which provides a constant source of infection for nurseries. The fungus first appears in necrotic spots on leaves and, if conditions are favorable, spores produced on the leaf spots will infect stems later in the season. High seedling density and poor air circulation under the foliage promote development of stem cankers which can completely girdle seedlings. Detection, prevention, and/or control of this disease are essential in order to avoid serious losses. Effective control has been demonstrated with monthly application of benomyl (Frankel 1990), although the registered form of this chemical (Benlate,® DuPont) is no longer available. Other fungicides with a similar mode of action are being tested. Preliminary results indicate that monthly application of Cleary 3336-F (Topsin M, Grace Sierra, Inc.) provides effective control of *Septoria* (S. Frankel, U.S. Forest Service, Pacific Southwest Region, pers. com.). Maintaining lower seedbed densities and better ventilation also reduces problems with this disease.

*Botrytis* top-kill has also been common on nursery crops of alder. Lower seedbed densities and adequate air circulation generally prevent serious damage. As with *Septoria,* benomyl was an effective treatment.

Serious problems with damping-off fungi are often prevented with fumigation of seedbeds, although these fungi still can be a problem with alder, because of the need for frequent irrigation during germination and establishment. Close monitoring of surface moisture and careful regulation of irrigation may be necessary to provide adequate moisture without promoting development of damping-off fungi. Measures that may reduce the occurrence of damping-off include irrigation in the morning (vs. afternoon), maintenance of slightly acid pH and low ratios of carbon to nitrogen, and avoidance of poorly rotted manures (Williams and Hanks 1976).

## Protection from Freezing

Significant damage and losses in seedling yield have occurred as a result of frost or extreme cold. A large proportion of the 1990 crop of alder seedlings was damaged or killed by freezing at several nurseries (J. McGrath, U.S. Forest Service, Wind River Nursery, pers. com.; A. Dobkowski, Weyerhaeuser Co., pers. com.).

In general, effective protection from frost may be achieved with overhead sprinkler irrigation systems. Little information is available concerning specific applications for protecting red alder. Regan (1988) provided a good review of key factors to consider when using irrigation to protect nursery stock from cold. Key variables include the timing, rate, and duration of irrigation; appropriate settings depend on the expected temperature and wind conditions. It is also important to achieve uniform coverage of all seedlings in the nursery bed. Improper irrigation can increase the level of damage to nursery stock.

Effective systems for frost protection may be essential to avoid excessive damage to red alder. In extreme cases, nursery sites with high risks of unseasonable frost should be avoided. It may be possible to prevent frost damage through development of nursery practices that induce cold hardiness early in the fall.

## Lifting and Storage

An optimal lifting date of mid-January has been recommended on the basis of work with seedlings grown in southwest Washington (Ahrens et al. 1992; Y. Tanaka, Weyerhaeuser Co., pers. com.). Quantitative relationships, however, between chilling degree-days, dormancy, and seedling performance attributes are not available. Such information is needed to assess optimal lifting dates in abnormal years or at different latitudes, and to assess risks involved with early or late lifting necessitated by operational limitations.

Storage temperatures of either +2°C or -2°C are recommended. Testing has shown little difference in vigor between cooler versus freezer storage; however, storage below freezing ensures more complete dormancy and reduces the tendency for planted alder to break bud too quickly after planting. Alder should be stored in sealed bags to prevent desiccation and physical damage to exposed parts.

## Summary

Work on determining key attributes which define alder seedling quality needs to continue. Further, this information needs to be used to develop or modify existing nursery practices so that they consistently produce the best range of these attributes. Indicators of stockiness, such as height/diameter ratio or bud density, are sensitive to the substantial variation in average size or bed densities between nurseries and years. Effective grading of alder stock must include such indicators, in addition to absolute size standards. Further research is necessary to identify moisture and temperature regimes and lifting/storage schedules that ensure optimal physiological quality of seedlings that meet more easily observed morphological criteria.

Quality alder seedlings can be produced in the greenhouse or in nursery beds. Control over spacing and uniformity of spacing are essential for consistent production. More work is needed to develop accurate methods for predicting germination, establishment, and survival rates for different seedlots and nursery locations. In addition, aggressive pest prevention and control programs are recommended.

## Acknowledgment

Much of the information summarized in this chapter was obtained with the assistance and cooperation of various nursery personnel who have been growing red alder. J. Nelson (U.S. Forest Service, Humboldt Nursery) and J. McGrath (U.S. Forest Service, Wind River Nursery) contributed valuable observations and provided for a variety of nursery treatments and measurements. B. Fangen (Washington Department of Natural Resources, Webster Nursery) and P. Morgan (Oregon Department of Forestry, Elkton Nursery) contributed their observations and recommendations concerning red alder. Y. Tanaka (Weyerhaeuser Co.) provided technical review.

## Literature Cited

Ahrens, G. R., A. Dobkowski, and D. E. Hibbs. 1992. Red alder: guidelines for successful regeneration. Forest Research Laboratory, Oregon State Univ., Corvallis, Spec. Publ. 24.

Berry, A. M., and J. G. Torrey. 1985. Seed germination, seedling inoculation and establishment of *Alnus* spp. in containers in greenhouse trials. Plant and Soil 87(1):161-173.

Borden, J. H., and W. F. Dean. 1971. Observations on *Eriocampa ovata* L. infesting red alder in southwestern British Columbia. Can. Entomol. 101:870–878.

Bunting, W. R. 1980. Seedling quality: growth and development—soil relationships, seedling growth and development, density control relationships. *In* Proceedings, North American Forest Tree Nursery Soils Workshop, 28 July–1 Aug. 1980, Syracuse, New York. *Edited by* L. P. Abrahamson and D. H. Bickelhaupt. State Univ. of New York, Syracuse. 21–42.

Cleary, B. D., R. D. Greaves, and P. W. Owston. 1978. Seedlings. *In* Regenerating Oregon's forests. *Edited by* B. D. Cleary, R. D. Greaves, and R. K. Hermann. Oregon State Univ. Extension Service, Corvallis. 63–97.

Cole, D. W., J. Compton, H. Van Miegroet, and P. Homann. 1989. Changes in soil properties and site productivity caused by red alder. *In* IUFRO Symposium on Management of Nutrition in Forests Under Stress, 18–21 September 1989, Frieberg, Germany.

DeBell, D. S., and M. A. Radwan. 1979. Growth and nitrogen relations of coppiced black cottonwood and red alder in pure and mixed plantings. Bot. Gaz. (Suppl.) 140:97–101.

Dommergues, Y. 1982. Ensuring effective symbiosis in nitrogen-fixing trees. *In* Biological nitrogen fixation technology for tropical agriculture, Proceedings of a workshop, 9–13 March 1981, Cali, Colombia. *Edited by* P. H. Graham and S. Harris. CIAT Press, Cali, Colombia. 395–411.

Duryea, M. L. 1984. Nursery cultural practices: impacts on seedling quality. *In* Forest nursery manual: production of bareroot seedlings. *Edited by* M. L. Duryea and T. D. Landis. Martinus Nijhoff/Dr. W. Junk Publishers, The Hague/Boston/Lancaster, for Forest Research Laboratory, Oregon State Univ., Corvallis. 143–163.

Duryea, M. L., *ed.* 1985. Evaluating seedling quality: principles, procedures, and predictive abilities of major tests. Proceedings of a workshop, 16–18 October 1984, Corvallis, OR. Forest Research Laboratory, Oregon State Univ., Corvallis.

Duryea, M. L., and T. D. Landis, *eds.* 1984. Forest nursery manual: production of bareroot seedlings. Martinus Nijhoff/Dr. W. Junk Publishers, The Hague/Boston/Lancaster, for Forest Research Laboratory, Oregon State Univ., Corvallis.

Elliot, D. M., and I. E. P. Taylor. 1981. Germination of red alder (*Alnus rubra*) seed from several locations in its natural range. Can. J. For. Res. 11:517–521.

Farmer, R. E., Jr. 1975. Dormancy and root regeneration of northern red oak. Can. J. For. Res. 5:176–185.

Frankel, S. 1990. Evaluation of fungicides to control *Septoria* leaf spot on white alder at Humboldt nursery. USDA For. Serv., Pacific Southwest Reg., Rep. R90-02.

Gara, R. I., and L. L. Jaeck. 1978. Insect pests of red alder: potential problems. *In* Utilization and management of alder. *Compiled by* D. G. Briggs, D. S. DeBell, and W. A. Atkinson. USDA For. Serv., Gen. Tech. Rep. PNW-70. 265–269.

Gordon, J. C., C. T. Wheeler, and D. A. Perry, *eds.* 1979. Symbiotic nitrogen fixation in the management of temperate forests. Oregon State Univ. Press, Corvallis.

Haeussler, S. 1988. Germination and first-year survival of red alder seedlings in the central Coast Range of Oregon. M.S. thesis, Oregon State Univ., Corvallis.

Heilman, P. E. 1990. Growth of Douglas-fir and red alder on coal spoils in western Washington. Soil Sci. Soc. Am. J. 54:522–527.

Hibbs, D. E., and A. A. Ager. 1989. Red alder: guidelines for seed collection, handling, and storage. Forest Research Laboratory, Oregon State Univ., Corvallis, Spec. Publ. 18.

Hilger, A. B., Y. Tanaka, and D. D. Myrold. 1991. Inoculation of fumigated nursery soil increases nodulation and yield of bare-root red alder (*Alnus rubra* Bong.). New For. 5:35–42.

Hughes, D. R., S. P. Gessel, and R. B. Walker. 1968. Red alder deficiency symptoms and fertilizer trials. *In* Biology of alder. *Edited by* J. M. Trappe, J. F. Franklin, R. F. Tarrant, and G. M. Hansen. USDA For. Serv., PNW For. Range Exp. Sta., Portland, OR. 225-237.

Kenady, R. M. 1978. Regeneration of red alder. *In* Utilization and management of alder. *Compiled by* D. G. Briggs, D. S. DeBell, and W. A. Atkinson. USDA For. Serv., Gen. Tech. Rep. PNW-70. 183–192.

Koo, C. D. 1989. Water stress, fertilization, and light effects on the growth of nodulated mycorrhizal red alder seedlings. Ph.D. thesis, Oregon State Univ., Corvallis.

Lavender, D. P. 1989. Characterization and manipulation of the physiological quality of planting stock. *In* Physiology and genetics of reforestation. Proceedings 10th North Am. For. Biol. Workshop, 10–22 July 1988, Vancouver, B.C. *Edited by* J. Worrall, J. Loo-Dinkins, and D. P. Lester. Univ. of British Columbia, Vancouver.

Martin, K. J., Y. Tanaka, and D. D. Myrold. 1991. Peat carrier increases inoculation success with *Frankia* on red alder (*Alnus rubra* Bong.) in fumigated nursery beds. New For. 5:43-50.

McNeill, J. D., M. K. Hollingsworth, W. L. Mason, A. J. Moffat, L. J. Sheppard, and C. T. Wheeler. 1989. Inoculation of *Alnus rubra* seedlings to improve seedling growth and forest performance. Great Britain For. Comm. Res. Div., Res. Inf. Note 144.

Mexal, J. G., and T. D. Landis. 1990. Target seedling concepts: height and diameter. *In* Target seedling symposium. Proceedings, Combined Meeting of the Western Forest Nursery Associations, 13–17 August 1990, Roseburg, OR. *Edited by* R. Rose, S. J. Campbell, and T. D. Landis. USDA For. Serv., Gen. Tech. Rep. RM-200. 17-35.

Morris, O. N. 1969. Susceptibility of several forest insects of British Columbia to commercially produced *Bacillus thuringiensis*. II. Laboratory and field pathogenicity tests. J. Invertebr. Pathol. 13(2):285—295.

Perinet, P., J. G. Brouillette, J. A. Fortin, and M. LaLonde. 1985. Large scale inoculation of actinorhizal plants with *Frankia*. Plant and Soil 87:175—183.

Pezeshki, S. R., and T. M. Hinckley. 1982. The stomatal response of red alder and black cottonwood to changing water status. Can J. For. Res 12:761–771.

———. 1988. Water relations characteristics of *Alnus rubra* and *Populus trichocarpa:* responses to field drought. Can. J. For. Res. 18:1159–1166.

Puttonen, P. 1989. Criteria for using seedling performance potential tests. New For. 3:67—87.

Radwan, M. A. 1987. Effects of fertilization on growth and foliar nutrients of red alder seedlings. USDA For. Serv., Res. Pap. PNW-375.

Radwan, M. A., and D. S. DeBell. 1981. Germination of red alder seed. USDA For. Serv., Res. Note PNW-370.

Radwan, M. A., Y. Tanaka, A. Dobkowski, and W. Fangen. 1992. Production and assessment of red alder planting stock. USDA For. Serv., Res. Pap. PNW-450.

Regan, R. 1988. Sprinkler salvation. Overhead sprinkler systems can provide an added measure of cold protection for nursery stock. American Nurseryman 168(5):70—77.

Rietveld, W. J., and R. D. Williams. 1978. Dormancy and root regeneration of black walnut seedlings: effects of chilling. USDA For. Serv., Res. Note NC-244.

Ritchie, G. A. 1984. Assessing seedling quality. *In* Forest nursery manual: production of bareroot seedlings. *Edited by* M. L. Duryea and T. D. Landis. Martinus Nijhoff/Dr. W. Junk Publishers, The Hague/Boston/Lancaster, for Forest Research Laboratory, Oregon State Univ., Corvallis. 243-257.

Roller, K. J. 1977. Suggested minimum standards for containerized seedlings in Nova Scotia. Can. For. Serv., Dept. Environ. Inf. Rep. M-X-69.

Rose, R., S. J. Campbell, and T. D. Landis, *eds.* 1990a. Target seedling symposium. Proceedings, Combined Meeting of the Western Forest Nursery Associations, 13–17 August 1990, Roseburg, OR. USDA For. Serv., Gen. Tech. Rep. RM-200.

Rose, R., W. C. Carlson, and P. Morgan. 1990b. The target seedling concept. *In* Target seedling symposium. Proceedings, Combined Meeting of the Western Forest Nursery Associations, 13–17 August 1990, Roseburg, OR. *Edited by* R. Rose, S. J. Campbell, and T. D. Landis. USDA For. Serv., Gen. Tech. Rep. RM-200. 1–8.

Schroeder, P., *ed.* 1984. Cultural treatment of selected species for woody biomass production in the Pacific Northwest. Final report. Seattle City Light, Seattle, WA.

Stowers, M. D., and J. E. Smith. 1985. Inoculation and production of container-grown red alder seedlings. Plant and Soil 87:153–160.

Sutton, R. F. 1979. Planting stock quality and grading. For. Ecol. Manage. 2:123–132.

Tanaka, Y., P. J. Brotherton, A. Dobkowski, and P. C. Cameron. 1991. Germination of stratified and non-stratified seeds of red alder at two germination temperatures. New For. 5:67–75.

Tarrant, R. F. 1961. Stand development and soil fertility in a Douglas-fir—red alder plantation. For. Sci. 7:238-246.

Thompson, B. E. 1985. Seedling morphological evaluation— what you can tell by looking. *In* Evaluating seedling quality: principles, procedures, and predictive abilities of major tests. Proceedings of a workshop, 16–18 October 1984, Corvallis, OR. *Edited by* M. L. Duryea. Forest Research Laboratory, Oregon State Univ., Corvallis. 59–71.

Webb, D. P., and F. W. von Althen. 1980. Storage of hardwood planting stock: effects of various storage regimes and packaging methods on root growth and physiological quality. New Zealand J. For. Sci. 10(1):83–96.

Williams, R. D., and S. H. Hanks. 1976. Hardwood nurseryman's guide. USDA For. Serv., Agric. Handb. 473. Washington, D.C.

Worthington, N. P., R. H. Ruth, and E. E. Matson. 1962. Red alder: its management and utilization. USDA For. Serv., Misc. Publ. 881.

# 13

# Red Alder Plantation Establishment

ALEXANDER DOBKOWSKI, PAUL F. FIGUEROA,
& YASUOMI TANAKA

Red alder (*Alnus rubra* Bong.) is one of the few quality hardwoods that can be grown to a high value commodity with a relatively short rotation of 35 to 40 years. In addition to its lumber value (Helmuth 1988), red alder is a desirable pulpwood species (Hrutfiord 1978). These facts coupled with a projected declining supply of alder and a likely increase in product demand give some land owners an optimistic view of the value of dedicating land to red alder production.

Red alder is a pioneer hardwood that can readily invade a recently logged area if the requirements for seedling establishment are met. Successful natural regeneration of red alder is dependent upon having a bare mineral soil seedbed, a well-distributed seed source, and adequate spring rainfall (Kenady 1978; Haeussler 1988; Ahrens et al. 1992). These conditions can vary considerably within and between logged areas, as well as over time, making it difficult to achieve uniform regeneration with certainty. In addition, past management practices may have excluded red alder seed sources on many areas otherwise suitable for alder production.

As a result of these limitations, natural seeding generally is not reliable for producing well-stocked and evenly distributed stands of red alder (even if scarification is used to expose mineral soil and prepare the seedbed). On a commercial scale, artificial regeneration through the planting of seedlings enables a land manager to achieve stands with desirable uniformity on sites best suited for red alder production.

Successful red alder plantation establishment is dependent upon an understanding of the requisites of the species for survival and growth. Three relationships of alder performance with site and environmental factors have been identified (as cited in Haeussler et al. 1990):

1. Naturally occurring red alder is generally tolerant of poorly drained soils (Krajina et al. 1982).

2. Young red alder can be adversely affected by drought stress (Chan et al. 1988; Pezeshki and Hinckley 1988) and spring frost (Peeler and DeBell 1987)

3. Red alder has low shade tolerance (Krajina et al. 1982; Chan et al. 1988).

These factors need to be carefully considered when selecting the appropriate regeneration strategy.

Over the past five years, the experimentation and operational experience of Weyerhaeuser Company, the USDA Forest Service, Oregon State University Department of Forest Science, and others have resulted in considerable progress in understanding the prerequisites for red alder plantation establishment. Proper site selection, quality seedlings, thorough site preparation, and outplant timing have been shown to be the keys to a successful red alder plantation.

## Site Selection

Red alder occupies sites with a range of soil and physiographic conditions (Harrington 1986; Hibbs and DeBell, Chapter 14). Well-drained upland or alluvium sites tend to be more productive than poorly drained, frost-prone, or droughty sites (Newton et al. 1968; Fergerson et al. 1978; Harrington 1986). Furthermore, the risk of plantation failure can be very high on poorly drained, frost-prone, or droughty sites.

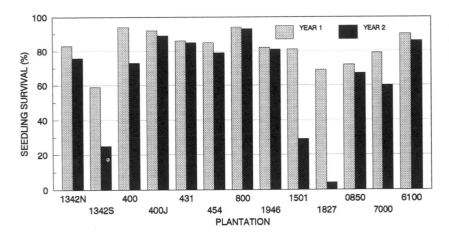

*Figure 1. Red alder seedling survival for the first and second years of plantation establishment on Weyerhaeuser Company ownership in western Washington (A. Dobkowski, unpub. data 1990).*

## Drainage

Although naturally occurring red alder can tolerate poorly drained soils (Krajina et al. 1982), careful examination shows that it occupies only the better drained microsites within the irregular topography of a wet site (see also Harrington et al., Chapter 1). The establishment of a well-stocked plantation is significantly hindered on wet sites because suitable microsites occur infrequently and are poorly distributed and because newly planted seedlings are adversely affected by poor drainage. Where saturated soils persist into the growing season, poor drainage induces seedling mortality and also severely restricts root growth of those seedlings that survive periodic soil saturation. The diminished root system can predispose newly planted seedlings to later summer drought stress. Given the heavy herbaceous weed communities that can develop on these sites and limited site preparation options, drought stress effects can be compounded, resulting in considerable seedling mortality.

A series of alder plantations established near Toledo, Washington, exemplify this point (Fig. 1). Plantations 800, 1946, 1501, and 1827 were established on soils of the same association that differed primarily by soil drainage. Two years after planting, the plantations on the upland soils (800 and 1946) showed survival rates to 80 and 90 percent, respectively. The sites with poor drainage and impervious clay subsoil showed survival rates of 5 and 30 percent, respectively. Summer drought stress was the primary cause of mortality on these units (Fig. 2).

## Frost

Areas of severe frost hazard should not be regenerated to red alder. These sites are associated with topographic features having a high probability of cold air drainage from higher elevations in the spring and fall seasons. Both late spring and early fall frosts can be detrimental to newly established alder plantations.

The adverse effects of frost on a first-year red alder plantation have been documented for a site near Doty, Washington, which was planted on 30 March 1989 with bare-root seedlings (plantation 7000; Figs. 1,

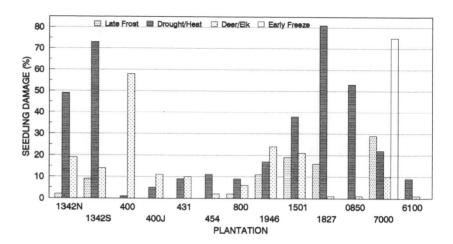

*Figure 2. Incidence of four types of damage to red alder seedlings in first-year plantations established on Weyerhaeuser Company ownership in western Washington (A. Dobkowski, unpub. data 1990).*

2). Cold hardiness testing of the stock at the end of the storage period showed LT10 and LT50 (lethal temperature for 10 and 50 percent of the sample, respectively) to be -9.7°C and -14.4°C, respectively. By 17 April 1989, 100 percent of the planted seedlings had broken bud; average leaf length was 5 to 10 mm. A spring frost occurred on 13 May 1989 (air temperature -2.2 to -1.7 °C for 5 hours) when seedlings were actively growing; 94 percent of them showed an average leaf length of 2.5 to 4 cm. Considerable damage resulted, with 50 percent of the trees in the plantation showing evidence of foliage loss and top-kill. Recovery from the event, however, was good. An assessment made on 28 September 1989 showed 80 percent of the trees free-to-grow with an average tree height of 1 m. Bud set had not yet occurred. On 2 October 1989, an early frost (-6.7 to 2.8°C for 8 hours) severely damaged 75 percent of the stand. The effects of the October freeze event were worsened by subsequent winter weather. By spring, stem-kill to ground level was observed on many trees 1 to 1.5 m in height at the end of the previous growing season. Mortality attributable to freeze damage was 20 percent. Careful examination of the topographic features surrounding the Doty site would lead one to suspect a high probability of cold air drainage from higher elevations. Environmental monitoring at other red alder plantations in the vicinity showed temperatures not falling below freezing for either of these events, indicating that the Doty site was frost-prone.

Other plantings of alder in frost pockets have shown severe damage (complete shoot die-back or plant mortality) with damage incidence rates of 50 to 90 percent (Peeler and DeBell 1987; A. Dobkowski and Y. Tanaka, unpub. data 1991).

Vegetation condition and topographic features can be used to assess the likelihood that a site is in a cold air drainage. Red alder and other naturally occurring woody vegetation can show evidence of previous severe frost events on a given site. Valleys exhibiting a gentle gradient from high elevation areas or topographic features that form blockages tend to slow the flow of cold air and be more frost-prone.

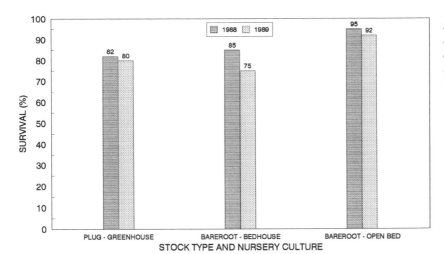

Figure 3. Survival after two years for red alder seedlings produced with different propagation technologies (A. Dobkowski and Y. Tanaka, unpub. data 1990).

Resprouting from root systems after complete stem-kill may provide acceptable survival, but the accumulation of effects from frost events can result in a stand with very poor stem form. When terminal buds are damaged on young red alder, stem sprouts proliferate and lateral branches develop into multiple dominant stems. It is unlikely that normal plantation spacing will inhibit the multistemmed form.

### Drought and Heat Stress

Summer drought and heat stress contribute significantly to reduced performance of newly planted red alder seedlings (Fig. 2). Regeneration difficulties have been particularly noted on droughty sites typified by south-southwest aspects, steep slopes, and coarser textured soils. Red alder test plantations 1342S and 0850 are examples of units with these characteristics (Figs. 1, 2). At plantation age 2, alder seedling survival was 25 and 65 percent respectively in these two plantations.

### Animal Damage

The density of black-tail deer (*Odocoileus hemionus*) and elk (*Cervus elapus*) populations for a locale is also a factor to consider in the site selection process. Red alder provides significant browse for deer and elk from midsummer through early fall (Leslie et al. 1984). Browse damage during the first growing season at plantation 0400 resulted in poor plantation performance (Figs. 1, 2). Top-kill from severe browsing resulted in the loss of seedling apical dominance. Seedlings were overtopped by weeds before regaining good height growth and were then negatively affected by competition. The 400J plantation was the portion of the planting unit not heavily affected by big-game damage and showed excellent performance. Stem barking, trampling, and top breakage (from feeding) by deer and elk also have a negative impact on red alder log quality by promoting multiple stems and probably increasing the incidence of decay. The hazard from stem barking can exist well past the period of plantation establishment.

*Figure 4. Height through two years for red alder seedlings produced with different propagation technologies (A. Dobkowski and Y. Tanaka, unpub. data 1990).*

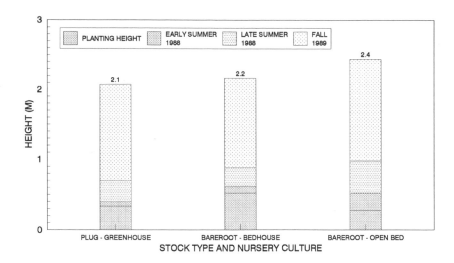

Mountain beaver (*Aplodontia rufa*) and fur beaver (*Castor canadensis*) have been observed damaging red alder plantations. Avoidance of this problem through site selection is possible, although both animals can be controlled with preventative trapping (see also Harrington et al., Chapter 1, for a discussion of animal damage).

Site evaluation based upon soil properties, topographic characteristics, condition of existing woody vegetation, and past experience can be applied to avoid locations that possess a high risk for plantation failure or poor stand quality.

## Planting Stock

As with any reforestation program, quality planting stock is essential if a successful plantation is to be established. Several propagation technologies are available to produce operational quantities of planting stock (Ahrens et al. 1992; Radwan et al. 1992; Ahrens, Chapter 12):

1. Plug-seedlings; greenhouse grown in plastic-foam blocks (82 or 131 cm$^3$) or in single plastic cells (164 cm$^3$).

2. Bare-root bed-house seedlings; seed sown in the nursery bed and grown under a transparent, tentlike covering to facilitate germination and provide shading.

3. Bare-root open-bed seedlings; seed sown in the nursery bed and grown without protective cover.

4. Plug-transplant; a small (33 cm$^3$) plug grown in the greenhouse from March until transplanted in June into a nursery bed where it remains for the rest of the growing season.

Although these technologies differ in production cost, all yield seedlings of suitable quality for outplanting in one growing season.

Morphological characteristics of a seedling that contribute to better survival, growth, and resistance to field rigor are listed in Ahrens (Chapter 12). Although seedlings 30 to 100 cm in height are recommended for superior stock, a larger seedling (60 to 75 cm in height) possessing the attributes of quality stock facilitates lower planting

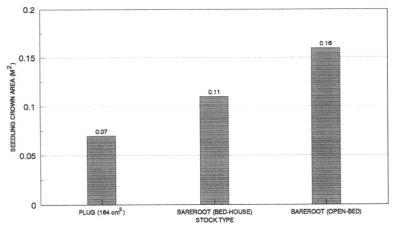

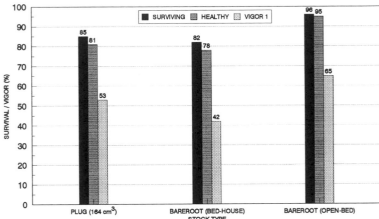

*Figure 5 (above). Crown area after one year for red alder seedlings produced with different propagation technologies (A. Dobkowski and Y. Tanaka, unpub. data 1990). Cross-sectional crown area calculated from crown diameter measurements.*

*Figure 6 (below). Survival and vigor after one year for red alder seedlings produced with different propagation technologies (A. Dobkowski and Y. Tanaka, unpub. data 1990). Healthy seedlings are likely to survive to the free-to-grow stage or are free-to-grow. Vigor 1 seedlings are free-to-grow.*

costs and more uniform plantation stocking. The taller stock allows the planter to readily see planted seedlings enabling the planting line to be followed more effectively. Seedlings with a height greater than 100 cm are difficult to handle and prone to breakage.

Greenhouse- and nursery-grown red alder seedlings are vulnerable to certain diseases. Considerable fall-down in crop yield has been attributed to *Septoria alnifolia* (a leaf-spot fungus that can develop stem cankers) and *Botrytis* species (a gray mold that results in leaf mortality and causes top-kill). Foresters need to recognize the symptoms of these diseases and not accept infected seedlings from the nursery for outplanting. See Ahrens (Chapter 12) for a discussion of seedling diseases in the nursery.

The seedling grading process needs to include an assessment of root systems, stem and root breakage, and overall seedling health. Seedlings with damage to roots that are greater than 2 mm in diameter should be excluded from pack. First-year plantation survival can be increased 10 to 15 percent by controlling culling operations to exclude obviously defective seedlings (A. Dobkowski, unpub. data 1988).

All the stock types listed above have been used to successfully establish red alder plantations; however, bare-root (open-bed) seedlings perform on average, across all site conditions, better than seedlings

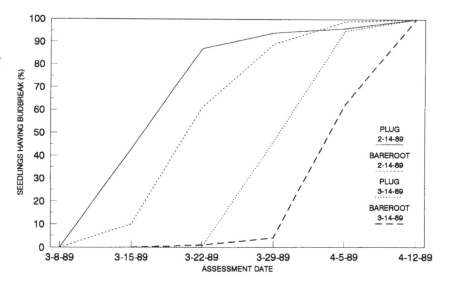

*Figure 7. Budbreak phenology of plug and bare-root red alder seedlings planted mid-February and mid-March (A. Dobkowski and Y. Tanaka, unpub. data 1990). Mean of four replications with a 25-seedling-cluster sample per replication.*

produced by greenhouse and bed-house technologies. Field performance tests at six locations in southwestern Washington showed survival after two years at 92 percent for open-bed bare-root stock as compared to 80 percent for styro-5 plugs and 75 percent for bed-house seedlings (Fig. 3). Height growth for the period was 2.1 m for the open-bed bare-root stock, 1.8 m for plug stock, and 1.6 m for the bed-house bare-roots (Fig. 4). The growing conditions in an open nursery bed, established at a proper seedling density, produce seedlings with healthy buds along nearly the entire length of the stem which then exhibit very rapid first-year growth after outplanting. As a result of this good health, open-bed seedlings showed a rapid accumulation of foliar biomass in the first growing season and better seedling health than other stock types (Figs. 5, 6). Bare-root seedlings tend to deharden and break bud after outplanting at a slower rate than plug seedlings (Fig. 7). This delay in bud break could make bare-root stock less vulnerable to damage from late frosts. Plug and bed-house stock show a greater incidence and severity of drought stress, sun-scald, and heat girdle than open-bed stock (A. Dobkowski and Y. Tanaka, unpub. data 1990).

Initial evaluations of the performance of plug-transplants to plug, open-bed, and bed-house stock types showed some advantages of the plug-transplant over the other stock types (Radwan et al. 1992). First-year height growth was superior for the transplants—from 81 to 89 cm compared to 42, 59, and 35 cm for the plug and other two stock types, respectively. Survival after two years was excellent for all stock types, 89 to 99 percent. Subsequent stock comparison experiments were designed to test field performance of plug-transplant to bare-root stock types for a range of site preparation methods, on sites with contrasting severity, and in the absence of herbaceous weed control. The results did not show a clear advantage in survival of the plug-transplant over the grade 1 bare-root (Table 1); however, there was

*Table 1. First-year plantation survival (%) of red alder plug-transplant and bare-root seedlings outplanted on sites in southwestern Washington with different site preparation methods (A. Dobkowski and Y. Tanaka, unpub. data 1991). Means with the same letter within a plantation are not significantly different (p = 0.05; arc-sine transformation of percentages; Tukeys Method).*

| Preparation[2] | Plantation | Stock Type and Grade[1] | | |
| | | Transplant #1 | Bare-root #1 | Bare-root #2 |
|---|---|---|---|---|
| None | 0447 | 89 a | 86 a | 71 b |
| | 9200 | 90 a | 91 a | 81 a |
| | 1844 | 80 a | 73 ab | 69 b |
| Scarified (Shovel) | 0460 | 96 a | 90 ab | 83 ab |
| | 0471 | 92 a | 84 a | 66 b |
| | 0019 | 84 a | 74 a | 75 a |
| | 9312 | 79 a | 74 a | 55 b |
| | 1827 | 83 a | 84 a | 77 a |
| Burned | 0002 | 62 a | 51 a | 46 a |
| | 0026 | 68 a | 50 a | 62 a |
| | 9400 | 79 a | 83 a | 65 b |
| | 1948 | 81 a | 74 a | 66 a |

[1]Transplants are grown in a greenhouse as small plugs and then transplanted into nursery beds in early summer. Bare-root seedlings are produced by sowing seed in nursery beds and allowing the seedling to develop unprotected. Both stock types are ready for outplanting in one growing season. The standard for grade 1 stock is height greater than 30 cm and stem caliper, 2.5 cm above root collar, greater than 4 mm. Grade 2 stock has height between 20 to 30 cm and stem caliper between 3.5 to 4 mm. Both grades of stock were damage- and disease-free.
[2]No pre-plant herbicides were used to control weed competition.

an advantage of grade 1 bare-root stock (height greater than 30 cm and stem caliper, 2.5 cm above root collar, greater than 4 mm) over grade 2 bare-root stock (height between 20 to 30 cm and stem caliper between 3.5 and 4 cm).

The selection of proper stock is dependent upon the regeneration risks associated with a particular site weighted against site preparation and stock cost. All stock types are vulnerable to a degree to the effects of herbaceous weed competition. Taking into consideration stock cost and field performance, bare-root seedlings produced with open-bed nursery technology are the preferred stock type for reforesting most alder sites.

## Site Preparation

Red alder's intolerance to drought stress, shade, and weed competition had been shown to hinder natural alder regeneration (Worthington et al. 1962; Kenady 1978). A key element of red alder stand management is the control of weed competition. There are currently no broadcast, aerial applied, selective herbicide treatments available for red alder release. All broadcast vegetation control measures must be taken prior to planting.

*Table 2. First-year plantation effects of selected herbicide site preparation treatments for bare-root[1] and plug[2] red alder seedling stock. (Adapted from Figueroa 1988).*

| Herbicide treatment | Weed cover (%) | Survival (%) | | Increase over check (%) Height | | Basal caliper | |
|---|---|---|---|---|---|---|---|
| | | Bare-root | Plugs | Bare-root | Plugs | Bare-root | Plugs |
| Glyphosate/ 2,4-D | 60 | 95 | 80 | 7 | 48 | 18 | 58 |
| Glyphosate/ atrazine | 63 | 93 | 75 | 10 | 14 | 10 | 6 |
| Hexazinone | 65 | 90 | 63 | -2 | 9 | 0 | 2 |
| Glyphosate | 90 | 88 | 58 | -13 | 30 | -1 | 34 |
| Check— no treatment | 166 | 73 | 50 | — | — | — | — |

[1]Bed-house bare-root stock; average seedling height of 50 cm at time of planting. One-year-old planting stock.

[2]Plug seedling (65 cm$^3$ container); average seedling height 23 cm at time of planting. One-year-old planting stock.

The earliest operational trials included tests of mechanical scarification with and without herbicides and tests to evaluate broadcast burning with and without herbicides. These tests were conducted in 1987 on a droughty site in southwestern Washington owned by Weyerhaeuser Company.

Results from these scarification tests showed that survival after one year with no site preparation was 27 percent. Scarification alone and scarification with glyphosate/ 2, 4-D as a spring pre-plant treatment increased survival to 64 percent. Broadcast burning with the pre-plant herbicides showed 89 percent survival. Scarification with or without herbicides failed to achieve adequate survival. The scarification created an ideal seedbed for the invasion of grasses and forbs. The glyphosate/ 2,4-D pre-plant herbicide treatment only controlled vegetation present at the time of application. The burn-plus-herbicide treatment had lower levels of grass competition than the scarified area and was a likely reason for the higher survival.

Results from the broadcast burned site preparation tests showed significant survival and growth losses associated with levels of weed control (Figueroa 1988). Weed cover at the end of the first growing season ranged between 60 and 90 percent for the burned-plus-herbicide treatments, while the burned-only treatment had 166 percent weed cover (Table 2). The effects of reduced weed competition increased survival by 30 percent for bare-root seedlings and 60 percent for plug seedlings. First-year total height and basal caliper showed improvement depending on the level of weed reduction and stock type. These early trials indicated that the level of weed competition was an important factor in determining the survival and growth of red alder seedlings and that stock type was an important link in successful regeneration.

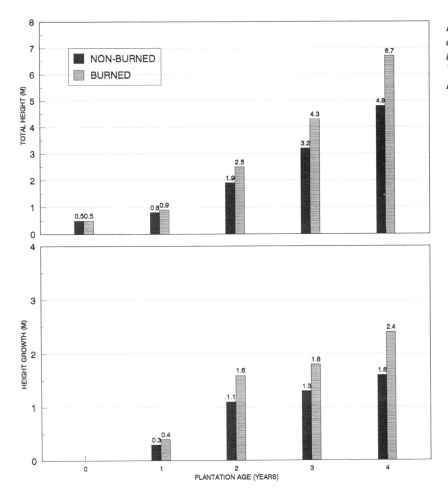

*Figure 8. Red alder total height and height growth for a burned versus non-burned site in the Washington coast range (P. Figueroa, unpub. data 1991).*

Burning alone has been shown to be an effective form of site preparation on sites not prone to drought stress or without significant later weed invasion. Three-year results evaluating alder seedling growth on a comparable burned and non-burned site showed trees on the burned site to be 38 percent larger in basal diameter and 26 percent taller in height (P. Figueroa, unpub. data 1991). Growth trajectories suggest a continuance of divergence in growth over time (Fig. 8). Broadcast burning, however, can promote a good seedbed for the rapid establishment of forbs and grasses on some sites, in which case herbaceous weed competition in the first year has resulted in plantation failures.

Scarification, burning, and site preparation chemicals are methods that can effectively reduce weed communities that exist in the understory of the harvested stand. A major concern of plantation establishment is weed invasion by forbs, grasses, and woody shrubs into newly harvested areas. Primary methods of site preparation, such as scarification and burning, can create favorable seedbeds. Without secondary weed control, competing plants can rapidly invade these areas.

Following the 1987 site preparation trial, a series of trials to assess and define the effects of weed competition on red alder plantation performance were established as part of Weyerhaeuser Company's

*Table 3. First-year vegetation ground cover and second-year mean stand statistics for red alder plantations with and without pre-plant herbicide treatments (P. Figueroa and A. Dobkowski 1990, P. Figueroa 1991, A. Dobkowski and R. Meade 1992,; Weyerhaeuser Company, unpub. data).*

| Unit | Site preparation[A] | Cover[B] (%) | Stocking (tph) | Dbh (cm) | Height (m) | High vigor[C] (%) |
|------|---------------------|--------------|----------------|----------|------------|-------------------|
| 0850 | Burn | 39 a[D] | 899 a | 1.0 a | 2.2 a | 79 a |
|      | Burn & Atrazine/2,4-D | 12 b | 766 a | 1.0 a | 2.1 a | 73 a |
| 1510 | Scarified/Burned Piles | 56 a | 1023 a | 2.3 a | 1.2 a | 92 a |
|      | Scarified/Burned Piles & Atrazine/Glyphosate/2,4-D | 42 a | 988 a | 2.5 a | 1.3 a | 91 a |
| 1080 | Scarified/Burned Piles | 68 a | 852 a | 1.8 a | 0.9 a | 73 a |
|      | Scarified/Burned Piles & Atrazine/Glyphosate/2,4-D | 45 a | 889 a | 2.0 a | 1.0 a | 70 a |
| 6100 | Scarified | 87 a | 820 a | 0.9 a | 2.0 a | 78 a |
|      | Scarified & Atrazine/2,4-D | 30 b | 877 a | 1.3 b | 2.3 b | 90 b |
| 0306 | None | 88 a | 1120 a | 1.7 a | 0.6 a | 98 a |
|      | None & Atrazine/2,4-D/Triclopyr | 71 a | 980 a | 1.5 a | 0.5 b | 92 a |
| 0431 | Burn | 94 a | 1517 a | 1.2 a | 2.4 a | 98 a |
|      | Burn & Atrazine/Glyphosate | 6 b | 1599 a | 1.4 b | 2.7 b | 100 a |
| 0312 | Burn | 96 a | 892 a | 1.8 a | 0.8 a | 95 a |
|      | Burn & Atrazine/2,4-D | 53 b | 1030 a | 2.3 b | 1.1 b | 100 a |
| 1946 | Burn | 110 a | 791 a | 0.7 a | 1.8 a | 74 a |
|      | Burn & Atrazine/2,4-D | 17 b | 1013 b | 0.9 a | 2.1 b | 93 b |
| 1300 | Scarified/Burned Piles | 112 a | 700 a | 1.6 a | 0.6 a | 72 a |
|      | Scarified/Burned Piles & Atrazine/Glyphosate/2,4-D | 48 b | 889 b | 2.2 b | 1.0 b | 93 b |
| 0800 | Burn | 122 a | 297 a | 0.2 a | 1.1 a | 30 a |
|      | Burn & Atrazine/2,4-D | 17 b | 1161 b | 0.9 b | 2.1 b | 95 b |
| 7000 | Scarified | 123 a | 912 a | 0.3 a | 1.2 a | 72 a |
|      | Scarified & Atrazine/Dalapon/2,4-D | 66 b | 694 b | 0.3 a | 1.2 a | 52 a |
| 1501 | Burn | 140 a | 25 a | 0.0 a | 0.1 a | 0 a |
|      | Burn & Atrazine/2,4-D | 82 b | 445 b | 0.1 a | 0.6 b | 30 b |
| 1827 | Scarified | 156 a | 0 | 0.0 | 0.0 | 0 |
|      | Scarified & Glyphosate | 100 b | 49 | 0.0 | 0.1 | 0 |

[A]Herbicides were applied at labeled rates, in water, 2 to 3 weeks prior to planting.
[B]Cumulative percentage ground cover for the following classes of vegetation: grass/sedge; forbs (excluding ferns); woody plants.
[C]High-vigor seedlings show good health and were essentially free-to-grow.
[D]Within a unit, treatment means with the same letter are not significantly different (p = 0.10; t-test).

operational plantation program (Table 3). Results from these plantations suggest a weed competition threshold above which the survival and growth of red alder can be adversely effected. This threshold appears to be 90 to 100 percent ground cover during the first-year. Survival begins to be negatively impacted when weed ground cover is greater than 110 percent, and weed competition that exceeds 125 percent can cause plantation failure. On those sites where severe weed competition does not cause appreciable mortality, it can severely retard alder growth. Delayed site occupancy resulting from slowed early growth may allow subsequent invasion of woody plants, thereby increasing weed competition. Weed ground cover levels below 75 percent have minimal effects on red alder stand development; not until levels are reduced to below 30 percent does a growth response occur. The presence of some weed cover may actually enhance seedling performance on some sites by providing shade to the seedlings lower stem, thereby decreasing the incidence and severity of sunscald or frost damage.

Sites with an expectation of low to moderate weed competition in the first few years can be adequately regenerated with minimal to no site preparation. An example of a harvested unit with low vegetation competition potential is a dense stand of western hemlock (*Tsuga heterophylla*) with little to no vegetation surviving in the

*Table 4. Second-year plantation survival and height growth for red alder plug and bed-house bare-root seedlings out-planted on different dates at sites in southwestern Washington (A. Dobkowski and Y. Tanaka, unpub. data 1989).*

| | | Planting date[A] | | | | | | |
|---|---|---|---|---|---|---|---|---|
| Plantation | Year | Nov. | Dec. | Jan. | Feb. | Mar. | Apr. | May |
| A. Survival (%) | | | | | | | | |
| 1342 | 1987 | — | — | — | 30 c | 65 a | 58 a | 48 b |
| 0431 | 1987-88 | 87 a[B] | — | — | 87 a | 82 a | 90 a | — |
| 1946 | 1987-88 | 78 ab | — | — | 67 b | 80 a | 83 a | — |
| 6100 | 1988-89 | 82 ab | 74 bc | 67 c | 82 ab | 82 ab | 87 a | 77 bc |
| B. Height growth (cm) | | | | | | | | |
| 1342 | 1987 | — | — | — | 51 c | 82 a | 76 a | 62 b |
| 0431 | 1987-88 | 165 b | — | — | 189 a | 172 ab | 174 ab | — |
| 1946 | 1987-88 | 207 b | — | — | 238 a | 242 a | 236 a | — |
| 6100 | 1988-89 | 126 c | 126 c | 123 c | 156 a | 149 a | 148 a | 131 c |

[A]Means with the same letter within a plantation are not significantly different (p = 0.10; arc-sine transformation of percentages; Tukeys Method).

[B]Seedlings were planted mid-month. Stock for the Nov. to Jan. plantings was taken directly from the nursery bed or greenhouse holding area. Stock planted Feb. to May was lifted in mid-January and freezer stored until planted. Frozen stock was thawed in the cooler prior to outplanting.

understory. After harvest, the reduced weed seed and sprout bank would result in a low probability of weed re-invasion and subsequent competition. Site preparation may only be needed if a reduction in logging slash is necessary to facilitate quality planting.

Prescribing a site preparation treatment includes an assessment of the risk of weed competition from both existing plant communities occupying the site and those communities that may invade the site. A judgment by foresters with experience in weed development in young plantations, coupled with vegetation survey data prior to planting, is the key to site preparation planning. Preference for alder planting should be given to sites which have a low risk of weed encroachment during the first-year of the plantation. The specific vegetation at a site will determine the exact herbicide, rate, and timing of entry or entries necessary to achieve effective control. Herbicides that act to provide pre-emergent control and have a residual effect are a part of effective treatments (Figueroa 1988).

On many units, two site-preparation treatments may be necessary: burning or scarification as a primary method of control followed by a pre-plant herbicide treatment. Effective herbaceous weed control can often be the difference between plantation success and failure.

There is undoubtedly an interaction between planting stock quality and weed competition thresholds. Table 2 shows performance differences between plug and bare-root seedlings with and without weed control. Transplant and bare-root seedlings, produced by the current open-bed nursery technologies, may be able to tolerate higher levels of weed competition than greenhouse or bed-house produced stock. The rapid first-year height growth and vigorous crown development indicative of transplant and open-bed bare-root stock may allow alder site capture more quickly than when other stock types are used. Though there may be a difference in performance between plug and bed-house stock as compared to open-bed and plug-transplant stock, the performance of plug-transplant and open-bed bare-root stock is very similar in the presence of first-year weed competition (Table 1).

## Outplant Timing

Seedlings are usually lifted from nursery beds in the middle of January and freezer stored until outplanting in early spring (Ahrens et al. 1992;

*Table 5. The effects of dehardening on the cold hardiness of one-year-old red alder plug seedlings (Y. Tanaka, unpub. data 1988).*

| Dehardening treatments | Leaf length (cm)[1] | Bud break (%)[1] | $LT_{10}$ (°C) | $LT_{50}$ (°C) |
|---|---|---|---|---|
| Control | — | — | -2.2 | -19.8 |
| 9 days in greenhouse[2] | 2.1 (0.1) | 33 (4) | -0.9 | -4.3 |
| 16 days in greenhouse[2] | 6.0 (0.4) | 45 (2) | -0.5 | -2.5 |

[1]Mean and (standard error).
[2]Warm greenhouse; 21°C.

Hibbs and DeBell, Chapter 14). The timing of the planting date can be very important. Results of trials to determine the optimum planting date are summarized in Table 4. Planting in November through January can result in serious winter freeze damage, which can result in top-kill and diminished root growth potential. Greenhouse studies have shown that red alder seedlings can deharden very quickly upon exposure to warm temperatures (Table 5). Several days of warm weather can bring on budbreak of newly planted seedlings. Seedlings planted in February can deharden and break bud in early March at a time when the risk of frost damage is still high (Fig. 9). Seedlings planted in mid-April and mid-May break bud very rapidly. Planting in late April to mid-May may not allow enough time for an adequate root system to develop before the onset of summer drought stress. Figure 10 shows the number of new roots for seedlings excavated from the field in early-summer relative to planting date. Seedlings planted in mid-March can have two and four times more new roots than seedlings planted in mid-April or mid-May, respectively. The depressed number of new roots in the November and February plantings could be attributed in part to freezing winter temperatures that occurred immediately following planting. Depending upon local site conditions and expected weather trends, a planting date should be selected to balance the risks of freeze damage and drought stress.

These planting date investigations support a recommended planting window of mid- March to mid-April for western Washington, at elevations less than 300 m. The key to extending these findings to other areas is to plant in the spring as opposed to the fall or winter. The spring planting period begins when the probability of severe frost is low and ends before there is appreciable seasonal drying of the soil. Depending upon the scope of a particular reforestation project it may be necessary to expand the time available for planting. Given the sensitivity of red alder to drought stress, it may be advisable to begin planting in early March (at sites with minimal risk of spring frost) as opposed to extending planting into late April or early May.

## Other Considerations Regarding Planting

Red alder seedlings are brittle and prone to breakage. Planting crews accustomed to handling conifers need to be cognizant that alder seedlings require more care. Careless loading of seedlings into planting

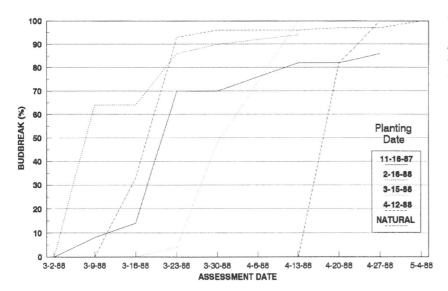

*Figure 9. Seasonal progression of budbreak from red alder seedlings planted at different dates in the southern Washington Cascades (A. Dobkowski and Y. Tanaka, unpub. data 1989).*

bags can result in considerable breakage to roots and stems. Care needs to be taken when closing the planting hole to assure that the stem is not wounded by the planter's boot.

Because red alder can deharden very rapidly when removed from cold storage, seedlings stored in the field on the day of planting need to be protected from heat to prevent premature dehardening. On-site daily storage in an insulated truck canopy or in the shade of standing timber covered with a heat shield (mylar) seedling protection tarp is recommended.

To partially offset the effects of heat and drought on newly planted seedlings, deep planting (ground level approximately 2 to 5 cm above the root collar) is recommended. Other species, including some less susceptible than red alder to heat stress, have been shown to benefit from deep planting to protect the root collar (Stroempl 1990). Heat girdling can be reduced by minimizing the scalping of surface debris during the planting process; exposed mineral soil at the base of the stem acts as a heat-sink, and the thin bark of alder is readily damaged.

## Conclusion

Key considerations for successful red alder plantation establishment are as follows:

1. Select sites with a low risk of regeneration failure. Poor soil drainage, frost, drought, competing vegetation, and big-game activity can hinder successful plantation establishment.

2. Plant only high-quality seedlings. Bare-root seedlings grown with the open-bed nursery technology can provide planting stock with the attributes necessary to regenerate most sites suitable for red alder production.

3. Reduce risk of poor plantation performance with thorough site preparation to control competing vegetation. Herbaceous weed competition in the first-year has been shown to be detrimental to red

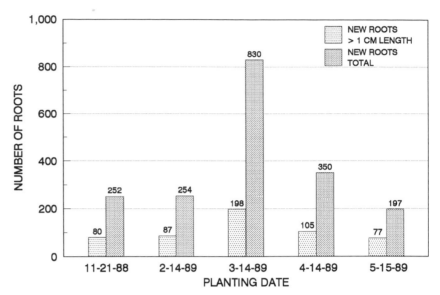

*Figure 10. Early summer assessment of new root growth for red alder seedlings planted at monthly intervals (A. Dobkowski and Y. Tanaka, unpub. data 1990). Assessed 6/15/89. Mean of 30 seedlings.*

alder seedling performance. Since there are no acceptable methods for broadcast release of red alder from weed competition, site preparation strategies need to consider herbicide treatments for controlling first-year weeds below the threshold level.

4. Depending upon local site conditions and expected weather trends, select a planting date to balance the risks of early season freeze damage and early summer drought stress. The recommended planting window is mid-March to mid-April in western Washington at elevations less than 300 m; seedlings planted in mid-February can deharden and break bud while the risk of frost is still high, and those planted in May may not develop an adequate root system before the onset of summer drought.

By applying the knowledge gained relative to site selection, seedling propagation, site preparation, and outplanting, successful red alder plantation establishment is predictable.

## Acknowledgment

Much of the information presented here is the result of work by a team of Weyerhaeuser Company scientists, nurserymen, and foresters. The authors would like to recognize Thomas S. Stevens, Jerry Barnes, Mark E. Triebwasser, Willis Littke, William Scott, Thomas Terry, Heinz J. Hohendorf, and John Keatley for their contributions.

# Literature Cited

Ahrens, G. R., A. Dobkowski, and D. E. Hibbs. 1992. Red alder: guidelines for successful regeneration. Forest Research Laboratory, Oregon State Univ., Corvallis, Spec. Publ. 24.

Chan, S. S., S. R. Radosevich, and J. D. Walstad. 1988. Physiological, morphological and carbon allocation strategies of Douglas-fir (*Pseudotsuga menziesii* {Mirb.} and Franco) and red alder (*Alnus rubra* Bong.) under varied light and soil moisture levels: third year results. Weed Sci. Soc. of Amer. Abstr. 28:60.

Fergerson, W., G. Hoyer, M. Newton, and D. R. M. Scott. 1978. A comparison of red alder, Douglas-fir, and western hemlock productivities as related to site: a panel discussion. *In* Utilization and management of alder. *Compiled by* D. G. Briggs, D. S. DeBell, and W. A. Atkinson. USDA For. Serv., Gen. Tech. Rep. PNW-70. 175-182.

Figueroa, P. F. 1988. First-year results of a herbicide screening trial in a newly established red alder plantation with 1+0 bare-root and plug seedling stock. Proceedings of 1988 Western Society of Weed Sci. Mtg. Weed Sci. 41:108-124.

Haeussler, S. 1988. Germination and first-year survival of red alder seedlings in the central Coast Range of Oregon. M.S. thesis, Oregon State Univ., Corvallis.

Haeussler, S., D. Coates, and J. Mather. 1990. Autecology of common plants in British Columbia: a literature review. Canada—British Columbia Forest Resource Development Agreement, Victoria, B.C., FRDA Rep. 158.

Harrington, C. A. 1986. A method of site quality evaluation for red alder. USDA For. Serv., Gen. Tech. Rep. PNW-192.

Helmuth, R. 1988. Red alder: Opportunities for better utilization of a resource. Forest Research Laboratory, Oregon State Univ., Corvallis, Spec. Publ. 16.

Hrutfiord, B. F. 1978. Red alder as a pulpwood species. *In* Utilization and management of alder. *Compiled by* D. G. Briggs, D. S. DeBell, and W. A. Atkinson. USDA For. Serv., Gen. Tech. Rep. PNW-70. 135-138.

Kenady, R. M. 1978. Regeneration of red alder. *In* Utilization and management of alder. *Compiled by* D. G. Briggs, D. S. DeBell, and W. A. Atkinson. USDA For. Serv., Gen. Tech. Rep. PNW-70. 183-192.

Krajina, V. J., K. Klinka, and J. Worral. 1982. Distribution and ecological characteristics of trees and shrubs of British Columbia. Univ. of British Columbia, Vancouver.

Leslie, D. M., Jr., E. E. Starkey, and M. Vavra. 1984. Elk and deer diets in old-growth forests in western Washington. J. Wildl. Manage. 48(3):762-775.

Newton, M., B. A. El Hassan, and J. Zavitkovski. 1968. Role of red alder in western Oregon forest succession. *In* Biology of alder. *Edited by* J. M. Trappe, J. F. Franklin, R. F. Tarrant, and G. M. Hansen. USDA For. Serv., PNW For. Range Exp. Sta., Portland, OR. 73-84.

Peeler, K. C., and D. S. DeBell. 1987. Variation in damage from growing season frosts among open-pollinated families of red alder. USDA For. Serv., Res. Note PNW-464.

Pezeshki, S. R., and T. M. Hinckley. 1988. Water relations characteristics of *Alnus rubra* and *Populus trichocarpa*: response to field drought. Can. J. For. Res. 18:1159-1166.

Radwan, M. A., Y. Tanaka, A. Dobkowski, and W. Fangen. 1992. Production and assessment of red alder planting stock. USDA For. Serv., Res. Pap. PNW-450. In press.)

Stroempl, G. 1990. Deeper planting of seedlings and transplant increase plantation survival. Tree Planters' Notes 41(4):17-21.

Worthington, N. P., R. H. Ruth, and E. E. Matson. 1962. Red alder: its management and utilization. USDA For. Serv., Misc. Publ. 881.

# 14

# Management of Young Red Alder
DAVID E. HIBBS & DEAN S. DEBELL

Management principles and strategies for young red alder stands are discussed in this chapter, which is divided into three major sections: (1) stand management principles, (2) strategies of plantation management, and (3) management of mixed plantations of alder with other species. Our objective is to provide an up-to-date synthesis of research results and management experience. Rather than give specific recommendations, we try to provide general information pertinent to decision making in any specific case. Because of the dynamic nature of the subject matter, however, information presented here will continue to expand and improve.

## Stand Management Principles
Principles of stand management are reviewed in this section according to the chronology of plantation management. Discussions of topics covered in depth in other chapters in this volume are referenced, but not duplicated. The focus is limited to biological information; application of this information rests in an economic and social context.

### Site Selection
Site selection for red alder must be made with regard to a large number of factors, including nutrition, air and water drainage, frost, drought, heat stress, and potential for heat damage. These growth-related factors are addressed in recent publications (Harrington 1986; Ahrens et al. 1991) and by Dobkowski et al. (Chapter 13). Other important considerations are operability (e.g., slope and soil strength), management constraints caused by other resource or social values, and proximity to markets and conversion plants.

### Stand Establishment
Although many issues of stand establishment are addressed in this volume (see Ahrens, Chapter 12, and Dobkowski et al., Chapter 13), selection of initial density requires additional consideration. Choice of initial planting density varies considerably with management goals. For example, densities as high as 1500 trees/ha (600 tpa) may be appropriate for pulpwood or biomass production, whereas relatively lower densities, down to 500 trees/ha (200 tpa), are used in saw-log production.

If stands are not established by planting, the alternative is to regenerate the stand from seed. Experience in the deliberate establishment of alder stands from seed is limited and is not encouraging. Test plots under a variety of growing conditions have shown that far less than 1 percent of the sown seed survives as seedlings at the end of the first growing season (Haeussler 1988). Thus, successful regeneration from seed requires high seed densities. In addition, efforts to establish red alder from naturally or artificially distributed seed must emphasize uniform or even spacing. When spacing is uneven, the number of trees with substantial lean or sweep is high. Evenness of establishment requires an evenly distributed mineral seedbed and even distribution of seed. Because regeneration from seed requires creation of extensive areas of exposed mineral soil, care must be taken to prevent erosion. Although seeding uniformly distributed and prepared spots by hand could resolve such problems, we suspect that planting would be more cost effective and successful.

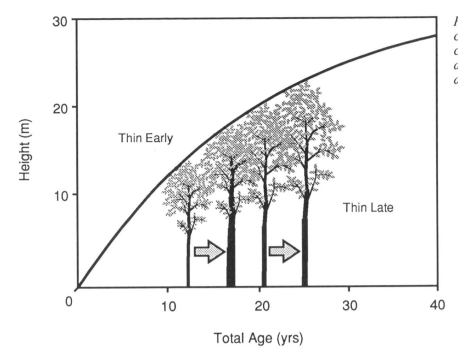

Figure 1. The effect of age on crown response to thinning. The curve presents stand height with an $SI_{20}$ = 20 (meters, Harrington and Curtis 1986).

*Density Management*

Few natural stands of red alder exist in the age range appropriate for density management (less than twenty years). Those stands which do exist tend to occur as small pockets in conifer plantations. In general, they can be treated as discussed in the following sections for plantations with recognition of the differences associated with their initial high density and uneven spacing.

Density management in alder carries the common objectives of capturing mortality, regulating rate of growth, and allocating growth among and within the remaining trees. Density management also provides opportunities to improve stem quality of residual trees, via selection of better stems for retention, and to create more uniform stands. Uniform spacing is essential for growing straight trees. Most of the poor form (sweep and lean) seen in natural stands is the result of growth at uneven spacings. Other factors, including site quality and tree age, however, interact with the level of competition to produce the actual level of growth.

**Timing of thinning.** Thinning in red alder should be done at a young age. A first thinning done after about age 20 produces little or no response in natural or otherwise high-density stands. A height growth curve ($SI_{20}$ = 20 m; Fig. 1) illustrates one of the reasons for this outcome. A thinning done at age 12 when the trees are 14 m tall produces a strong thinning response, because the trees still have the height-growth potential and crown vigor with which to build a large crown. In contrast, trees of twenty years have a much slower rate of height growth, and, therefore, a much slower response to thinning. In managed stands or otherwise low-density stands, the level of crown vigor to be expected in trees twenty years old and older is unknown;

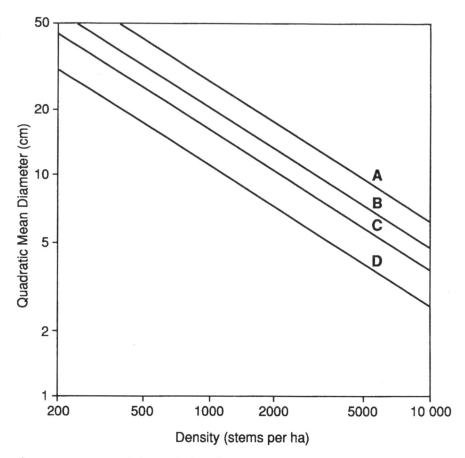

*Figure 2. Stand density management diagram from K. J. Puettmann et al. (1993). The A line is the boundary of all data (biological maximum). The B line represents the average maximum density. The C line is the lower boundary for competition-caused mortality (operating maximum), whereas the D line is the upper boundary for maximum individual tree growth and approximates crown closure (competition threshold).*

thus, more research is needed to determine the thinning response for older, managed stands. In addition, the potential benefits of pruning and fertilization in stands more than twenty years old need to be examined.

**Density management tools.** A common tool in density management is the stand density diagram, a graphical representation of the relationship between mean stand diameter and stem density. Several stand density diagrams have been prepared for red alder; the most recent (Fig. 2; Puettmann et al. 1993) contains four parallel lines that describe critical stages of stand development. The A line is a boundary to all existing data and represents current knowledge of the biological maximum, which in this case indicates that the maximum stocking level of red alder is considerably below that of most conifers in the region. The B line represents the average maximum, or the degree of stocking that most natural stands approach and then follow as a result of self-thinning. The average maximum has a relative density (RD; relative to the biological maximum) of 65 percent. The C line represents the operating maximum, or the relative density at which mortality becomes important. In this example, relative density is 45 percent. Above this density, mortality rates increase rapidly. Thus, the operating maximum generally represents the upper density limit in stand management. Finally, the D line represents the competition threshold, or the density below which competition between trees is slight. It is the lower limit for density management and approximates the point of canopy closure. In the example given, relative density is 25 percent.

A number of sources describe characteristics of natural, unmanaged stands and may be used to predict their development (Worthington et al. 1960; Chambers 1974; Arney 1985); these sources are reviewed in Puettmann (Chapter 16). Initial spacing and thinning in red alder may result in variations in a number of these characteristics.

**Responses to thinning.** As would be expected, thinning generally results in an increase in diameter growth over the prethinning level in the stand. Both an immediate increase in growth after thinning and a continued further increase can occur for five to ten years as crowns grow to occupy the vacant canopy space. The new level of diameter growth can persist at least until stand density begins to approach that at which mortality occurs (C line; Fig. 2). Examples of prolonged average stand diameter growth rates up to 1.3 cm yr$^{-1}$ have been found five to ten years after thinning on good sites (D. E. Hibbs et al., in review), but this growth rate is near the maximum to be found in managed stands (see also Hibbs et al. 1989 for a general review). Observed rates vary with site quality, tree age, time since thinning, and other aspects of stand history.

Diameter growth rates in thinned stands with a density between the operating and average maximum are uncertain. In one example that appears to contradict some of the generally assumed principles underlying a stand density diagram, D. E. Hibbs et al. (in review) found that diameter growth increased as stands grew from the competition threshold to the operating maximum (D to C lines; Fig. 2) and that crown reclosure occurred at or a little above the operating maximum twelve years after thinning. Thus, crown closure occurred at a relative density at which mortality is usually seen in unmanaged stands. As the stands in this example develop further (i.e., grow toward the average maximum), crown classes will begin to differentiate and diameter growth reductions will appear in these lower crown classes. The rate at which these changes will occur is unknown. Until competition reduces crown size, diameter growth could be maintained. Longer-term studies are

needed to provide data on actual growth rates in these types of stands at these densities.

Thinning red alder has frequently resulted in an increase in stand basal area growth. A basal area growth rate of 1.25 m$^2$ yr$^{-1}$ was found by D. E. Hibbs et al. (in review) after thinning. This rate is higher than that observed in normal stands (Worthington et. al. 1960) and almost twice the rate of unthinned stands at the same location. The review of alder thinning studies in Hibbs et al. (1989) shows that the increase in stand basal area growth is general, except when thinning intensity is very high. After severe thinning, basal area growth may eventually recover and exceed the prethinning level.

Thinning has commonly resulted in a temporary loss of height growth when natural stands up to 21 years old have been thinned. Height growth recovers in about five years. Whether or not a height growth loss occurs when low-density stands (below the operating maximum; C line, Fig. 2) are thinned remains unknown.

Some uncertainty is associated with estimation of tree volume in managed stands. Existing volume equations and tables have been derived from natural stands (see Puettmann, Chapter 16). Thinning alters relative rates of height and diameter growth, however, and thus changes tree form. Therefore, use of equations derived from natural stands incorporates an unknown degree of error into volume estimates. Because increasing proportions of the alder resource in the Pacific Northwest will be managed, the applicability of existing tables and equations to managed stands needs to be assessed; new tools may be needed.

Thinning red alder does not appear to increase the number of epicormic branches, but research on this topic is limited (Lloyd 1955; Berntsen 1961a; Warrack 1964; Hibbs et al. 1989). Thinning may increase the size of existing branches (Hibbs et al. 1989), but the long-term product consequence of this is unknown. Research on other hardwood species suggests that many epicormic branches may die following reclosure of the tree crowns (Erdmann and Peterson 1972); our casual observations have also shown epicormic branch loss following crown reclosure in red alder. Windthrow and stem breakage have been observed after thinning young (less than five years old), dense stands, but have been rare or nonexistent when older stands were thinned.

**Mechanics of thinning.** The usual thinning tool is a chain saw. Some tests have also been conducted on chemical thinning (Hibbs et al. 1989; W. Emmingham, unpub. data). Stem injection of full strength 2,4-D in June resulted in severe flashback (injury) to trees within 1.5 m of the treated trees. Although the flashback resulted in growth loss and some mortality, stem injection required much less time than cutting and was much easier to execute because there was no immediate debris on the ground.

A combined manual treatment of trees close to crop trees and chemical injection of more distant trees may be possible. A manual technique of girdling the alder close to crop trees with the same hatchet used for chemical injection has been used successfully (W. Emmingham, unpub. data). The tip of the hatchet was slipped under the bark and around the tree to remove a large section of bark. A few chops to the trunk were added to reduce stem strength. Almost 80 percent of these trees broke at the girdle the following winter. For stem injection, W. Emmingham (unpub. data) found that half-strength 2,4-D also worked well.

Alder trunks are damaged easily and are very susceptible to secondary fungal infection. Thus, thinning activities must take place with care to avoid damage to leave (crop) trees. Avoiding the spring season, when the bark is weakly attached to the wood, will also help minimize damage.

**Protection.** Issues of protection are covered in Chapters 12 and 13. Most wildlife problems are associated with the establishment phase (Harrington et al., Chapter 1; Dobkowski et al., Chapter 13). Further, many disease problems have their source in the nursery (Ahrens, Chapter 12), and are best treated there.

Although there are three common defoliators of red alder (Gara and Jaeck 1978), no assessments have been made of their effects on tree growth. A list of common insect pests of red alder is given in Harrington et al. (Chapter 1).

## Pruning

For alder, as for most hardwoods, the economic value differential between lumber with knots and lumber without knots is large. Thus, management practices that encourage the natural branch pruning process or actively remove branches should be considered.

Four branch management strategies are currently being discussed or employed. First, a high initial stand density encourages branch mortality and can result in a small knotty core. Our field observations, however, have shown that whereas branches do die and break off, the break is frequently several centimeters out from the trunk. Many years of growth are required to grow over these stubs, and they produce a loose knot in lumber. A second strategy involves actively removing dead branches in combination with stand thinning. Falling trees are first used to knock dead branches from leave trees, and a work crew then can quickly remove the remaining branches and branch stubs. In the third strategy, live and dead branches are removed by saws; this is the most labor-intensive method.

A fourth strategy to encourage alder branch pruning utilizes western redcedar (*Thuja plicata*) as a companion species in a mixed-species plantation. In contrast to what is described in a later section as the general strategy in mixed species culture, the objective here is to create a two-tiered forest. Redcedar is sufficiently shade tolerant to grow well under an alder canopy. Thus, both species are managed at nearly the full number of stems. The cedar shades and physically abrades lower alder branches. With careful thinning and eventual harvesting of the alder, the cedar is released to grow to maturity.

## Fertilization

Radwan and DeBell (Chapter 15) review fertilization effects in red alder. Phosphorus appears to be the most generally beneficial nutrient, especially on wet sites. Because nitrogen fixation is an energy-consuming process (Binkley et al., Chapter 4), fertilization with nitrogen on some sites may also result in an increase in growth.

## Rotation Length

Rotation length for managed stands of red alder is relatively short by regional standards. Small saw logs can be produced in 25 to 30 years on good sites. Perhaps a diameter of 40 cm can be grown in less than forty years with intensive management. Pulp wood rotation length, where cubic volume production is the objective, may be as short as ten to fifteen years.

## Plantation Management

The many strategies of plantation management with choices of initial density, intensity of thinning, and timing of thinning make it difficult to provide specific recommendations. Therefore, we use examples of specific strategies to illustrate the range of possibilities and discuss their advantages and disadvantages. The selection of a particular strategy, however, depends on management objectives, cash flow considerations, and product markets.

Although we have made assumptions about growth rates and patterns, the principles illustrated in the examples are valid, irrespective of small errors in the assumptions. The specific sizes and ages given are likely to differ as growth rates differ. In the examples, we assume a moderately good site, $SI_{20}$ = 20 m (Harrington and Curtis 1986), and a diameter growth rate of 1.0 cm $yr^{-1}$ for intraspecific competition levels up to relative densities of 25 to 30 percent in unmanaged stands and up to 45 to 50 percent in thinned stands (Worthington et al. 1960; Hibbs et al., in review). Assuming a higher growth rate would result in a shorter time to a target size or density; a lower rate would increase the time. The assumed rate is near the maximum observed in current managed stands and varies with age, time since thinning, density, and site quality.

In the thinning examples, mortality after thinning is assumed to be minimal and little reduction is assumed to occur in diameter growth until the stand reaches average maximum density (B line; Fig. 2). Even though this assumption is not strictly true (i.e., growth will be reduced before the stand reaches this density; see previous discussion on responses to thinning), a simplifying assumption about growth rate was needed to estimate final size and rotation lengths. Thus, the act of thinning is

assumed to increase average stand diameter by 10 percent. To the extent that this assumption is wrong, rotation lengths to a final product size will be lengthened.

We have ended the rotation at the average maximum density because of increased uncertainty about subsequent growth patterns. In practice, a target size is chosen at the beginning of a rotation, and the rotation is terminated when the target size is reached.

Clear bole length of stands with a density higher than the operating maximum is assumed to equal about two-thirds the total tree height. This approximation is only useful near the middle of the red alder density range (i.e., about 500 to 2000 trees/ha).

### Example 1: High Density

Trajectory 1 in Figure 3 represents the development of a naturally regenerated or planted stand with an initial density of 5000 trees/ha (2000 tpa). This density falls at the high end of the range being investigated in biomass research, and is representative of densities that often occur in young, natural alder thickets. Early stages of stand development are characterized by diameter growth and low mortality: the trajectory is vertical. Crown closure and diameter growth reductions begin at a relative density of about 25 percent, the competition threshold (D line; Fig. 2); the stand is about four years old and has an average diameter of nearly 4 cm at this stage. Density-dependent mortality becomes important as the relative density approaches 45 percent, the operating maximum (C line; Fig. 2); individual tree diameter growth continues to decrease as this relative density is approached and is less than 0.5 cm $yr^{-1}$ at the operating maximum. As trees continue to grow, the mortality rate increases to the point that the trajectory curves to the left and approaches the relative density of 65 percent, the average maximum (B line; Fig. 2). When mortality is episodic, the trajectory may oscillate around the average maximum. By the time density has dropped to 3000 stems/ha, average diameter growth has dropped to 0.25 cm $yr^{-1}$ (Worthington et al. 1960).

This management strategy is attractive in its simplicity. The high initial density results in rapid

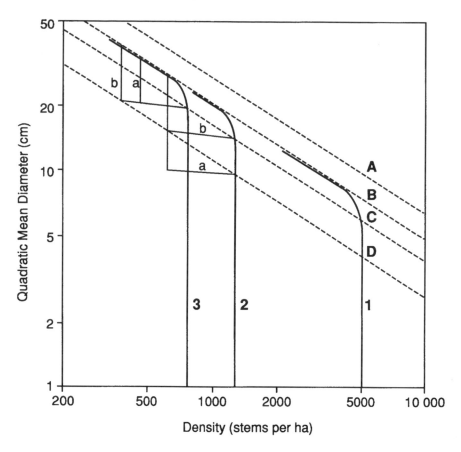

*Figure 3. Trajectories for stand development at three different initial densities. Trajectories 1, 2, and 3, respectively, illustrate the effects of high-density planting with no thinning, timing of thinning, and intensity of thinning.*

site occupancy, which minimizes weed control concerns. No thinning is performed; therefore, the only cost, which may be very high, is in stand establishment. Clear wood production and additions of nitrogen and organic matter to the forest floor and soil begin early. In addition to high establishment cost, other drawbacks occur when saw-log production is the management objective. Saw-log production requires long rotation lengths, and it is not certain that large sizes can ever be achieved. Moreover, stem form may be inferior because of uneven competition associated with the mortality pattern.

Thinning could be done in this stand. At age four (when the stand reaches the competition threshold), thinning would act to maintain diameter growth while still affording the advantages of early site capture with high initial density. In a stand with an initial density this high, thinning at the usually recommended relative density (i.e., when the stand reaches the operating maximum) results in a diameter growth loss before thinning, a slower thinning response, and a height growth loss. After the first thinning, the stand response would be similar to that in the following medium density example.

### Example 2: Intermediate Density

A plan for intermediate density thinning might involve establishing 1250 to 1500 trees/ha (500 to 600 tpa), a density which, as in the preceding example, gives rapid site occupancy and thus control of competing vegetation. We assume an initial density of 1375 trees/ha (550 tpa) in this example (trajectory 2; Fig. 3).

It takes about ten years for an alder tree to reach a diameter of almost 10 cm (10 cm/1.0 cm yr$^{-1}$ = 10 yrs) and cross the competition threshold (25 percent RD). These trees have a dominant height of nearly 12 m and a clear bole length of about 7 m. At fourteen years, the stand has an average diameter of about 14 cm as it nears the operating maximum (C line; Fig. 2), mortality is beginning, and diameter growth is decreasing.

To illustrate the effect of the timing of thinning, we have charted the consequences of thinning at age 10 and of waiting an additional four years. A longer delay is not considered because of the impending onset of density-dependent mortality, at which time crowns will lose vigor and the potential thinning response will be reduced. Two options are considered. In the first option trees are thinned to a density of 600 trees/ha (240 tpa) at age 10 (trajectory 2a; Fig. 3), and stand conditions are as described previously. In the second option thinning occurs at age 14 (trajectory 2b), when the stand has a dominant height of 16 m, a clear bole length of about 11 m, and an average diameter of 14 cm. If left unthinned, diameter growth declines and mortality increases as the stand nears and curves to the left to track the average maximum (B line; Fig. 2).

Both thinning options result in a final stand diameter (i.e., reach the average maximum) of 28 cm. Option one (trajectory 2a) reaches this diameter in a little over 28 years, whereas option two (trajectory 2b) results in slower diameter growth after crossing the competition threshold (D line) at ten years and has a reduced thinning response because of crown loss after age 10. Thus, option two requires two to five more years than does option one to reach the final diameter. The delay in thinning also results in an increase in clear bole length and an increase in size at time of thinning. This increase in size may be economically beneficial or harmful, depending on whether or not thinned material is removed for sale.

## Example 3: Low Density
A low initial density might be chosen to reduce early stand costs in planting and thinning, but this advantage comes with an increase in the diameter of the knotty core, a slower rate of site occupancy, and a potential for weed control problems. In this example (trajectory 3; Fig. 3), we have an initial density of 750 trees/ha (300 tpa). When the stand reaches the operating maximum (i.e., when relative density is 45 percent), the average diameter is 19 cm and the age is about nineteen years. The stand has a dominant height of 19 m, and a clear bole length of 13 m. The age is estimated at nineteen years with the untested assumption that the diameter growth rate does not drop much before the stand reaches the operating maximum. In making this assumption, we have drawn on a parallel between growth rate of this initially low density stand and the growth rate of thinned stands in this relative density zone and near this absolute density.

If left unthinned, the stand soon exceeds the operating maximum, the rate of diameter growth begins to decline, and the mortality rate increases. As a result of these changes, the stand trajectory curves left and follows the average maximum (B line; Fig. 2).

To examine the effects of intensity of thinning, we have reduced stand density to 450 trees/ha (180 tpa) and 363 trees/ha (145 tpa) (trajectories 3a and 3b; Fig. 3). Although the absolute difference in density between these two treatments is relatively small, the impact on the stands at the end of the rotation would be large. The lighter thinning can result in a rotation length of 34 years, a tree height of 26 m, and a diameter of 34 cm. The heavier thinning can result in a rotation length of 39 years, a tree height of 27 m, and a diameter of 39 cm. Clear bole length does not differ between the stands. Thus, when large final diameters are sought, heavy thinning is required.

## Summary
In a comparison of the three thinning examples, higher initial densities contribute to wood quality by minimizing the diameter of the knotty core and by encouraging rapid natural pruning. Higher densities are also a tool for early weed control. If not carefully tended, however, high density stands can develop an uneven spacing and thus reduce lumber recovery. Lower initial densities or multiple

thinnings aid in the development of large diameter trees. The final choice of management strategy will depend on site quality, pre-existing shrub competition, markets, and social and economic factors.

## Mixed-species Stands

Mixed-species forests have been managed around the world for hundreds of years. These forests are attractive to some managers because they offer the possibility of multiple harvests and markets and, thus, a form of economic security through diversity. Although recent silvicultural practices in many areas of the world have tended to move away from mixed-species stands, concerns over biological diversity and long-term productivity are resulting in renewed interest in mixed-species stands, particularly in North America and Europe.

Tree species such as red alder, which can fix atmospheric nitrogen, are of special interest in mixed-species stands. On soils where nitrogen is a limiting factor in plant growth, spatial or temporal mixtures of nitrogen-fixing plants can result in an increase in ecosystem productivity. In designing and managing a mixed-species system, however, care must be taken to prevent the detrimental effects of unequal competition from out-weighing the beneficial effects of nitrogen fixation.

Six factors require consideration in the design of a mixed-species system. These factors are site quality, species choice (here, one is always red alder), species proportion, total plant density, spatial pattern, and timing of management activities. We will review each of these factors and develop two examples of mixed-species stands.

The rotation or alternation of species is the simplest management option and, therefore, will probably be the most used system. A crop-rotation system would utilize one rotation of pure red alder followed by one or more rotations of another species. The red alder rotation yields a timber crop, builds soil fertility, and contributes to landscape-level diversity; the subsequent rotation(s) of another species will return fertility to near the pre-alder level. In addition to such normal monoculture considerations as competition control, the issue of site quality must be addressed.

Newton (1978) also described a management system that involves minimal direct interactions between species. This is a system in which young conifers and red alder could be grown together initially until the latter approached the point of suppressing the former. At that time, the conifers can be released from the alder. This approach is likely to produce good potential for supplying nitrogen on deficient soils if the population density of alder is high, if nutritional conditions for fixation are good (both condition must be met), and if economical means of conifer release are available. Release should occur before significant replacement of sun-foliage occurs on the conifers. This approach will not lead to intermediate yields of alder, but may increase intermediate yields of conifers.

### Site Quality

There are two major site quality issues. First, is the site appropriate for red alder? Second, do the site and companion species have the capacity to derive a benefit from nitrogen fixation? The first issue is addressed in Dobkowski et al. (Chapter 13) on plantation establishment.

The key to a successful mixed-species stand is identifying and managing the limiting resource(s). When soil nitrogen is a limiting resource, then a mixture may be beneficial. The commonly used index of productivity, site class, gives a preliminary indication of limiting factors. Site class I obviously has little that limits growth (McArdle et al. 1961). Site Classes II, III, and IV, however, need to be examined carefully to identify the limiting factor(s). If soil nitrogen is a predominant factor, then the location is a good candidate for mixed-species management. Site Classes V and below probably have several important limiting factors and may be poor candidates for mixed-species management.

### Species Choice

Among the Pacific Northwest conifers, only Douglas-fir (Pseudotsuga menziesii) has received much attention in mixtures (Berntsen 1961b; Tarrant 1961; Miller and Murray 1978; Tarrant et al. 1983; Binkley et al. 1992; Newton and Cole, Chapter 7). Western redcedar, western hemlock (Tsuga heterophylla), Sitka spruce (Picea sitchensis), and grand fir (Abies grandis) also appear to have potential, because their site requirements overlap those of red alder. Short rotation mixtures with black cottonwood (*Populus trichocarpa*) have also been explored (DeBell and Radwan 1979; Pezeshki and Oliver 1985; Heilman and Stettler 1985), but results have varied widely with site conditions and relative growth rates of the two species.

Initial height growth of red alder is greater than that of any conifer species considered here for mixture, but height growth of all species varies with growing conditions. Thus, careful choice of a partner for alder may allow a more even matching of growth rates under a particular set of growing conditions.

All the potential conifer partners are able to sustain height growth far longer than can red alder. Initial growth rates may prove difficult to match for a particular location; in this case, choice of a shade-tolerant species may allow the partner to grow through the alder canopy. In the event that height growth rates and shade tolerance are incompatible, alder can be planted several years after the first species (Newton et al. 1968), although such a delay can create new animal damage and weed-control requirements.

The sustained height growth of the conifer partners means that they would probably be managed on a longer rotation than alder. Thus, whether they are managed for pulpwood, saw timber, or both, the alder may provide an earlier, intermediate crop in such mixtures. With cottonwood, alder could be managed on an equal rotation length for pulp or a longer rotation for saw timber. If an intermediate harvest is planned, some additional care in species choice is required. Species like western hemlock and western redcedar are quite susceptible to decays following logging-induced injuries.

In addition to matching ecological characteristics, there may be other reasons to select a particular partner for a particular site. Redcedar may fit into a branch pruning strategy as described earlier, or it may be sought for its ability to buffer alder's soil acidifying tendency (Alban 1969; DeBell et al. 1983; Cole et al. 1990; Bormann et al., Chapter 3). A desired wildlife species or community may be favored by a particular combination.

## Species Proportion

Given full-site occupancy, the proportion of red alder in a stand is the major determinant of the amount of nitrogen fixed and the severity of interspecific competition. The relationship between alder proportion and nitrogen fixed as a proportion of the potential fixation on the site (Binkley et al., Chapter 4) is probably not linear. At low alder proportions, individual trees have larger crowns and capture a disproportionate share of site resources compared to the more even distribution of resources at high proportions. Thus, an alder proportion of 10 to 25 percent may fix 15 to 40 percent as much nitrogen as a fully stocked alder stand. The data in Binkley et al. (Chapter 4) suggest that 100 to 150 kg ha$^{-1}$yr$^{-1}$ is a common rate of fixation for fully stocked alder stands. The proportion of alder that will best balance competitive and beneficial effects probably will fall within the large range of 15 to 50 percent alder. Miller and Murray (1978) suggested that 50 to 100 alder trees/ha (20 to 40 tpa) are sufficient. The basis for any recommendation is weak; additional research is needed.

## Density

Initial densities commonly recommended for pure species stands of alder or conifer species fall between 750 and 1250 trees/ha (300 to 500 tpa). The total stem density of mixed stands generally lies within the range of densities found in monoculture and is based on site-specific management objectives. When the two species considered have different monoculture densities, the mixed-species density should be similar to that of the less dense species. In our examples this species is red alder.

In understanding competitive relationships in mixed stands, the plantation can be viewed as an area with a fixed number of planting spots that are being allocated among the species rather than as a fully stocked stand of a given species with alder added on top. This view helps to separate the effects of density and proportion.

Competitive effects of alder increase with the number of alder trees present. As long as alder density remains less than about 250 evenly spaced alder/ha (100 tpa), alder will not close crown until mean diameter exceeds about 28 cm. This number

Top View　　　　　　　　　Side View

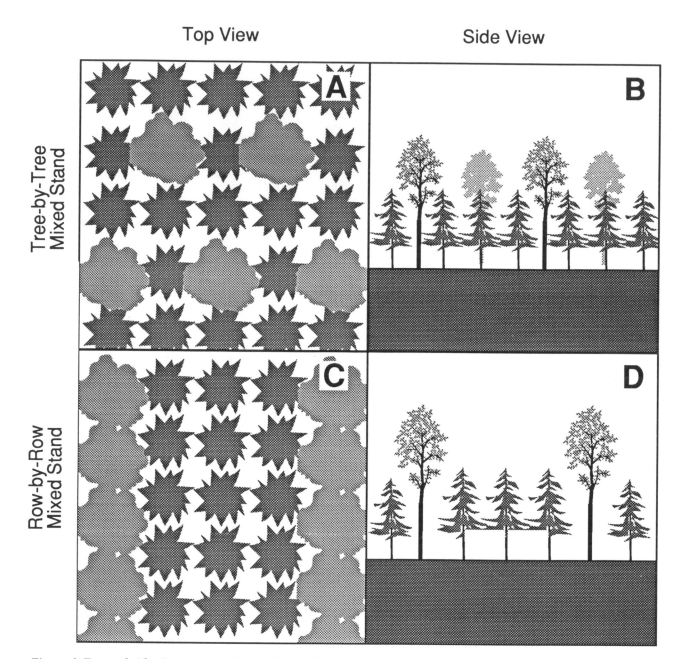

Figure 4. Top and side views, respectively, of (A and B) tree-by-tree and (C and D) row-by-row mixed stands from 5 to 20 years in age.

comes from the competition threshold of the density management diagram (D line; Fig. 2), which represents our best approximation of diameter-density relationships at the point of canopy closure. Thus, understory conifers will receive some direct sunlight, which minimizes conifer mortality and helps to ensure eventual conifer dominance. Maximum density without crown closure can be estimated for other target sizes via Figure 2.

### Spatial Pattern

Species mixtures involve a systematic design that balances two conflicting factors. First, the nonalder species needs to be near enough to the alder to benefit from improved soil fertility. Second, the two species need to be separated sufficiently to minimize the shading effects of the alder on the other species. This can be done by regulating density and proportion in a tree-by-tree mix. A row-by-row mix may offer an advantage in systems that will be thinned commercially.

The tree-by-tree mix disperses the nitrogen resource and the competitive effects evenly. When the alder proportion is low, the competitive effects on any one associate will be minimized. In Figure 4, A and B illustrate an evenly mixed stand of 800 trees/ha, 25 percent of which are alder. This places the alder trees about 7 m apart, within the effective dispersal distance for the nitrogen fixed by adjacent alder. Early in stand development, the alder will stand above the conifer, but spacing will ensure that light is adequate for lower story growth for many years. The large-crowned alder will be a vigorous nitrogen fixer; the conifer lower story will help prune the alder. By about thirty years, the mean diameter of alder could be about 30 cm, the conifer height will equal that of alder, and rates of nitrogen fixation will begin to decline. A commercial thinning could then remove the alder and some conifer. The stand then would grow to final harvest as a conifer monoculture.

A row-by-row mixture would probably utilize one row of alder for every two or three rows of an associated species. In Figure 4, C and D illustrate the 3:1 mix. Except at very low densities, a 1:1 mix would probably result in suppression of the associated species. A row-by-row mix allows row harvesting of the alder in a commercial thinning

and provides access to the associated species for thinning.

### Timing of Activities

The differences in growth rates and ages to maturity of the species in a mixture require careful planning and timing in the execution of a silvicultural plan. Some of the important options have been mentioned already. First, one technique for compensating for a species with a slow initial growth rate is to plant it several years before the alder. Newton et al. (1968) produced a set of graphs that define the delays needed in alder planting when mixed with Douglas-fir to prevent overtopping. The delays range from four to eight years, depending on site conditions. Delayed alder planting requires a program to control competing vegetation, both at the time of planting Douglas-fir and at the time of planting red alder.

Precommercial thinning is commonly practiced in conifer plantations. It allows a manager to use a high alder density for a few years, which gives a pulse of nitrogen to a site. Most of the alder is then removed in an early precommercial thinning before it has a significant competitive effect on the conifer. This leaves a mixed stand as described under the preceding section.

Red alder commonly seeds into conifer plantations. This opens the door to the option of leaving some alder in a precommercial thinning. There is no cost to leaving well-spaced alder where conifer mortality has occurred. In addition, a manager should examine the potential for benefit from nitrogen fixation at the site and decide whether or not the benefit is sufficient to warrant allowing some alder to replace conifers. Current management plans on the Siuslaw National Forest in western Oregon, for example, call for leaving some red alder in Douglas-fir plantations growing on soils low in nitrogen (Turpin 1981).

### Conclusions

Considerable research and practical experience during the past fifteen years have demonstrated that vigorous plantations can be established by planting. Reliable techniques to control weed, insect, and animal damage exist, and the methodology of their application has been developed. The value of density management in improving growth, shortening rotations, and improving tree form and wood quality has been demonstrated, and some estimates of the growth responses to density management have been made.

The primary limitation in the current state of red alder management information is the lack of extensive testing; most studies have been conducted over a limited area and for relatively short time periods. More extensive tests of many of these management principles and practices are underway and will extend the generality of our current knowledge. Major questions related to wood quality, repeated rotations, and sustained resource availability remain to be addressed. These matters notwithstanding, the information available for managing red alder is impressive and suggests a potential and options considerably beyond those envisioned when the first book on alder, *Biology of Alder* (Trappe et al. 1968), was assembled.

## Acknowledgments

This is paper 2893 of the Forest Research Laboratory, Oregon State University, Corvalis. This paper draws on the research and ideas of many forest managers and scientists, to whom we are indebted for contributions to the knowledge of red alder and silviculture in general. The first author received sabbatical support from the Institut National de la Recherche Agronomique of France and research support under USDA grant 84-CRCR-1-1434 during the writing of this paper. The second author has been supported in part by the Short Rotation Woody Crops Program (now Biofuels Feedstock Development Program) of the U.S. Department of Energy under interagency agreement No. DE-A105-310R20914.

## Literature Cited

Ahrens, G. R., A. Dobkowski, and D. E. Hibbs. 1991. Red alder: guidelines for successful regeneration. Forest Research Laboratory, Oregon State Univ., Corvallis, Spec. Publ. 24.

Alban, D. H. 1969. The influence of western hemlock and western redcedar on soil properties. Soil Sci. Soc. Am. Proc. 33:453-457.

Arney, J. D. 1985. SPS: stand projection system for mini- and micro-computers. J. For. 83:378.

Berntsen, C. M. 1961a. Pruning and epicormic branching in red alder. J. For. 59(6):675-676.

———. 1961b. Growth and development of red alder compared with conifers in 30-year-old stands. USDA For. Serv., Res. Pap. PNW-38.

Binkley, D., R. Bell, and P. Sollins. 1992. Soil nitrogen transformations in adjacent conifer and alder/conifer forests. Can. J. For. Res. (In press.)

Chambers, C. J. 1974. Empirical yield tables for predominantly alder stands in western Washington. Washington State Dept. of Natural Resources, DNR Rep. 31. Olympia, WA.

Cole, D. W., J. Compton, H. Van Miegroet, and P. Homann. 1990. Changes in soil properties and site productivity caused by red alder. Water, Air, Soil Pollut. 54:231-246.

DeBell, D. S., and M. A. Radwan. 1979. Growth and nitrogen relations of coppiced black cottonwood and red alder in pure and mixed plantings. Bot. Gaz. (Suppl.) 140:97-101.

DeBell, D. S., M. A. Radwan, and J. M. Kraft. 1983. Influence of red alder on chemical properties of a clay-loam soil in western Washington. USDA For. Serv., Res. Pap. PNW-313.

Erdmann, G. G., and R. M. Peterson, Jr. 1972. Crown release increases diameter growth and bole sprouting of pole-size yellow birch. USDA For. Serv., Res. Note NC-130.

Gara, R. I., and L. L. Jaeck. 1978. Insect pests of red alder: potential problems. *In* Utilization and management of alder. *Compiled by* D. G. Briggs, D. S. DeBell, and W. A. Atkinson. USDA For. Serv., Gen. Tech. Rep. PNW-70. 265-269.

Haeussler, S. 1988. Germination and first-year survival of red alder seedlings in the central Coast Range of Oregon. M.S. thesis, Oregon State Univ., Corvallis.

Harrington, C. A. 1986. A method of site quality evaluation for red alder. USDA For. Serv., Gen. Tech. Rep. PNW-192.

Harrington, C. A., and R. O. Curtis. 1986. Height growth and site index curves for red alder. USDA For. Serv., Res. Pap. PNW-358.

Heilman, P., and R. F. Stettler. 1985. Mixed, short-rotation culture of red alder and black cottonwood: growth, coppicing, nitrogen fixation, and allelopathy. For. Sci. 31:607-616.

Hibbs, D. E., W. H. Emmingham, and M. C. Bondi. 1989. Thinning red alder: effects of method and spacing. For. Sci. 35:16-29.

Hibbs, D. E., W. H. Emmingham, and M. C. Bondi. Thinning response of red alder. Western J. Applied Forestry (in review).

Lloyd, W. J. 1955. Alder thinning—progress report. USDA Soil Conservation Serv., West Area Woodland Conservation Tech. Note 3. Portland, OR.

McArdle, R. E., W. H. Meyer, and D. Bruce. 1961. The yield of Douglas-fir in the Pacific Northwest. USDA, Tech. Bull. 201.

Miller, R. E., and M. D. Murray. 1978. The effects of red alder on the growth of Douglas-fir growth. *In* Utilization and management of alder. *Compiled by* D. G. Briggs, D. S. DeBell, and W. A. Atkinson. USDA For. Serv., Gen. Tech. Rep. PNW-70. 283-306.

Newton, M. 1978. Herbicides in alder management and control. *In* Utilization and management of alder. *Compiled by* D. G. Briggs, D. S. DeBell, and W. A. Atkinson. USDA For. Serv., Gen. Tech. Rep. PNW-70. 223-230.

Newton, M., B. A. El Hassan, and J. Zavitkovski. 1968. Role of red alder in western forest succession. *In* Biology of alder. *Edited by* J. M. Trappe, J. F. Franklin, R. F. Tarrant, and G. M. Hansen. USDA For. Serv., PNW For. Range Exp. Sta., Portland, OR. 73-83.

Pezeshki, S. R., and C. D. Oliver, C.D. 1985. Early growth patterns of red alder and black cottonwood in mixed species plantations. For. Sci. 31:190-200.

Puettmann, K. J., D. S. DeBell, and D. E. Hibbs. 1993. Density management guide for red alder. Forest Research Laboratory, Oregon State Univ., Corvallis, Res. Contrib. 2.

Tarrant, R. F. 1961. Stand development and soil fertility in a Douglas-fir-red alder plantation. For. Sci. 7:238-246.

Tarrant, R. F., B. T. Bormann, D. S. DeBell, and W. A. Atkinson. 1983. Managing red alder in the Douglas-fir region: some possibilities. J. For. 81(12):787-792.

Trappe, J. M., J. F. Franklin, R. F. Tarrant, and G. M. Hansen, *eds*. 1968. Biology of alder. USDA For. Serv., PNW For. Range Exp. Sta., Portland, OR.

Turpin, T.C. 1981. Managing red alder on the Siuslaw National Forest. *In* Proceedings, National Silviculture Workshop—Hardwood Management, 1-5 June 1981, Roanoke, VA. USDA For. Serv., Timber Management, Washington, D.C. 268-272.

Warrack, G. C. 1964. Thinning effects in red alder. British Columbia For. Serv., Res. Div., Victoria.

Worthington, N. P., F. A. Johnson, G. R. Staebler, and W. J. Lloyd. 1960. Normal yield tables for red alder. USDA For. Serv., Res. Pap. PNW-36.

## 15

# Fertilization and Nutrition of Red Alder
M. A. RADWAN & D. S. DEBELL

## The Need for Fertilization in Red Alder Management

Red alder (*Alnus rubra* Bong.) is well known for its ability to increase soil nitrogen (N) and organic matter through biological atmospheric-nitrogen ($N_2$) fixation and production of litterfall (e.g., Tarrant and Miller 1963, DeBell et al. 1983, Radwan et al. 1984). Because of these attributes, red alder has been proposed for reclamation of coal mine spoils and as a source of biological nitrogen to manage in mixture or in crop rotation with Northwestern conifers (Tarrant and Trappe 1971).

Potential benefits from red alder, however, should not obscure the fact that the species has its own nutritional requirements for growth and development. Good growth of and benefits from red alder, therefore, cannot be accomplished unless the soil has or is provided with sufficient amounts of nutrients other than nitrogen to meet the trees' requirements for healthy growth (Radwan et al. 1984).

Red alder has also been known to have some undesirable effects on the soil. It has been reported to acidify the soil (Bollen et al. 1967; Franklin et al. 1968), accelerate leaching of important bases, such as calcium and magnesium (Bollen et al. 1967), and decrease available phosphorus (Comp- ton and Cole 1989). Moreover, red alder is a relatively short lived tree (Smith 1968), and declining growth with age may be related to limiting supplies of some essential elements (DeBell and Radwan 1984).

Clearly, red alder may require and benefit from fertilization. Application of fertilizers starts in the nursery where production of quality planting stock requires large amounts of nutrients, including nitrogen (Radwan et al. 1992). Fertilizers may also be applied to young stands to improve productivity on some sites and, perhaps, to older stands to stave off early aging and deterioration.

## Growth Responses to Fertilizer

The exact nutritional requirements for normal growth and development of red alder have not been determined. Also, the literature contains little on fertilization of the species. This is not surprising. Most forest managers have considered red alder a weed tree. Few have appreciated the species, and herbicides and a variety of mechanical methods have been used to kill the trees in conifer stands, along forest roads, and in power-line rights-of-way.

Few fertilization tests have been conducted with red alder, and most were with potted seedlings. Documented fertilization work, both published and unpublished, is detailed below.

### Potted Seedlings

Early fertilization tests with red alder seedlings started with experiments by Hughes et al. (1968) at the University of Washington. They grew red alder seedlings for 20 weeks in Lyndon loam soil mixed with various fertilizers as follows: N, P, N+P, N+P+S, N+P+S+K, N+P+S+K+Ca, and N+P+S+K+Ca+Mg. Nitrogen, as ammonium nitrate, significantly depressed growth (-76 percent). Without nitrogen, the phosphorus fertilizer improved growth (57 percent), but the N+P treatment resulted in significantly smaller seedlings (-52 percent). In presence of nitrogen, calcium increased growth, potassium and sulfur fertilizers were without effect, and magnesium resulted in very poor growth. Seedlings suffered from poor drainage and inadequate aeration, however, and no recommendations were made for field fertilization.

*Figure 1. Red alder seedlings grown in Wishkah (left) and Grove soils without fertilizer.*

More pot tests were carried out in the 1980s. Two experiments were conducted at the USDA Forest Service Forestry Sciences Laboratory in Olympia, Washington. The objective was to determine effects of phosphorus, potassium, calcium, magnesium, sulfur, cobalt, and molybdenum fertilizers on red alder seedlings grown in three different soils in a lathhouse (Radwan 1987). Soil series used were Grove and Bunker in experiment 1 and Grove and Wishkah in experiment 2. Experiment 1 compared untreated controls with five different fertilizers, which were applied singly at low and high rates equivalent to the following in kg/ha: 150 and 300 P, 100 and 200 K, 1200 and 2400 Ca, 75 and 150 Mg, and 75 and 150 S. In experiment 2, comparisons were made between controls, a single fertilizer (phosphorus), and six different fertilizer mixtures (P+Ca, P+Ca+Mg, P+Ca+Mg+K, P+Ca+Mg+K+S, P+Ca+Mg+K+S+Co, and P+Ca+Mg+K+S+Co+Mo). Rates of application in kg/ha were as follows: 300 P, 1800 Ca, 75 Mg, 100 K, 75 S, 0.1 Co, and 0.5 Mo. Briefly, the results showed that growth in the unamended soil was much better in Grove than in Bunker or Wishkah soils; the latter two soils were especially low in extractable phosphorus. Among fertilizer treatments, growth in Grove soil was enhanced most by phosphorus, calcium, and P+Ca+Mg+K+S. Growth in Bunker soil was increased only by the phosphorus treatment. In Wishkah soil, best growth was obtained when phosphorus was used alone; all fertilizer mixtures produced significantly less growth than that obtained with phosphorus alone. Differences between red alder seedlings grown in Wishkah (low phosphorus) and Grove (higher phosphorus) soils without fertilizer are illustrated in Figure 1. The stimulating effect of phosphorus fertilizer on growth of red alder seedlings is shown in Figure 2. Figures 3 and 4 show percent response of seedling dry weight to various fertilizers used in experiments 1 and 2.

*Figure 2. Red alder seedlings grown in Wishkah soil, with (left) and without phosphorus fertilizer.*

## New Plantations

In 1986, the authors participated in installation of a comprehensive screening trial of nitrogen, phosphorus, potassium, and lime amendments in a new red alder plantation near Yelm in western Washington. The study involved two low-elevation half-sib red alder families (Nisqually and John's River), two irrigation levels (high and low), and selected fertilization treatments of the four amendments. Irrigation treatments were applied soon thereafter to main plots containing 100 measurement trees and a surrounding buffer row. Families were planted in alternating rows within the plots, and fertilizer treatments were applied to four-tree subplots. Each treatment plot included tests of two nutrients (i.e., either nitrogen and phosphorus or potassium and lime) applied at all combinations of five levels each. Fertilizers and rates of application tested were as follows: triple superphosphate at 0, 100, 300, 500, and 1,000 kg P/ha; ammonium nitrate at 0, 20, 50, 100, and 200 kg N/ha; muriate of potash at 0, 100, 200, 400, and 1,000 kg K/ha; and lime at 0, 1, 2, 5, and 10 tons/ha. The potassium and lime plots also received a blanket application of triple superphosphate at 300 kg P/ha.

After four years, our unpublished results showed that height and diameter growth differed by family, irrigation regime, and rate of fertilizer application. In both irrigation regimes, growth of both families responded positively to added phosphorus fertilizer. The 300 kg P/ha treatment appeared optimum for growth of the Nisqually under low irrigation. The 500 kg P/ha treatment, however, was associated with best growth of both families under high irrigation and with that of the John's River family under low irrigation. Low amounts of nitrogen fertilizer (50 kg/ha) also stimulated height growth, but beneficial effects were limited to the John's River family and were greatest under low irrigation. Growth was not affected by lime additions, and growth response to potassium fertilizer was inconsistent.

A spacing study also established by the authors in 1986 at the same location produced similar findings. Nineteen half-sib families were distributed randomly within plots spaced at 0.5 x 0.5 m, 1.0 x 1.0 m, and 2.0 x 2.0 m. Twelve plots were established at each spacing; six each were irrigated at high and low levels; and within each irrigation level, three plots were fertilized with triple superphosphate at 300 kg P/ha and three plots were not fertilized. After five years, no differences were associated with phosphorus fertilizer in spacings of 0.5 m and 1.0 m irrigated at the high level. Tree growth in plots receiving irrigation at the low level, however, was stimulated by phosphorus; response was 10 to 12 percent in both height and diameter. Height growth in the widest-spaced (2.0 m) plots was increased about 5 percent by phosphorus in plots irrigated at both high and low levels, but the fertilizer had little or no effect on diameter growth.

*Established Young Plantations*

In 1983, scientists of the Olympia Forestry Sciences Laboratory installed a fertilizer test in a 4-year-old red alder plantation near McCleary, Washington. The primary soil is Olympic, but inclusions of other soil series are also present. The experiment consisted of three fertilizer treatments applied to plots in two blocks. The fertilization treatments included (1) unfertilized control, (2) triple superphosphate (200 kg P and 125 kg Ca/ha), and (3) triple superphosphate plus potassium sulfate (100 kg P + 62 kg Ca + 100 kg K + 41 kg S/ha). One block was well drained and the other was on a poorly drained site, sometimes covered with standing water. Within each block, treatments were assigned to two plots each, and fertilizers were broadcast by hand in March. Height and diameter of 25 trees in each plot were measured before the experiment was installed and two years after fertilization. Soil samples collected in summer 1984 were analyzed for selected chemical properties. The experiment was terminated in spring 1985 when the study area was accidentally sprayed with herbicides by the landowner. Results were assessed by covariance analysis, using pretreatment size as the covariate.

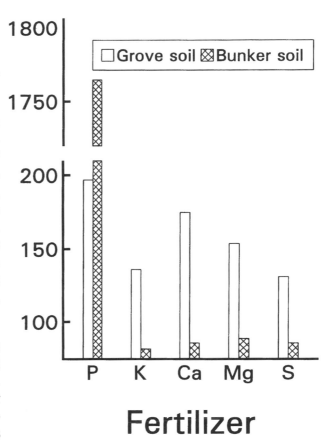

Figure 3. Effect of different fertilizers on seedling dry weight of red alder. Fertilizers were applied singly. Fertilizer sources and application rates are given in the text. Values are averages of two application rates.

*Figure 4. Effect of different fertilization treatments on seedling dry weight of red alder. Fertilizers were applied in the order shown in an additive manner (i.e., P, P + Ca, P + Ca + Mg, etc.). Fertilizer sources and application rates are given in the text.*

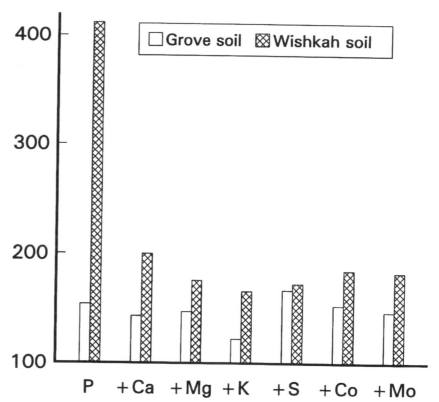

## Seedling Dry Weight
## (percent of control)

## Fertilizer

The unpublished results showed that fertilization affected height more than diameter growth. On the well-drained plots, effects on growth were negligible or slightly negative with triple superphosphate and decidedly negative (i.e., -16 percent for height growth) with the superphosphate plus potassium sulfate treatment. On the poorly drained plots, however, effects of the triple superphosphate treatment were substantial and positive (i.e., 15 percent for diameter and 29 percent for height); response to the superphosphate plus potassium sulfate treatment was much lower (i.e., 4 percent for diameter and 9 percent for height). The favorable effects of phosphorus fertilizer on alder in poorly drained soil may have resulted, at least in part, from the much lower extractable phosphorus in that soil compared with levels in the well-drained soil (9 vs. 83 ppm). The same soils were used later in a pot test to further explore interactions between phosphorus and soil waterlogging as discussed in a subsequent section.

## Natural Stands

In 1961, the University of Washington installed a fertilizer test in a 17-year-old red alder stand at the Pilchuck Tree Farm, near Arlington, Washington (S. P. Gessel, University of Washington, Seattle, pers. com.). There were twelve 0.04-ha plots, and each plot received one of twelve fertilization treatments, including an unfertilized control. Fertilization treatments supplied nitrogen, phosphorus, potassium, calcium, magnesium, sulfur, boron, zinc, and copper in various combinations and at different rates. Fertilizers used were ammonium nitrate, triple superphosphate, muriate of potash, limestone flour, dolomite, borax, zinc sulfate, copper sulfate, and fritted trace elements. The trees were remeasured in 1965, and basal-area growth was calculated. Unpublished results showed that some nutrient elements, such as potassium, seemed to have increased growth of red alder at that site. Results, however, were inconclusive. No further assessment of treatment effects was possible because the experiment was terminated when the landowner harvested the trees. We know of no other fertilizer tests in natural stands of red alder.

## Planting Stock

Quality red alder planting stock cannot be raised without fertilizers containing various nutrients, including nitrogen (see also Ahrens, Chapter 12). Both manure and synthetic fertilizers are used, and different nurseries follow different regimes. The Webster State Forest Nursery, near Olympia, Washington, has had much experience in producing various types of quality red alder planting stock. Following are examples of the successful fertilization regimes used at that nursery (W. Fangen, Webster State Forest Nursery, pers. com.).

Containerized seedlings (plugs) are started from seed sown in 1:1 (v/v) mixture of peat moss and vermiculite inoculated with red alder endophytes. Usually, fertilization with a dilute solution of a starter fertilizer, such as 7-40-17, is started after seedling emergence and continues to be used twice a week until the seedlings are 1 inch tall. A dilute solution of 20-20-20 fertilizer is then used twice a week until mid-August when a finisher fertilizer, such as 4-25-35, is applied once a week until mid-October.

For production of bed-house and open-bed stock, seed is sown in nursery beds which have been inoculated with red alder endophytes and fertilized with 10-20-20 fertilizer at 100 lb/acre, 0-45-0 fertilizer at 200 lb/acre, and dolomite at 2,000 lb/acre. The same fertilization regime is used when containerized seedlings, raised in the greenhouse, are transferred in early summer to open nursery beds to provide transplant stock (plug + 1/2) by the end of the year.

## Phosphorus and Growth in Waterlogged Soil

Application of phosphorus fertilizer to improve growth of southern pines on poorly drained sites in the lower Coastal Plain is well known (Pritchett 1979). Results from our fertilization trial at the McCleary plantation reported above suggested that phosphorus may also improve growth of red alder on wet sites. Before testing phosphorus in the field, we conducted a test on potted seedlings to compare performance of red alder under drained and waterlogged conditions, with and without phosphorus fertilizer.

Test soils were collected from the control plots of the McCleary site. One soil (S-I) was collected from a wet, waterlogged plot, and the other soil (S-II) was obtained from a well-drained plot. One-year-old red alder containerized seedlings were planted individually in pots filled with the different soils. Seedling performance in the two soils, under drained and waterlogged conditions, with and without triple superphosphate fertilizer (0, 200, and 500 kg P/ha) was compared. Some of our unpublished results are shown in Figures 5—7. Briefly, the results showed that under drained conditions and without fertilizer, seedlings grew much better in S-II than in S-I. Without phosphorus fertilizer, waterlogging reduced growth in both soils, but reductions were much greater for seedlings grown in S-I than for those grown in S-II. Application of phosphorus enhanced shoot and root growth in drained S-I but not in drained S-II and also improved growth of the waterlogged seedlings in both soils; the fertilizer was more effective at 200 kg P/ha than at 500 kg P/ha, but effects differed by soil. In S-I, phosphorus greatly reduced the adverse effects of waterlogging on seedling growth. In S-II,

*Figure 5. Growth of red alder seedlings in drained (left) and waterlogged soil I (a naturally wet soil) without fertilizer.*

the combination of fertilizer and waterlogging resulted in greater height growth and seedling weight than was obtained in any of the other treatments. Although many questions remain, results of this study show clearly that phosphorus fertilization of red alder can increase productivity of the species on both well-drained and waterlogged sites. Poorly drained and waterlogged sites are most likely to be used for commercial production of alder because they are inherently less productive for conifers, provided that such management is not precluded by wetland regulations.

## Nutrient Concentrations of Unfertilized and Fertilized Trees

The literature on foliar nutrient concentrations of unfertilized and fertilized red alder, especially in natural stands, is limited. Available information, both published and unpublished, comes mainly from work by our group in Olympia (Table 1). As expected, nutrient concentrations in foliage of unfertilized trees vary with tree age, soil/site conditions, time of the year, and so forth. In general, nutrient concentrations in red alder leaves are within the range of values reported in the literature for other deciduous tree species, including green alder (*Alnus crispa* [Ait.] Pursh) (Henry 1973; Grigal et al. 1979). Table 1 shows the following concentration ranges for the species:

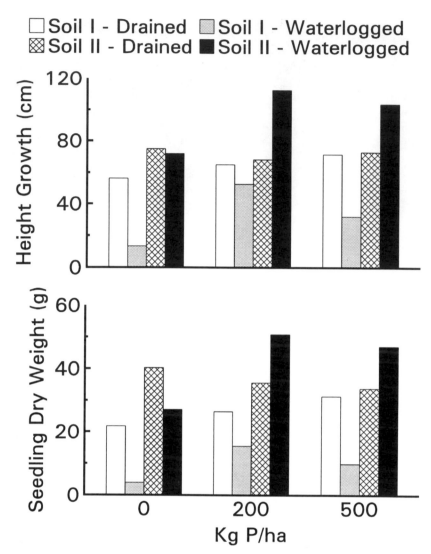

*Figure 7. Effect of waterlogging and phosphorus fertilizer on height growth (A) and dry weight (B) of red alder seedlings grown in two different soils.*

nitrogen = 1.65 to 2.96 percent; phosphorus = 0.11 to 0.23 percent; potassium = 0.80 to 1.25 percent; calcium = 0.35 to 1.07 percent; magnesium = 0.13 to 0.34 percent; sulfur = 0.14 to 0.20 percent; iron = 46 to 500 ppm; manganese = 100 to 9500 ppm; copper = 8 to 13 ppm; and zinc = 30 to 132 ppm. There is no information on foliar concentrations of boron or molybdenum, the other two micronutrients considered essential for normal growth of green plants. Molybdenum could be especially important because of its known influence on biological fixation of atmospheric nitrogen.

Fertilization may affect foliar nutrient concentrations in red alder, but trends vary by nutrient, fertilizer(s) used, and rate of application (Table 1). Important factors which may ultimately influence fertilizer effect(s) on foliar nutrient concentrations include dilution by increased growth following fertilization, changes in soil pH due to fertilizer action, and nutrient interactions.

As with most forest tree species, there is no information on optimum concentrations of foliar nutrients required for maximum growth

Table 1. *Foliar nutrient concentrations of unfertilized and fertilized red alder.*

| Tree age yr | Fertilizer | Time of year | N % | P % | K ppm | Ca | Mg | S | Fe | Mn | Cu | Zn | Reference |
|---|---|---|---|---|---|---|---|---|---|---|---|---|---|
| **A. Potted Seedlings** | | | | | | | | | | | | | |
| 1 | 0 | | 1.95 | 0.13 | 1.25 | 1.07 | 0.23 | -- | 500 | 9,500 | -- | -- | Hughes et al. 1968 |
| 1.5 | 0 | Sept. | 2.45 | 0.18 | 1.17 | 0.97 | 0.23 | 0.20 | 359 | 1,367 | 11 | 89 | Radwan 1987 |
| 1.5 | 0 | Sept. | 2.34 | 0.12 | 0.90 | 0.83 | 0.19 | 0.17 | 196 | 1,558 | 9 | 79 | |
| 1.5 | 0 | Sept. | 2.40 | 0.12 | 1.09 | 0.42 | 0.34 | 0.17 | 151 | 350 | 13 | 132 | |
| 1.5 | P | Sept. | 2.69 | 0.17 | 0.86 | 0.87 | 0.23 | 0.18 | 133 | 294 | 8 | 47 | |
| 1.5 | K | Sept. | 2.19 | 0.18 | 1.46 | 0.89 | 0.22 | 0.17 | 306 | 1,880 | 10 | 92 | |
| 1.5 | P+Ca+Mg | Sept. | 2.47 | 0.15 | 0.87 | 1.01 | 0.35 | 0.18 | 191 | 1,198 | 9 | 72 | |
| **B. Young Plantations** | | | | | | | | | | | | | |
| 3 | 0 | Aug. | | 0.16 | 0.80 | 0.66 | 0.17 | -- | -- | -- | -- | -- | Radwan, DeBell, and Kraft, unpublished |
| 3 | L+K | Aug. | | 0.15 | 1.04 | 0.62 | 0.15 | -- | -- | -- | -- | -- | |
| 3 | N+P | Aug. | | 0.15 | 0.67 | 0.73 | 0.18 | -- | -- | -- | -- | -- | |
| 5 | 0 | Aug. | 2.76 | 0.18 | 0.80 | 0.67 | 0.21 | 0.18 | 88 | 122 | 12 | 36 | |
| 5 | P+Ca+K+S | Aug. | 2.83 | 0.20 | 0.72 | 0.76 | 0.22 | 0.18 | 88 | 86 | 9 | 33 | |
| **C. Natural Stands** | | | | | | | | | | | | | |
| -- | 0 | -- | 2.96 | 0.23 | 0.85 | 0.64 | 0.19 | -- | 100 | 100 | -- | -- | Hughes et al. 1968 |
| 36 | 0 | Aug. | 2.55 | 0.11 | 1.00 | 0.92 | 0.19 | -- | -- | 170 | -- | -- | Turner et al. 1976 |
| 9 | 0 | July | 2.45 | 0.17 | 0.96 | 0.44 | 0.15 | 0.18 | 58 | 173 | 9 | 33 | DeBell and Radwan 1984 |
| 28 | 0 | July | 1.79 | 0.15 | 0.99 | 0.38 | 0.13 | 0.14 | 46 | 109 | 7 | 30 | |
| 38 | 0 | July | 1.82 | 0.14 | 0.94 | 0.35 | 0.15 | 0.15 | 46 | 103 | 12 | 35 | |
| 45 | 0 | July | 1.65 | 0.14 | 0.97 | 0.40 | 0.16 | 0.14 | 46 | 186 | 11 | 42 | |

of red alder. Such information would be useful in development of fertilization guidelines for red alder in plantations or natural stands.

## Relationships Between Foliar Nutrients and Stand Age

Red alder grows rapidly the first quarter century of its life; growth slows substantially thereafter. Some stands begin to show die-back and other symptoms of aging as early as forty years (Smith 1968). Reasons for the growth decline and deterioration are unknown, but such problems are observed at earlier ages on sites of low quality than on sites of high quality (Smith 1968). It is possible, therefore, that nutritional factors may be involved in the growth decline and premature aging of red alder.

The nutritional status of trees is often determined by assessing levels of foliar nutrients. In 1979, therefore, we started a study to determine correlations between foliar concentrations of important nutrients and age of red alder stands. Six stands of natural origin, ranging in age from 9 to 45 years, were selected for the study. The stands were located on sites of similar soil and topographic conditions. We determined stand age; collected foliar samples from each stand; determined foliar concentrations of nitrogen, phosphorus, potassium, calcium, magnesium, sulfur, iron, manganese, copper, and zinc; and calculated correlation coefficients (r) between concentrations of each nutrient and stand age from mean values for each stand.

Results showed that foliar concentrations of the nutrients determined varied among the different stands (DeBell and Radwan 1984). Highest concentrations of nitrogen, phosphorus, calcium, sulfur, and iron were found in trees of the youngest stand (9 years), and concentrations of four of these elements (i.e., phosphorus, nitrogen, sulfur, and iron) showed significant negative correlations with

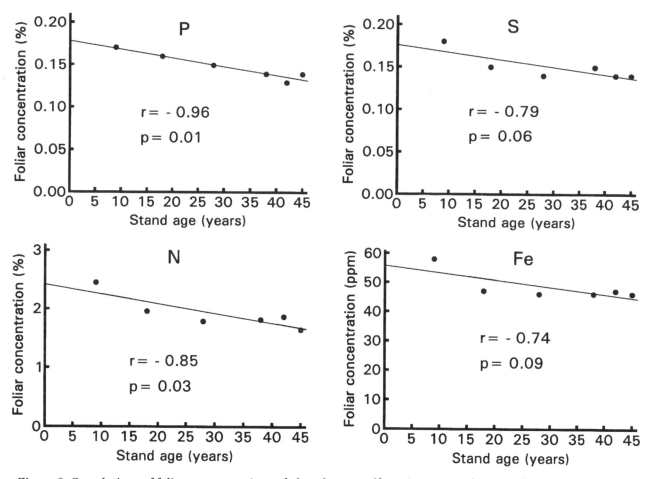

*Figure 8. Correlations of foliar concentrations of phosphorus, sulfur, nitrogen, and iron with age of red alder stands.*

stand age (Fig. 8). Results also indicated that foliar concentrations of phosphorus, nitrogen, and sulfur were usually lowest in the 42- or 45-year-old stands.

Findings from this study suggested that supplies of one or more important nutrients, such as phosphorus, nitrogen, and sulfur, may become limiting in red alder stands more than forty years old. Nutrient deficiencies may, therefore, be involved in the growth decline and deterioration of red alder stands. Moreover, application of phosphorus, nitrogen, and/or sulfur fertilizers may slow down or even prevent premature aging and enhance red alder production.

## Effects on Soil Chemistry Corrected by Fertilizers

It is well known that a crop of red alder produces some important changes in soil chemistry, including increased nitrate levels and decreased pH, concentrations of calcium, magnesium, potassium, and available phosphorus (Bollen et al. 1967; Franklin et al. 1968; Compton and Cole 1989). These changes may affect future site productivity, and a decline in productivity of second-rotation red alder has been reported on one site (Compton and Cole 1989). It is possible, therefore, that application of various fertilizers to increase available phosphorus and important bases, such as calcium and magnesium, may help restore site productivity for second-rotation red alder or other species.

*Figure 9. Growth of red alder in soils high (left) and very low in available phosphorus. The very small seedlings (right) died within one year.*

## Nutrient Requirements and Use of Fertilizer Amendments

Nutrient demands by red alder are high, but the exact nutrient requirements of the species have not been determined. It is assumed that red alder requires the same essential elements as other higher green plants for normal growth and development. However, an additional element, cobalt, shown to be required for fixation of atmospheric nitrogen by *Alnus glutinosa* (Bond and Hewitt 1962), may similarly be required by red alder (Russell et al. 1968). Root nodules and the high-energy demand of the atmospheric-nitrogen fixation process also require relatively large amounts of molybdenum and available phosphorus in the soil (Richards 1974); Figure 9 shows extremely poor growth of red alder in low-phosphorus soil. In addition, large amounts of lime are usually required whenever alder is to be successfully used for revegetation of mine spoils which tend to be high in acidity and low in available calcium. In a small pot test, we obtained the best growth of red alder seedlings in an acid spoil from a coal mine near Centralia, Washington, by adding lime at 60 mT/ha (Radwan, unpub. data). Also, application of lime and phosphorus enhanced growth of *A. glutinosa* seedlings planted in two acid mine spoils (Seiler and McCormick 1982). Clearly, soil tests followed by addition of the required fertilizer amendment(s) can correct special nutrient problems and enhance growth of alder whenever such problems occur.

## Summary and Conclusions

Although management interest in red alder has increased significantly during the 1980s in the Pacific Northwest, research in the nutrition and fertilization of the species has been very limited. Fertilization tests have been few and they have been conducted primarily with potted plants. Results of these tests indicate that fertilization with nitrogen and other nutrients is essential to production of quality planting stock of red alder. They further suggest that phosphorus fertilizer can be effectively used to improve growth of red alder in new plantations and young established stands, particularly on soils that are low in available phosphorus, poorly drained, or both, and on droughty sites. Large amounts of lime should also enhance growth of red alder on acid mine spoils.

Much basic and applied research on nutrition is still needed to optimize productivity of the species under various conditions. Basic research should include the following: (1) determination of the nutritional requirements for normal growth and development, and (2) estimation of the optimum concentrations of foliar nutrients to help guide future fertilization programs. Applied research should be directed at formulating effective fertilization prescriptions for the species under different environmental conditions. Studies should also be conducted to explore the usefulness of fertilizers containing phosphorus, nitrogen, and/or sulfur to slow down premature aging in alder. In addition, effects of fertilizers should be assessed in future studies of possible decline in productivity of second-rotation red alder.

As with other tree species, prospects for better management of red alder in the Pacific Northwest hinge on the amount and success of the biological research to be conducted in the near future. A major part of this research, we believe, should be devoted to work on nutrition and fertilization.

## Literature Cited

Bollen, W. B., C. S. Chen, K. C. Lu, and R. F. Tarrant. 1967. Influence of red alder on fertility of a forest soil. Microbial and chemical effects. Forest Research Laboratory, Oregon State Univ., Corvallis, Res. Bull. 12.

Bond, G., and E. J. Hewitt. 1962. Cobalt and the fixation of nitrogen by root nodules of *Alnus* and *Casuarina*. Nature 195:94-95.

Compton, J. E., and D. W. Cole. 1989. Growth and nutrition of second rotation red alder. Amer. Soc. Agronomy. Agron. Abstracts 1989:300.

DeBell, D. S., and M. A. Radwan. 1984. Foliar chemical concentrations in red alder stands at various ages. Plant and Soil 77:391-394.

DeBell, D. S., M. A. Radwan, and J. M. Kraft. 1983. Influence of red alder on chemical properties of a clay loam soil in western Washington. USDA For. Serv., Res. Pap. PNW-313.

Franklin, J. F., C. T. Dyrness, D. G. Moore, and R. F. Tarrant. 1968. Chemical soil properties under coastal Oregon stands of alder and conifers. *In* Biology of alder. *Edited by* J. M. Trappe, J. F. Franklin, R. F. Tarrant, and G. M. Hansen. USDA For. Serv., PNW For. Range Exp. Sta., Portland, OR. 157-172.

Grigal, D. F., L. F. Ohman, and N. R. Moody. 1979. Nutrient content of some tall shrubs from northwestern Minnesota. USDA For. Serv., Res. Pap. NC-168.

Henry, D. G. 1973. Foliar nutrient concentrations of some Minnesota forest species. Univ. of Minnesota, St. Paul. Minnesota For. Res. Note 241.

Hughes, D. R., S. P. Gessel, and R. B. Walker. 1968. Red alder deficiency symptoms and fertilizer trials. *In* Biology of alder. *Edited by* J. M. Trappe, J. F. Franklin, R. F. Tarrant, and G. M. Hansen. USDA For. Serv., PNW For. Range Exp. Sta., Portland, OR. 225-237.

Pritchett, W. L. 1979. Properties and management of forest soils. Wiley, New York.

Radwan, M. A. 1987. Effects of fertilization on growth and foliar nutrients of red alder seedlings. USDA For. Serv., Res. Pap. PNW-375.

Radwan, M. A., C. A. Harrington, and J. M. Kraft. 1984. Litterfall and nutrient returns in red alder stands in western Washington. Plant and Soil 79:343-351.

Radwan, M. A., Y. Tanaka, A. Dobkowski, and W. Fangen. 1992. Production and assessment of red alder planting stock. USDA For. Serv., Res. Pap. PNW-450.

Richards, B. N. 1974. Introduction to the soil ecosystem. Longman, New York.

Russell, S. A., H. J. Evans, and P. Mayeux. 1968. The effect of cobalt and certain other trace metals on the growth and vitamin $B_{12}$ content of *Alnus rubra*. *In* Biology of alder. *Edited by* J. M. Trappe, J. F. Franklin, R. F. Tarrant, and G. M. Hansen. USDA For. Serv., PNW For. Range Exp. Sta., Portland, OR. 259-271.

Seiler, J. R., and L. H. McCormick. 1982. Effects of soil acidity and phosphorus on the growth and nodule development of black alder. Can. J. For. Res. 12:576-581.

Smith, J. H. G. 1968. Growth and yield of red alder in British Columbia. *In* Biology of alder. *Edited by* J. M. Trappe, J. F. Franklin, R. F. Tarrant, and G. M. Hansen. USDA For. Serv., PNW For. Range Exp. Sta., Portland, OR. 273-286.

Tarrant, R. F., and R. E. Miller. 1963. Accumulation of organic matter and soil nitrogen beneath a plantation of red alder and Douglas-fir. Soil Sci. Soc. Am. Proc. 27:231-234.

Tarrant, R. F., and J. M. Trappe. 1971. The role of *Alnus* in improving the forest environment. Plant and Soil (spec. vol.) 1971:335-348.

Turner, J., D. W. Cole, and S. P. Gessel. 1976. Mineral nutrient accumulation and cycling in a stand of red alder (*Alnus rubra*). J. Ecol. 64:965-974.

# 16

# Growth and Yield of Red Alder

Klaus J. Puettmann

The increased demand for red alder as a commercial species has put a major focus on predicting alder growth and yield. As prices increase, new management options also become of interest. The evaluation of these options requires knowledge of their effect on wood quality, individual tree growth, and stand growth and yield.

This chapter discusses growth and yield aspects of red alder. The discussion is limited to wood yield, with the major emphasis on sawtimber production. Yield aspects such as nitrogen input are discussed by Binkley et al. (Chapter 4).

Growth and yield are of special interest in newly established stands, either plantations or naturally regenerated, and in existing unmanaged stands. The majority of the latter stands are 20 years and older and started from natural regeneration after harvesting operations or other disturbances. The interest in young stands derives from their flexibility in management; the interest in older stands is on the optimal transition to a managed stand condition. Most available information on growth and yield comes from unmanaged stands.

No recent summary of growth and yield information exists. Information about managed stands is especially rare. What exists is sketchy and must be assembled from various studies which vary in objectives and design. The major sources of growth and yield information include spacing studies, thinning studies, and natural stand studies. Since no stand was monitored for a full rotation, however, yield estimates have to be based on chronosequences (an age sequence of similar stands) and through extrapolation of short-term studies.

The following is an overview of the current status of growth and yield information and research, starting with a review of the existing growth and yield tools. Special emphasis is placed on tools developed in the 1980s, their strengths and shortcomings, and the need for future work. Next follows a review of spacing and thinning studies, which for ease of explanation is broken down into height growth, diameter growth, and mortality. This information is then combined in a discussion of the potential yield of managed stands.

## Overview of Current Growth and Yield Tools

A number of tools are available to predict growth and yield of individual trees and whole stands. This section reviews the current status of these tools. The discussion includes site index and height growth equations, volume equations and tables, yield tables and growth models, and stand density management guides.

### Site Index and Height Growth Equations

Until recently, the commonly used method of indexing site was based on the work of Bishop et al. (1958) as presented by Johnson and Worthington (1963) and used in the Normal Yield Table for Predominantly Red Alder Stands by Worthington et al. (1960). This index was based on the top height at 50 years of age, reflecting the length of then-common rotations. Since that time, however, the expected rotation length has been reduced to 25 to 45 years. This drop in rotation length has resulted in attempts to index current stands which fall at the lower age limit of the data base used to develop the site index curves. Since any site indexing based on top height is more accurate when the measurement age is close to index age, using an index age of 50 years to determine the site index in young stands results in decreased accuracy.

Responding to this concern, Harrington and Curtis (1986) developed a site index based on top height at age 20 ($SI_{20}$). They included Johnson and Worthington's (1963) data and measured additional trees, emphasizing the age range below 20 years. This should lead to increased accuracy for site index estimation in younger stands. During the first decade of stand growth, however, the influence of site index on height growth is hard to separate from other factors (DeBell and Giordano, Chapter 8; Newton and Cole, Chapter 7). Site estimation in stands younger than 10 years should not be based on height. Other measures, like height growth, relative height growth, or soil-site factors (Harrington and Courtin, Chapter 10) might be more useful for estimating site index in young stands. Switching measurement systems will lead to some confusion during the conversion period, but since an extension of the red alder rotation for managed stands beyond 45 years is not foreseeable, the use of $SI_{20}$ in future site-quality estimations is recommended.

In many red alder stands, top height (and thus site index) cannot be directly measured because alder grows in mixture or the effect of past treatments on height growth is not known. To obtain site index information on these sites and on those sites currently not stocked with red alder Harrington (1986; Harrington and Courtin, Chapter 10) developed an indexing method based on soil and site properties. Information about the geographic and topographic position, soil fertility, and physical condition as well as soil moisture and aeration during the growing season was related to the site requirement of red alder. This method is of special interest to landowners who want to expand red alder plantations to sites that currently have no alder stands.

The initial validation by Harrington (1986) showed extremely good fit between the estimated and measured site index (n = 15 $r^2$ = 0.96), but Harrington placed special emphasis on applying this method to homogeneous sites. This may be a serious problem, because red alder is often found on very heterogenous sites.

Utilizing the site conditions and site requirements of a species for establishment of site index has an intrinsic attraction over measuring top height because it links the biological requirements of a species to environmental conditions. This has several advantages. First, site index estimation is not influenced by past stand conditions (overstory, weeds, density, etc.). Several environmental variables can be linked to biological requirements of several species, allowing a conversion of site index from one species to another. Finally, the changes in biological requirements (e.g., by genetic improvement) or environmental conditions (e.g., fertilization) can be incorporated into estimation of site quality. Nonetheless, application of this method has been very limited and further validation, especially of intensively managed forest, is needed to establish confidence in this method.

The most frequently used height growth model in the past was based on the site index equation developed by Bishop et al. (1958) and presented in Worthington et al. (1960). These equations were developed to predict site index from stand height and, therefore, are not the best equations to predict stand height (for detailed discussion see Curtis et al. 1974). Also, as pointed out in the discussion about site index, the age range in the data reflected the contemporary rotation length and the equations use the minimum age of 10 years. Harrington and Curtis (1986) expanded the data base to include younger stands and developed height growth equations as distinct from site index equations for red alder. Because of their broader age range in the data base, the proper choice of the variable in the analysis, and, most important, the polymorphic curve shape, these equations are an improvement over the earlier work. Differences between these curves are shown in Figure 1 (Figure 3 from Harrington and Curtis 1986). Growth and yield tools that utilize height to predict stand growth (e.g., Stand Projection System [SPS], Arney 1985) should use the new height growth equation in the prediction process.

Worthington et al. (1960) assumed that red alder would reach breast height in 3 years regardless of site quality. Harrington and Curtis (1986) had more young trees in their data base and their analysis indicated that the time to reach breast height varies with site quality. On higher (SI > 25 m), medium (15 m < SI < 25m), and lower quality sites

## Red Alder Height Growth

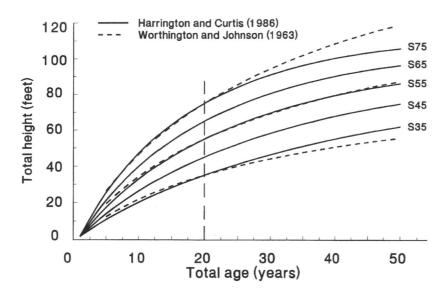

*Figure 1. Height growth curves derived by Harrington and Curtis (1986) and Johnson and Worthington (1963) (Figure 3 from Harrington and Curtis 1986).*

(SI < 15 m) 1, 2, and 3 to 4 years, respectively, are necessary for red alder to reach breast height.

In summary, the 1980s have shown a transition in application of site index and height growth equations from unmanaged natural mature stands to managed stands. While the new tables and equations are still based on data collected from unmanaged stands, the age range and index age reflects the changed need for information about younger stands.

### Volume Tables and Equations

Volume tables and equations are a vital link between the easily measured height and diameter (or basal area) information and the estimation of tree and stand yields. Nonetheless, any volume equation has to compromise the need to be simple while including enough information to be able to adjust to a range of stem forms and tree sizes. This compromise and the realization that plant growth patterns vary in different environments have led to development of different volume equations for British Columbia and for Washington and northern Oregon.

Smith (1968) reviewed the development of volume equations in British Columbia. The currently used individual tree volume equations are based on diameter and height of a tree (Smith and Breadon 1964). Munro (1967) expanded these equations to predict stand volume based on basal area and average tree height. Smith (1972) analyzed a data set consisting largely of young trees and developed equations for individual trees.

Several equations are in use in Washington and Oregon. The Normal Yield tables (Worthington et al. 1960), which are frequently used for stand volume estimations are based on volume tables generated by Johnson (1955). The Normal Yield Tables were constructed for natural stands; minimum diameter in the board foot table is 10 inches. The cubic and Scribner board foot volumes in the Empirical Yield Tables for Predominantly Red Alder Stands in western Washington (Chambers 1973) are based on tariff information generated by Turnbull et al. (1963), while the stock tables use the tables generated by Curtis et al. (1968). Other equations used today (e.g., Hibbs et al. 1989) were developed around the same time. Skinner's (1959) data set had a minimum height of 16 m and a minimum diameter of 15 cm. He then extrapolated the volume relationship to smaller trees. No subsequent verification of this extrapolation has been made and use of this table for small trees should be avoided.

The above studies were developed when intensive management of red alder stands for timber production was not foreseeable. Consequently, their focus was on mature, harvestable natural stands and did not include young or intensively managed stands. Thus caution is necessary when these equations are used to evaluate thinning or spacing studies, especially when ages of stands are beyond the data range used in the development of the equations.

In addition, the discussion about height and diameter growth (this chapter) indicates that these growth components do not respond in parallel fashion to density manipulations (Hibbs and DeBell, Chapter 14). Management can strongly influence taper and tree volume, yet the currently available equations do not reflect these changes. This complicates any interpretation of yield in thinning or spacing studies. Most emphasis should, therefore, be put on diameter and height growth responses because volume growth is confounded with errors in volume calculations.

In summary, volume tables and equations have been neglected since the 1960s. Research is needed to make the transition from mature, unmanaged to young, intensively managed stands. While lumber recovery (Plank et al. 1990) and allometric studies (Cooke 1987) have investigated tree form, thorough study of the effects of management practices is needed to develop volume equations that can help the land managers in their decision. Because management can increase the range of growing conditions for the individual trees, the allometric variation will probably increase. It can be hypothesized that a volume equation based solely on diameter at breast height (dbh) and height is not sufficient to characterize the full spectrum. Additional measurements (e.g., a second diameter measurement at another height) might be necessary to characterize the stem form and improve accuracy for volume prediction. A series of local volume equations may be an effective way to deal with regional differences in stem form.

## Yield Tables and Growth Models

Currently, most yield estimates are based on the Normal Yield Tables for Red Alder (Worthington et al. 1960), the Empirical Yield Table for Predominantly Alder Stands in Western Washington (Chambers 1973), and the Stand Projection System (SPS) (Arney 1985). All three prediction tools are derived from data collected in unmanaged stands.

To evaluate the accuracy of these tools, Puettmann and Hibbs (1988) compared the yield table and growth model predictions with growth records from actual stands. The stand growth data were obtained from research and inventory plots in natural stands in southwestern British Columbia, western Washington and northwestern Oregon. The 46 permanent plots included a total of 212 measurement periods. Only directly measured variables such as quadratic mean diameter ($D_q$), trees per acre (tpa), and basal area were compared to avoid the confounding effects of volume equations.

The Normal Yield Tables for Red Alder (Worthington et al. 1960) predicts stand characteristics of normal stands for a given age and site index. The average difference between predicted and actual values are presented in Table 1. For a stand with a minimum dbh of 5.5 inches, the average $D_q$ of the yield table is very close to the $D_q$ from the measured stands. The average density is underestimated by 47 trees per acre.

Using a minimum dbh of 9.5 inches resulted in less accurate predictions. The average underestimation of the $D_q$ is notable. Also, the difference between predicted and actual trees per acre and basal area per acre is substantially larger than for trees greater than 5.5 inches dbh.

For both diameter options, further investigation showed underestimation of $D_q$ in young stands and

Table 1. *Differences between values predicted (pred.) by the Normal Yield Table for Red Alder (Worthington et al. 1960) and the Empirical Yield Table for predominantly alder stands in western Washington (Chambers 1973) versus observed (obs.) plot values.*

| | Normal Yield Table (trees > 14 cm) | | | Normal Yield Table (trees > 24 cm) | | | Empirical Yield Table (trees > 18 cm) | | |
|---|---|---|---|---|---|---|---|---|---|
| | pred. | obs. | pred./obs. | pred. | obs. | pred./obs. | pred. | obs | pred./obs. |
| *Pure red alder stands* | | | | | | | | | |
| Basal area (m$^2$/ha) | 26.1 | 29.7 | 0.88 | 27.0 | 19.6 | 1.38 | N.A. | N.A. | N.A. |
| Density (tpha) | 521 | 637 | 0.82 | 519 | 257 | 2.02 | 533 | 487 | 1.10 |
| Quadratic mean diameter (cm) | 25.38 | 25.20 | 1.01 | 25.9 | 31.5 | 0.82 | 24.92 | 26.00 | 0.96 |
| *Red alder component of mixed red alder/ Douglas-fir stands* | | | | | | | | | |
| Basal area (m$^2$/ha) | 29.3 | 16.5 | 1.78 | 30.8 | 21.9 | 1.41 | N.A. | N.A. | N.A. |
| Density (tpha) | 457 | 319 | 1.4 | 442 | 296 | 1.5 | 232 | 259 | 0.90 |
| Quadratic mean diameter (cm) | 29.00 | 24.79 | 1.17 | 30.27 | 30.32 | 1.00 | 26.42 | 28.04 | 0.94 |

overestimation in older stands. Trees per acre showed the opposite trend (i.e., overestimation in young stands and underestimation in older stands). The most accurate predictions were in the age range of 30 to 50 years.

These results indicate that the diameter distribution underlying the yield tables is shifted toward higher diameters than observed in the actual stands. In today's market, the minimum diameter of usable trees is lower than 9.5 inches so the inaccuracies in the 9.5-inches-and-above table is less important now than it was when the tables were made.

The Empirical Yield Tables for Predominantly Alder Stands in Western Washington (Chambers 1973) require the measurement of basal area in addition to age and site index. This information allows calculation of percent normal basal area (PNBA), which is then used to adjust for stands with different densities. This procedure improves estimates of $D_q$ and tpa substantially (Table 1). Detailed investigation determined that $D_q$ tends to be underestimated at lower densities and overestimated at higher density. Since basal area was measured, trees per acres shows the reverse trend. Stand density was also overestimated for stands on lower sites. Overall, however, the added measurement of stand basal area increased the accuracy of the predicted stand characteristics significantly compared with Worthington et al. (1960).

SPS (Arney 1985) is a growth model, and therefore diameter growth, basal area growth, and mortality were evaluated. A stand table derived from the first measurement of every permanent plot was used as input in the model. This plot was projected by SPS to all ages at which subsequent measurements were made. Predicted values were then compared with actual stand values. The major shortcoming of SPS predictions was the substantial underestimation of mortality (Table 2), which is also reflected in overestimation of basal area growth. The underestimation of mortality becomes a major problem with increased projection length (average projection length in the study was 13 years). Most foresters will not use a diameter distribution table as input, and longer projection periods than were tested by this approach will lead to even less-accurate predictions.

Curiously, while SPS underpredicted mortality in these unmanaged high-density stands, it overestimated mortality in stands of lower densities. A new red alder plantation stand of 400 tpa (1000 tpha) will not experience density related mortality for a number of years (DeBell, unpub. data). Since SPS predicts mortality during the initial growing phases, it therefore is not recommended for prediction of either unmanaged or managed red alder stands.

To evaluate the predictive abilities of these tools for mixed-species stand characteristics, different modifications of the basal area proportion of red alder were used. No accurate predictions could be developed for the Normal Yield Tables for Red Alder (Worthington et al. 1960) (see Table 1). To utilize the Empirical Yield Table for Predominantly Alder Stands in Western Washington (Chambers

*Table 2. Differences between values predicted (pred.) by the Stand Projection System (Arney 1985) and observed (obs.) plot values.*

| | pred. | SPS obs. | pred./o bs. |
|---|---|---|---|
| *Pure red alder stands* | | | |
| Basal area (m²/ha) | 8.5 | 5.9 | 1.44 |
| Mortality (tpha) | 168 | 286 | 0.6 |
| Quadratic mean diameter growth (cm) | 3.53 | 4.11 | 0.86 |
| *Red alder component of mixed red alder/ Douglas-fir stands* | | | |
| Basal area (m²/ha) | 5.4 | 2.6 | 2.05 |
| Mortality (tpha) | 72 | 114 | 0.6 |
| Quadratic mean diameter growth (cm) | 5.94 | 4.77 | 1.24 |

1973), the PNBA was calculated only from the red alder proportion of mixed stands. While the average predicted values were very close, at lower red alder proportions (less than 40 percent) diameter and trees per acre of red alder were overestimated. A stand diameter table for each species was used as input for growth projections with SPS. As in pure stands, mortality was greatly underestimated. In combination with the overestimation of diameter growth, this also resulted in severe overestimation of red alder basal area, a problem that increased with the length of projection.

In summary, the Normal Yield Table is not sufficiently accurate for yield estimates of pure and mixed red alder stands, especially in stands below 30 and over 50 years of age. Measuring basal area allows usage of the Empirical Yield Table and is recommended as giving the most accurate yield estimates. At red alder proportion of 50 percent and greater, the Empirical Yield Table can also be used for the red alder component in mixed red alder/Douglas-fir stands. SPS cannot be recommended for growth projection of red alder stands, because, both in pure and mixed stands, the mortality estimates were too high and in mixed stands the diameter growth also was overestimated.

All three tools estimated trees per acre and mortality with less accuracy than did diameter estimates. Because spacing and thinning activities (rather than natural mortality) determine stand density in managed stands, growth and yield estimation for managed stands can be simplified and improved by assuming controlled pattern of mortality.

## Stand Density Diagram

The need for a red alder density diagram to provide general guidance in density management was recognized in the 1980s, and Hibbs (1987) undertook a first attempt. The diagram was based on sketchy data from natural stand thinning studies and a pot seedling study. As more data became available, Hibbs and Carlton (1989) compared red alder density diagrams based on volume and diameter as size variables. They found differences in application of both diagrams. While they hypothesized problems with basic assumptions or the methodology, the

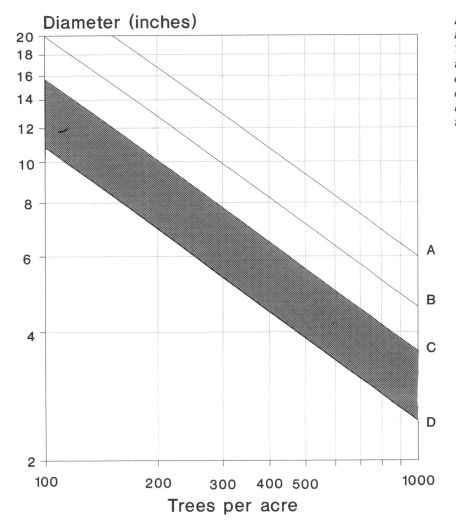

**Diameter (inches)**

**Trees per acre**

*Figure 2: Stand density diagram from (Puettmann et al. 1993a). The A line describes the biological maximum; the B line is the average self-thinning line; lines C and D are the upper and lower limit of the management zone, respectively.*

differences could simply be due to application of inaccurate volume calculations. Even though these publications included management aspects, the diagrams were not developed primarily as a management tool.

After pooling a broad data base that included mature natural stands as well as thinning and spacing studies, Puettmann et al. (1993a) developed a density management diagram for the use of land managers (Fig. 2). The diagram establishes a biological maximum (2540 tpha [1000 tpa] at 15-cm [6-inch]) diameter as the basis for relative density measure. The average self-thinning line that stands will approach is at 65 percent relative density. The suggested management zone is between 25 and 45 percent relative density, a range that assures full site occupancy while minimizing competition and mortality.

Several problem areas surfaced during the development of the diagram and need to be addressed in future work. Genetic improvement and stand treatments such as irrigation, fertilization, and weed control might increase the carrying capacity of a site and change the development pattern of stands (DeBell, unpub. data). Also, the effect

*Table 3. Summary of spacing studies of red alder.*

| Author | SI$_{50}$ | Current age | Initial density (tpha) | Current density (tpha) | Current height (m) | % of highest density | Current dbh (cm) | % of highest density |
|---|---|---|---|---|---|---|---|---|
| DeBell (unpubl.) | 30 | 12 | 11,918 | 3,528 | 14.6 | — | 9.5 | — |
| | 30 | 12 | 4,272 | 2,374 | 15.4 | 105 | 11.2 | 118 |
| | 30 | 12 | 1,272 | 1,137 | 15.6 | 107 | 14.7 | 155 |
| DeBell (unpubl.) | * | 5 | 40,000 | 13,600 | 7.5 | — | 4.2 | — |
| | * | 5 | 10,000 | 6,800 | 8.3 | 111 | 5.6 | 133 |
| | * | 5 | 2,500 | 2,450 | 8.4 | 112 | 7.0 | 167 |
| | ** | 5 | 40,000 | 21,600 | 4.9 | — | 2.6 | — |
| | ** | 5 | 10,000 | 9,400 | 6.1 | 124 | 3.8 | 146 |
| | ** | 5 | 2,500 | 2,475 | 6.4 | 130 | 5.4 | 08 |
| Hibbs (unpubl.) | 30 | 5 | 1,111 | 1,111 | 6.7 | — | 7.1 | — |
| | 30 | 5 | 567 | 567 | 7.2 | 107 | 8.0 | 113 |
| Hibbs (unpubl.) | 24 | 5 | **** | **** | 3.4 | — | 2.7 | — |
| Hook et al (1987) and DeBell (unpubl.) | 23 | 7 | 2,500 | 2,500 | 8.8 | — | 9.1 | — |
| Hook et al. (1990) | *** | 2 | 11,080 | 11,080 | 6.4 | — | 4.4 | — |
| Lester and DeBell (1989) | 24 | 15 | 2,150 | 1,075***** | 14.0 | — | 15.3 | — |

* high irrigation
** low irrigation
*** irrigation
**** density variable, replanted with wildlings
***** thinned to current density

of thinning on the size-density relationships needs to be investigated (see Hibbs and DeBell, Chapter 14).

In summary, the work on the size-density relationship has made significant progress since 1987. The density management diagram by Puettmann et al. (1993a) can be considered a comprehensive tool which cannot be improved upon until a substantial data base from managed stands is developed.

## Yield of Managed Stands

In past decades red alder growth and yield research has largely reflected the then-current management activities: the harvesting of naturally regenerated, unmanaged stands. Now that foresters are considering management of red alder, a new research effort has started to examine the potential yield of managed stands. The research addressing growth and yield in managed stands has focused on potential gain due to density management but also has

included genetic evaluations (DeBell and Wilson 1978; Agar and Stettler, Chapter 6). The following discussion utilizes available thinning and spacing studies to address height and diameter growth and mortality.

Since most of the spacing studies are fairly young and results have not been published, results and trends are discussed in general terms. A list of spacing and thinning studies is presented in Tables 3 and 4, respectively.

### Height Growth

Spacing and thinning studies indicate red alder height growth to be sensitive to spacing. Spacing studies covering a wide range of densities (DeBell and Giordano, Chapter 8; Giordano and Hibbs 1993; Hibbs, unpub. data; Newton, unpub. data) showed that during the initial years height growth was reduced in trees growing at very high densities. The same studies have also shown that trees grown at

| Author | Stand age at treatment | Site index | Duration of study | Stand density (tpha) | Average annual height growth (m) | % of highest density | Average annual dbh growth (cm) | % of highest density |
|---|---|---|---|---|---|---|---|---|
| Berntsen (1961) | 12 | 29 | 20 | 3342 | 0.58 | — | | |
| | 11 | 26 | 20 | 1811 | 0.69 | 118 | | |
| Berntsen (1962) | 21 | 29 | 20 | 1717 | | | 0.47 | — |
| | 21 | — | 20 | 692 | | | 0.56 | 119 |
| Bormann (1985) | 7 | 35 | 20 | 1235 | 25.6* | — | 17.8 ** | — |
| | 7 | 35 | 20 | 691 | 26.2* | 102 | 22.6 ** | 127 |
| | 7 | 35 | 20 | 395 | 25.8* | 101 | 28.5 ** | 160 |
| Hibbs et al. (in review) | 20 | 30 | 5 | 1482 | 0.58 | — | 0.38 | — |
| | 20 | 30 | 5 | 488 | 0.18 | 31 | 0.76 | 100 |
| | 20 | 30 | 5 | 425 | 0.28 | 68 | 0.76 | 100 |
| | 14 | 30 | 10 | 1697 | 0.51 | — | 0.59 | — |
| | 14 | 30 | 10 | 556 | 0.42 | 82 | 0.83 | 141 |
| | 14 | 30 | 10 | 304 | 0.40 | 78 | 0.85 | 144 |
| | 14 | 30 | 10 | 296 | 0.38 | 75 | 0.96 | 163 |
| Lloyd (1955) | 26 | — | 6 | 1013 | — | | 0.28 | — |
| | 26 | 35 | 6 | 417 | — | | 0.48 | 171 |
| Smith (1983) | 10 | 29 | 3 | 8018 | 0.9 | — | 0.63 | — |
| | 10 | 29 | 3 | 2550 | 0.5 | 56 | 0.69 | 110 |
| | 10 | 29 | 3 | 1110 | 0.6 | 67 | 0.97 | 154 |
| | 10 | 26 | 3 | 638 | 0.8 | 89 | 1.13 | 179 |
| | 10 | 28 | 3 | 406 | 0.3 | 33 | 1.17 | 186 |
| | 4 | | 3 | 45600 | 1.3 | — | 0.23 | — |
| | 4 | | 3 | 4524 | 1.3 | 100 | 0.87 | 378 |
| | 4 | | 3 | 2632 | 1.5 | 115 | 1.50 | 652 |
| | 4 | | 3 | 1266 | 1.4 | 108 | 1.67 | 726 |
| | 4 | | 3 | 725 | 1.5 | 115 | 1.53 | 665 |
| | 4 | | 3 | 389 | 1.2 | 92 | 1.97 | 857 |
| Warack (1964) | 21 | 35 | 8 | 963 | 0.49 | — | 0.98 | 390 |
| | 21 | 34 | 8 | 469 | 0.27 | 50 | 0.73 | 290 |
| | 21 | 35 | 8 | 272 | 0.27 | 50 | 0.25 | — |

\* final height (m)
\*\* final dbh (cm)

extremely wide spacings have reduced height growth.

Two important conclusions are drawn from these results. First, stand densities at the time of establishment probably need to be higher than 500 trees per hectare to avoid height growth loss due to too wide spacing (see DeBell and Giordano, Chapter 8; Hibbs and DeBell, Chapter 14). Second, because site index curves like that of Harrington and Curtis (1986) are based on natural stand growth and natural stands tend to have very high initial densities, applying them to trees grown in plantations may overestimate site quality.

A reduction in height growth after heavy thinning has also been found in a variety of thinning studies (Table 4), both in young stands (Smith 1983) and in older stands (Lloyd 1955). A reduction of height growth can lead to a substantial reduction in tree and stand volume. Because of complicating effects of thinning shock, more investigation is needed to determine the relationship between height growth and stand characteristics, but thinning high-density stands to relative densities below 25 percent seems to result in height growth reduction (Hibbs and DeBell, Chapter 14). In one study, Hibbs et al. (in review) found this effect to be temporary. The height growth reduction was observed in the first 5 years after thinning, but by the second measurement period (5 to 10 years after thinning), height growth in the thinned plots seemed to equal or exceed that in control plots. This might explain why studies which have not tracked height growth over time have not detected this phenomenon (see Bormann 1985).

The phenomena of reduced height growth at extreme low densities indicates the need to reevaluate earlier studies about red alder growth and yield. For example, Apsey (1961) compared open-grown alder with alder grown in closed forests. He interpreted the lower heights in the open-grown sample to reflect lower site qualities. In fact, this could have been an effect of reduced height growth because of the wider spacing. There is a need for the interpretation of the relationship between site quality and potential stand yield of red alder to also include information on stand density.

While initial height growth in some studies is extremely fast (Table 3), most studies confirm that height growth of natural and managed stands appears to follow the relationships presented by Harrington and Curtis (1986). Caution, however, has to be used when these equations are applied to stands that have been managed at extremely high or low densities, or to stands that have been thinned heavily. If kept in an "optimal" density range and thinned very carefully, height growth might be able to temporarily exceed the values predicted by Harrington and Curtis (1986).

### Diameter Growth

Diameter is one of the major factors determining individual tree volume and value. It is generally known to be strongly influenced by stand density and has therefore been investigated in several red alder spacing and thinning studies (Tables 3, 4). Red alder seems to follow the general rule that di-

ameter growth increases with decreased stand density. At extremely low densities, however, reduced diameter growth has been observed (DeBell and Giordano, Chapter 8; Hibbs, unpub. data). Specific relationships between thinning treatments and diameter growth response are hard to detect from Tables 3 and 4 because of the confounding effects of age, site quality, and thinning intensity among these unrelated and uncoordinated studies. Thinning response seems to last until stands reach high relative densities (40 to 45 percent); in the case of Hibbs et al. (in press) this elevated diameter growth lasted more than 10 years. Older stands (> 20 years) also have shown some ability to respond to reductions of stand density (Warrack 1964).

### Mortality

Stand mortality is a major factor in the dynamics of unmanaged stands and stands managed at higher densities. Hibbs and Carlton (1989) measured a variety of mature red alder stands in the Oregon Coast Range and found evidence of mortality in stands with a relative density of 50 percent and greater. Puettmann et al. (1993b), using the size-density relationship, analyzed mortality from repeated measurement of stands. This analysis was based heavily on young stands and indicated that mortality starts at the lower relative density of 30 percent. A mortality level of 20 percent was reached at a relative density of 45 percent (Puettmann et al. 1993a).

An analysis such as that used by Hibbs and Carlton (1989) can only detect mortality in the preceding few years, but in general, mortality is not a continuous process; stands show periods of higher and lower mortality. Mortality rates in red alder stands have been found to be less variable than those in Douglas-fir stands, and the average mortality rate over a 4- to 5-year period was around 2 to 3 percent (Puettmann et al. 1992). Comparison of mortality in control plots of thinning experiments confirmed these values (Hibbs et al. 1989; Hibbs et al., in review). With proper density management, natural mortality in stands should be limited to catastrophic events such as windstorms or herbivory.

The size-density relationship of red alder has been investigated in pure (Puettmann et al. 1993b) and mixed red alder/Douglas-fir stands (Puettmann et al. 1992). This information, a basic element for the integration of mortality information into growth and yield models, provides a good basis for development of a growth simulator.

## Stand Yield

The above studies give information about growth and yield responses of red alder to specific stand management practices. To get robust estimates of potential yield of red alder in managed stands requires integrating all this knowledge and extrapolating to other stand and site characteristics. Consequently, any estimate of potential yield in managed stands is based on interpretation and speculation and should therefore be viewed with caution. As discussed earlier in this chapter, taper and volume equations were developed from natural stands and might not be accurate for stands where spacing has been controlled. Because of this concern about volume estimates, the following evaluation of growth data will also include height and diameter growth estimates.

**Height.** The comparison of height growth curves (Harrington and Curtis 1986) and site index curves (Worthington et al. 1960) for medium to higher sites shows the new height growth curves to be lower in stands older than 20 years. While both curves are based on unmanaged stands, yield estimates based on the Normal Yield Table of Worthington et al. (1960) overestimated this yield component (e.g., DeBell et al. 1978; Arney 1985).

When considering thinning previously unmanaged stands, height growth response is hard to predict accurately. Thinning and spacing studies have shown that height growth is not necessarily optimal in density ranges that encourage maximum diameter growth. In addition, it can be assumed that height growth might be reduced at least temporarily when a stand is thinned heavily.

Spacing studies indicate that for most of the density range, the initial height growth rates are as high or higher than those projected by Harrington and Curtis (1986). If stand density is thereafter kept in optimal ranges, a thinning shock will probably have no, or only a minor, effect on height growth. Thus, for plantations the height growth projections used by Harrington and Curtis

(1986) seem achievable and might underestimate growth potential for managed stands.

Reduced height does not necessarily result in reduced stand yield, especially when yield is measured as board foot yield of sawtimber. In stands older than 30 to 40 years, height growth has slowed considerably (Harrington and Curtis 1986) and further growth increases crown length but has little influence on clear bole length. A height growth reduction at this age will thus reduce total tree volume but not result in reduced board foot yield from the clear bole.

**Diameter.** Diameter growth is more sensitive to density than is height growth. Thus, density management is very important to achieve full diameter growth potential. The stands in Tables 2 and 3 cover a wide range of site qualities and stand conditions, but mean annual diameter growth rates of 0.6 to 1.2 cm were found in young plantations as well as in older thinned stands on medium to higher quality sites. The duration of observation in these studies, however, is limited. There are individual trees in natural stands exhibiting these growth rates for 30 years or more (Puettmann, pers. obs.; Apsey 1961). Thus it can be assumed that, under optimal management regimes (i.e., with good planting stock, weed control, and density management), red alder stands also should be able to sustain these growth rates for 30 years or more. From this information it can be speculated that an average diameter of 38 cm can be reached after 25 to 30 years on sites with site indices (50 year) of 37 and 30 m ($SI_{20}$ = 23 and 20 m), respectively.

**Volume.** At 25 years the tree height on a site with $SI_{50}$ = 121 feet ($SI_{20}$ = 75 ft) is approximately 85 feet. On a site with a $SI_{50}$ = 100 feet($SI_{20}$ = 66 ft) the height at age 30 would be approximately 80 feet. Using the volume tables by Johnson (1955), the average tree cubic foot volume (4-inch top) for trees of 12 inches and 85 or 80 feet height would be 26 or 25 cubic feet (4-inch top), respectively. Thus a density of 150 tpa at harvest time (based on Figure 2) results in approximately 3900 and 3675 cubic foot (4-inch top) per acre.

For the same ages and equivalent site qualities ($SI_{50}$ = 120 and 100) Worthington et al. (1960) predicted 3500 and 3420 cubic foot per acre (4-inch top) for trees greater than 5.5 inches dbh. The

Empirical Yield Table (Chambers 1973) predicted 3986 and 3774 cubic foot (total stem, trees greater than 7 inches dbh) for the above stated conditions. Based on Brackett (1973), this is equivalent to 3600 and 3430 cubic foot (4-inch top). Using the above calculations, managed stands would yield approximately 10 percent more than unmanaged stands. In addition, the volume in managed stands is concentrated in fewer, bigger trees. The average diameter in managed stands at age 25 ($SI_{50}$ = 120 ft) and 30 ($SI_{50}$ = 100 ft) is 12 inches compared with 9.2 and 9.4 inches in unmanaged stands. Since diameter is important in determining the quality of products that can be recovered from logs (Plank et al. 1990; Plank and Willits, Chapter 17), management not only improves the yield volume, but also improves the quality of the yield.

While the diameter growth in managed stands derived by DeBell et al. (1978) is very similar to the one stated above, their projections are based on a higher stand density at harvest time (170 vs. 150 tpa). Consequently, the cubic foot volume yield per acre is higher than that predicted here. In addition, while DeBell et al. (1978) claimed their estimate to be conservative, thinning and spacing studies indicate that achieving these growth rates requires intensive management during the whole rotation. Further increase in yield might be gained through genetic improvement. (Agar and Stettler, Chapter 6).

## Summary and Outlook for the Future

Recent development of the red alder market has raised interest in management for timber. Because of the recent interest, no coherent information about growth and yield of managed stands is available. In the 1980s, height growth and site index equations, and a stand density management guide have been developed specifically addressing young, managed stands. Other tools, such as volume equations or yield tables, have not been improved significantly in that time period and thus are not addressing the special needs of managed stands.

The greatest potential gain over natural unmanaged stands can be expected through density management. A review of thinning and spacing studies indicated that, above all, there is a need for coordinated research in growth and yield. Thinning

will increase diameter growth, resulting in increased tree size and thus increased log value. Heavy thinning in older stands will reduce height growth and thus the total yield per acre. If, however, thinning is done after the clear trunk has developed, the loss in recoverable lumber is minimal. Density management is crucial to fully capture diameter and height growth potential.

Considering these scenarios and the results of thinning and spacing studies under optimal management regimes, a cubic foot yield (4-inch top) of 3900 and 3675 can be reached on high quality sites in 25 ($SI_{20}$ = 75 ft) and 30 years ($SI_{20}$ = 65 ft), respectively. This is approximately 10 percent higher than yield of unmanaged stands as predicted by the yield tables. In addition, the bigger trees yield a greater proportion of high-quality products.

The need for further investigation of red alder stand dynamics is reflected in the foundation of the Hardwood Silviculture Cooperative (HSC). To establish a data base that allows modeling of red alder growth and yield, the HSC started the Stand Management Study. By setting up initial spacing and thinning plots within a general framework, the HSC has taken advantage of the opportunity to control plot location, design, and layout, so that the overall data set can be used to established a regional growth and yield model. The first plantations were established in 1989, and the first results can be expected within the next decade.

The installment of research areas has to be complemented with work on modeling to develop a growth and yield simulator. The currently available simulator (SPS) is adapted from Douglas-fir. Evaluating the performance of this model reveals the need for an alder simulator. The monetary and time effort in establishing, maintaining, and measuring the research installation are only useful when complemented by intensive utilization of the data base.

Biodiversity and long-term site productivity become important in forest management, thus species mixtures of red alder with western red cedar or Douglas-fir must be included in research studies. Yield studies need to expand to include yield aspects reflecting multiple use. Cooperation with wildlife and wood products researchers, soil scientists, and ecosystem ecologists is necessary to address future aspects of red alder growth and yield.

## Acknowledgments

I thank Dean DeBell and especially David Hibbs for their assistance in the preparation of this chapter and for allowing me to use unpublished data. William Scott also gave a useful review. The Hardwood Silviculture Cooperative provided support for some of the work described herein.

## Literature Cited

Apsey, T. M. 1961. An evaluation of crown width of open-grown red alder as an aid to the prediction of growth and yield. B.S.F. thesis, Univ. of British Columbia, Vancouver.

Arney, J. D. 1985. SPS: stand projection system for mini- and micro-computers. J. For. 83:378.

Berntsen, C. M. 1961. Growth and development of red alder compared with conifers in 30-year-old stands. USDA For. Serv., Res. Pap. PNW-38.

———. 1962. A 20-year growth record for three stands of red alder. USDA For. Serv., Res. Note PNW-219.

Bishop, D. M., F. A. Johnson, and G. R. Staebler. 1958. Site curves for red alder. USDA For. Serv., Res. Note PNW-162.

Bormann, B. T. 1985. Early wide spacing in red alder (*Alnus rubra* Bong.): effects on stem form and stem growth. USDA For. Serv., Res. Note PNW-423.

Bormann, B. T., and J. C. Gordon. 1984. Stand density effects in young red alder plantations: productivity, photosynthate partitioning, and nitrogen fixation. Ecology 65:394-402.

Brackett, M. 1973. Notes on tree volume computation. Washington State Dept. of Natural Resources, Resource Manage. Rep. 24.

Chambers, C. J. 1973. Empirical yield tables for predominantly alder stands in western Washington. Washington State Dept. of Natural Resources, DNR Rep. 31. Olympia, WA.

Cooke, P. 1987. The influence of density and proportion on allometric determination of plant growth. M.S. thesis, Oregon State Univ., Corvallis.

Curtis, R. O., D. Bruce, and C. Van Coevering. 1968. Volume and taper tables for red alder. USDA For. Serv., Res. Pap. PNW-56.

Curtis, R. O., D. J. DeMars, and F. R. Herman. 1974. Which dependent variable in site index-height regression? For. Sci. 20:74-87.

DeBell, D. S., R. F. Strand, and D. L. Reukema. 1978. Short-rotation production of red alder: some options for future forest management. *In* Utilization and management of alder. *Compiled by* D. G. Briggs, D. S. DeBell, and W. A. Atkinson. USDA For. Serv., Gen. Tech. Rep. PNW-70. 231-244.

DeBell, D. S., and B. C. Wilson. 1978. Natural variation in red alder. *In* Utilization and management of alder. *Compiled by* D. G. Briggs, D. S. DeBell, and W. A. Atkinson. USDA For. Serv., Gen. Tech. Rep. PNW-70. 193-208.

Giordano, P. A., and D. E. Hibbs. 1993. Morphological response to competition in red alder: the role of water. Functional Ecology 7:462-468.

Harrington, C. A. 1986. A method of site quality evaluation for red alder. USDA For. Serv., Gen. Tech. Rep. PNW-192.

Harrington, C. A., and R. O. Curtis. 1986. Height growth and site index curves for red alder. USDA For. Serv., Res. Pap. PNW-358.

Hibbs, D. E. 1987. The self-thinning rule and red alder management. For. Ecol. Manage. 18:273-281.

Hibbs, D. E., and G. C. Carlton. 1989. A comparison of diameter- and volume-based stocking guides for red alder. West. J. Appl. For. 4(4):113-115.

Hibbs, D. E., W. H. Emmingham, and M. C. Bondi. 1989. Thinning red alder: effects of method and spacing. For. Sci. 35(1):16-29.

———. Thinning response of red alder. Western J. Applied Forestry (in review).

Hook, D. D., M. D. Murray, D. S. DeBell, and P. C. Wilson. 1987. Variation in growth of red alder families in relation to shallow water-table levels. For. Sci. 33(1):224-229.

Hook, D. D., D. S. DeBell, A. Ager, and D. Johnson. 1990. Dry weight partitioning among 36 open-pollinated red alder families. Biomass 21:11-25.

Johnson, F. A. 1955. Volume tables for Pacific Northwest trees. USDA For. Serv., Agric. Handb. 92. Washington, D.C.

Johnson, F. A., and N. P. Worthington. 1963. Procedure for developing a site index estimating system from stem analysis. USDA For. Serv., Res. Note PNW-170.

Lester, D. T., and D. S. DeBell. 1989. Geographic variation in red alder. USDA For. Serv., Res. Pap. PNW-409.

Lloyd, W. J. 1955. Alder thinning—progress report. USDA Soil Conservation Serv., West Area Woodland Conservation Tech. Note 3. Portland, OR.

Munro, D. D. 1967. Ratios of standard cubic-foot volume to basal area for the commercial tree species of British Columbia. Univ. of British Columbia. Faculty of Forestry. Litho.

Plank, M. E., T. A. Snellgrove, and S. Willits. 1990. Product values dispel "weed species" myth of red alder. For. Prod. J. 40(2):23-28.

Puettmann, K. J., D. S. DeBell, and D. E. Hibbs. 1993a. Density management guide for red alder. Forest Research Laboratory, Oregon State Univ., Corvallis, Res. Contrib. 2.

Puettmann, K. J., D. W. Hann, and D. E. Hibbs. 1993b. Development and comparison of the size-density trajectories for red alder and Douglas-fir stands. For. Sci. 39:7-27.

Puettmann, K. J., and D. E. Hibbs. 1988. Comparison of stand characteristics from long-term remeasured plots or red alder with yield table and growth model prediction. Report to the Hardwood Silviculture Cooperative.

Puettmann, K. J., D. E. Hibbs, and D. W. Hann. 1992. Extending the size-density analysis to mixed species populations and investigation of underlying dynamics using red alder and Douglas-fir stands. J. Ecol. 80:448-459.

Skinner, E. C. 1959. Cubic volume tables for red alder and sitka spruce. USDA For. Serv., Res. Note PNW-170.

Smith, J. H. G. 1968. Growth and yield of red alder in British Columbia. *In* Biology of alder. *Edited by* J. M. Trappe, J. F. Franklin, R. F. Tarrant, and G. M. Hansen. USDA For. Serv., PNW For. Range Exp. Sta., Portland, OR. 273-286.

———. 1972. Tree size and yield in juvenile red alder stands. Paper presented at the Forestry Section, Northwest Scientific Association. 23-25 March, Bellingham, WA.

———. 1983. Spacing of red alder increases tree size but reduces total yield. Univ. of British Columbia, Faculty of Forestry Paper.

Smith, J. H. G., and R. E. Breadon. 1964. Combined variable equations and volume basal area ratios for total cubic foot volume of commercial trees of British Columbia. For. Chron. 40:258-261.

Turnbull, K. J., G. R. Little, and G. Hoyer. 1963. Comprehensive tree-volume tariff tables. Washington State Dept. of Natural Resources, Olympia.

Warrack, G. C. 1964. Thinning effects in red alder. British Columbia For. Serv., Res. Div., Victoria.

Worthington, N. P., F. A. Johnson, G. R. Staebler, and W. S. Lloyd. 1960. Normal yield tables for red alder. USDA For. Serv., Res. Pap. PNW-36.

# 17

# Wood Quality, Product Yield, and Economic Potential

MARLIN E. PLANK & SUSAN WILLITS

Red alder (*Alnus rubra* Bong.) is the most commercially important hardwood species in the Pacific Northwest because of its abundance and marketability. Its wood is extremely popular in the furniture industry for its uniform color and ability to accept stains. Often alder is used as a substitute for birch because of its similar appearance.

## Wood Characteristics

When alder is compared to the other important North American hardwoods, its overall index of workability is among the best (Leney et al. 1978). In finishing, its ranking is similar to yellow birch (*Betula alleghaniensis* Britton), black cherry (*Prunus serotina* Ehrh.), and sugar maple (*Acer saccharum* Marsh.). Its color uniformity is second to none. In machining, alder is exceeded only by yellow birch, black cherry, and black walnut (*Juglans nigra* L.), and only the latter equals the sanding and polishing characteristic of alder. Red alder has superior gluing ability, ranking with silver maple (*Acer saccharinum* L.) and yellow poplar (*Liriodendren tulipifera* L.). Relative to other hardwoods, red alder has low specific gravity with only a few species, such as black cottonwood (*Populus trichocarpa* Torr. & Gray) or quaking aspen (*Populus tremuloides* Michx.), showing lower ratings; however, a positive trait is that volumetric shrinkage of red alder is also one of the lowest of all the hardwoods (USDA 1987).

Products resulting from primary manufacturing facilities include lumber, pallet stock, veneer, pulp chips, hog fuel, and firewood. Red alder logs yield a significant percentage of high-grade lumber that is used in the manufacture of secondary products including fine furniture, cabinets, and turned-wood novelties. Lower-grade lumber is used for upholstered furniture frames, interior furniture parts, and pallets. Face veneer is used extensively for cabinets, while core veneer is used in high-quality paneling. Red alder pallets have gained wide acceptance in the grocery industry, especially on the West Coast. Although other hardwood species may have greater strength, red alder exhibits superior qualities of low warpage, high nail-holding ability, and good dimensional stability.

## Basic Recovery

If the management, harvest, and manufacture of red alder are to be encouraged, accurate and reliable estimates of product recovery are essential pieces of information for forest managers and wood products manufacturers. A lack of this information led to two studies (Plank 1992) by the Timber Quality Research Unit of the USDA Forest Service, Pacific Northwest Research Station. In the first study, a sample of 159 trees was selected from six areas administered by the Bureau of Land Management and Siuslaw National Forest in the Coast Range of northwestern Oregon (Plank et al. 1990). The trees selected were representative of the type

*Table 1. Value and percentage of volume by lumber grade for red alder in Oregon and Washington.*

| Lumber grade | 1991 prices ($) | Oregon (%) | Washington (%) |
|---|---|---|---|
| Select | 915 | 25 | 27 |
| No. 1 shop | 500 | 24 | 30 |
| No. 2 shop | 280 | 9 | 19 |
| No. 3 shop | 200 | 3 | 9 |
| Frame | 300 | 6 | 14 |
| Pallet[1] | 240 | 33 | — |

[1]Pallets were not produced in the Washington study.

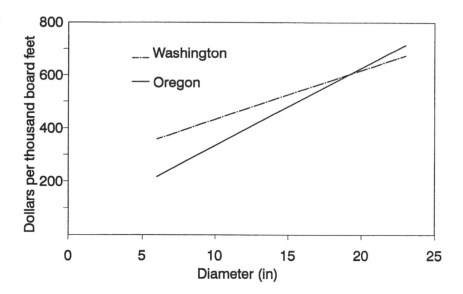

*Figure 1. Comparison of average lumber values for two red alder studies.*

of trees processed in alder mills. They averaged about 65 years of age and ranged in diameter from 12 to 28 inches.

In the second study, a sample of 153 trees was selected in northwestern Washington from lands administered by the Darrington Ranger District of the Mt. Baker-Snoqualmie National Forest and by the Washington State Department of Natural Resources. These trees were slightly younger but of a similar diameter range as those used in the Oregon study.

In both studies, logs were identified by tree number and log position, and that number was used to identify the logs and resulting lumber. Measurements of total height, tree age, log length, and log diameter were also taken during the logging phase. Logs were measured for Scribner and cubic scales at the sawmill both before and after bucking to sawmill lengths.

Both of the mills where study logs were sawn were typical of any mill that saws hardwoods. Each had a head saw, resaw, edger, and trimmer. The main difference in equipment between these two mills was that the Oregon mill had a circular saw headrig with a 0.325 in kerf, and an optimizing edger, whereas the Washington mill had a band saw headrig with a 0.175 in kerf, but only a board edger. Board-foot recovery for the Washington study was greater than for the Oregon study, primarily because of narrower saw kerf on the head saw. Because the Oregon mill had an optimizing edger, the resaw was used infrequently. The Washington mill used the resaw as the main piece of equipment to break down cants produced at the headrig.

Both mills also had pallet plants that normally are the destination for the center cants and for logs with a diameter less than 10 inches. Pallet stock was produced during the study at the Oregon mill, but not during the Washington study. Instead, center cants were sawn into lumber, and logs too small for sawing (diameter < 10 in) were chipped. This apparently had a major impact on the percentage of each lumber grade recovered. Table 1 shows the lumber grade distribution for

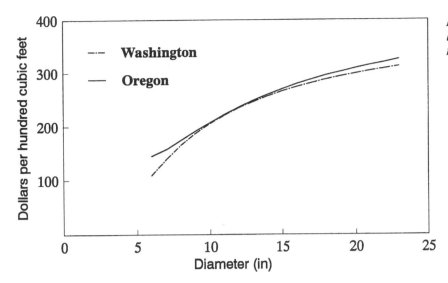

*Figure 2. Comparison of woods-length log values (lumber + chips) for two red alder studies.*

each study. Except for the pallet stock in the Oregon study, all lumber was sawn to a nominal 1-inch thickness and then surfaced to an actual thickness of 15/16 inch.

Current prices for each lumber grade, also shown in Table 1, were applied to the volume of lumber recovered from each log. The value per thousand board feet of lumber for the two studies reflects the difference in the lumber grade recovered for the sawn logs. This comparison is shown in Figure 1, plotted over log diameter, for mill-length logs. The higher values for the Washington study are the result of the lumber grade mix.

A different relation between the two studies develops when we compare value of the woods-length logs, where that value includes lumber plus chips. Lumber value was determined by using early 1991 lumber prices, and chip volumes were priced at $60 per bone-dry unit. In the Oregon study, all logs were sawn into either shop lumber or pallet material; edgings and slabs were chipped. In the Washington study, long logs were sawn into shop lumber, or part of the log was chipped if it was determined that subsequent bucking would result in mill-length logs with too small of a diameter (< 10 in). Apparently, the value of pallet stock produced in the Oregon study is more than enough to overcome the price of the higher chip volumes produced in the Washington study, thereby resulting in a merging of the values. This comparison, shown in Figure 2, indicates little difference between the two studies.

A comparison between young-growth Douglas-fir (< 100 years) and red alder demonstrates the high quality of the latter. Douglas-fir (Ernst and Fahey 1986) has a consistently higher recovery percentage than red alder because of the products manufactured, but the quality of its lumber, which is mostly Standard or No. 2 and Better, remains fairly constant as log diameter increases. A look at the value per hundred cubic feet for each species (Fig. 3) reveals that red alder increases in value, even on a log scale basis, much faster than does

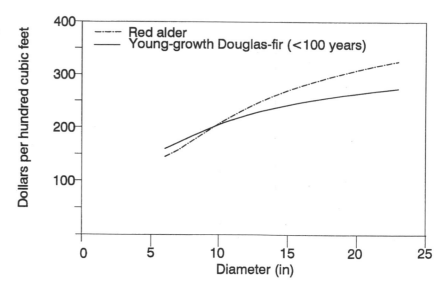

*Figure 3. Comparison of woods-length log values for red alder and young-growth Douglas-fir.*

Douglas-fir over similar diameter ranges. The greater value of alder lumber offsets its relatively low recovery percentage.

## Management and Silviculture Issues

Management and silvicultural aspects of red alder are appropriately addressed in other chapters of this book (Hibbs and DeBell, Chapter 14; Puettmann, Chapter 16); however, there are some considerations related to wood quality that need to be discussed here. Red alder is harvested from existing stands while trees are still quite small; thus, a large part of the volume goes to pulp. This early harvesting may occur because alder grows in mixture with softwoods, which tend to produce more volume per acre after 25 to 30 years than does red alder. Because yield of alder in Select and No. 1 Shop grades increases dramatically as tree diameter increases, the increase in value as the trees grow needs to be examined. Most fir softwood lumber from small trees is sold as Standard or No. 2 with an average value of $225 to $250/mbf. A significant portion of lumber produced from red alder logs of the same size can be worth anywhere from $250 to $900/mbf because red alder has the ability to produce clear wood on relatively small trees. Managers may need to carefully consider annual increases in value before harvesting.

In managed stands where the goal has been to maximize volume production, emphasis has been on increased growth through thinning or spacing treatments. Wood quality, however, is determined by size and spacing of limbs, with the highest value lumber coming from logs with small branches and wide spacing between them. Careful attention, then, must be given to ensure that treatments do not promote growth of branch size or inhibit natural pruning which will result in decreased lumber quality. Epicormic branches, however, do not seem to cause a significant reduction in quality as their influence is usually eliminated in the initial sawing cuts. Again, managers need to be aware that stand treatments made to increase wood volume may have a serious effect on the quality of the wood produced.

## Economic Considerations

In the mid 1980s, hardwood mills in Oregon and Washington sold products with an estimated value of nearly $100 million annually while directly employing about 1000 people with an annual payroll approaching $20 million (Beachy and McMahon 1987). Additional manufacturing added another $50 million. A conservatively estimated 85 percent of these dollars can be attributed to red alder products and, if multiplier effects of jobs created in associated service industries are counted, the economy is enhanced by considerably more dollars.

## Conclusion

With declining softwood harvest there is increased interest in using red alder and other hardwoods to maintain the manufacturing base in the rural communities of the Northwest and to help offset the decline in employment of the softwood industry. The studies described above demonstrate the product potential and the extremely high value of the products that are produced from red alder. Many of the final products require additional processing steps which also increase employment and add to the economy. Even though growth of red alder exceeds harvest by a margin of two to one, the availability of the resource for wood products is affected by land ownership, forest practices acts (i.e., riparian zone protection), habitat conservation areas, and land management principles. Therefore, it is extremely important in ecosystem management and other forest planning efforts to recognize not only the biological importance of the red alder resource, but also the economic opportunities available through the use of alder for wood products.

## Literature Cited

Beachy, D. L., and R. O. McMahon. 1987. Economic value of the Pacific Northwest hardwood industry in 1985. Forest Research Laboratory, Oregon State Univ., Corvallis.

Ernst, S., and T. D. Fahey. 1986. Changes in product recovery of Douglas-fir from old growth to young growth. *In* Proc. Douglas-fir stand management for the future, 18-20 June 1985. *Edited by* C. D. Oliver, D. P. Hanley, and J. A. Johnson. Univ. of Washington, Seattle. 103-107.

Leney, L., A. Jackson, and H. D. Erickson. 1978. Properties of red alder (*Alnus rubra* Bong.) and its comparison to other hardwoods. *In* Utilization and management of alder. *Compiled by* D. G. Briggs, D. S. DeBell, and W. A. Atkinson. USDA For. Serv., Gen. Tech. Rep. PNW-70. 25-33.

Plank, M. E. 1992. Red alder promising as commercial species. For. Ind. 119(2):18-19.

Plank, M. E., T. A. Snellgrove, and S. Willits. 1990. Product values dispel "weed species" myth of red alder. For. Prod. J. 40(2):23-28

USDA. 1987. Wood handbook: wood as an engineering material. USDA Forest Products Laboratory, Madison, WI. USDA For. Serv., Agric. Handb. 72. Washington, D.C.

# Whose Fat Shadow Nourisheth

DAVID E. HIBBS, DEAN S. DEBELL, & ROBERT F. TARRANT

Probably for the first time in the history of forestry, considerable biological information for a tree species has been developed in advance of its need in forest resource management. The quantity and quality of information on red alder, from basic biology to applied management issues, has increased tremendously since publication of *Utilization and Management of Alder* (Briggs et al. 1978). The expanded scope of red alder information provides a much broader current understanding of the interrelated processes that affect growth, development, and management of red alder—a synthesis on a level not possible in previous writings on red alder. Some important advances have been possible only because knowledge has reached, in some sense, a critical mass. Research has reached large geographic scales and has established functional, interdisciplinary connections between single discipline research projects. In this chapter, we review highlights of alder research since the 1970s and list some research needs that appear most critical.

## Below-ground Biology

Red alder is a partner in an important three-way symbiosis, a partnership whose interactions and ecosystem roles are becoming clear. Below ground, alder roots grow in an intimate and essential association with mycorrhizal fungi and the actinomycete *Frankia*. This three-way symbiosis contributes to the rapid growth rates of red alder, to its flexibility in site requirements, and to its large influence on soil structure and fertility. As host, alder provides energy to both symbionts as well as shelter to *Frankia*. The biology of the intimate mycorrhizal association as well as the biology of the formation and function of the *Frankia*-containing nodules has been studied extensively and fundamentals of the processes are known.

The *Frankia* of red alder can fix large amounts of nitrogen over a wide range of growing conditions. This fixation contributes more significantly to the rapid growth of red alder than the presence of mycorrhizal fungi, and fixation takes place even in the presence of high levels of soil nitrogen. In nature, colonization of red alder by *Frankia* appears to take place readily and very early in seedling development. It is not well known what kind of tradeoffs are made throughout the life cycle of red alder between fixing nitrogen, absorbing soil nitrogen (expanded root systems), and above-ground growth.

The few ectomycorrhizal fungal species of red alder are very host specific, yet seem to be present wherever alder is found growing, always arriving after *Frankia*. These mycorrhizae play a critical role in phosphorus acquisition; red alder uses and cycles large amounts of phosphorus.

Laboratory research on red alder and experience with other species has shown that *Frankia* strains and fungal species differ in growth characteristics, but these differences have not been systematically investigated in red alder symbioses. Tailoring nursery inoculations to specific field growing conditions may enhance alder survival and growth.

There has been considerable progress in understanding alder-soil interactions. Tools developed to predict alder growth from soil (and climatic) characteristics need to be expanded through the range of red alder and related to systems of soil classification. At the same time, the impacts of alder on soil nitrogen, phosphorus, organic matter, pH, and cation exchange capacity are increasingly studied. In addition to its obviously beneficial effects on soil nitrogen and organic matter, alder appears to increase soil acidity and decrease amounts and/or availability of calcium, magnesium, and phosphorus.

## The Tree

Red alder has adopted a stress-avoidance strategy in plant-water relations. Its roots explore a huge soil volume early in life. While stomatal conductance under high soil and atmospheric moisture conditions is much higher in red alder than in associated conifers, relatively small changes in plant moisture potential or vapor pressure deficit restrict stomatal conductance. At a plant water potential of -0.8 MPa, stomatal conductance begins to be reduced; stomata are closed at -1.5 MPa. So far, all studies of alder physiology have utilized seedlings or tree branches; now, research on whole, large trees is needed.

A variety of studies have provided a basic genetic characterization of red alder. Regional and elevational trends in characteristics have been documented and initial calculations of heritabilities of simple growth traits made. The most important research needs are assessment of the risks involved in seed transfer and of the benefits of a breeding program.

Growth studies have demonstrated the different seasonal growth patterns of height, diameter, and biomass, and described the roles of weather and soils in regulating these patterns. Tree age, stand density, and genetics have also been shown to play a role in regulating growth and allocation of dry matter.

## Natural Stands

Alder stands generally begin at such high densities that understories are nearly excluded. As stands grow, structural and species complexity increases as a succession of understory species takes place, eventually resulting in a forest with a high, open tree canopy, a tall shrub layer, and a simple herb layer. Unless large conifer logs are present on the ground, colonization by any tree seedlings is rare in many parts of alder's range. The details of this process vary with site conditions, but the direction of this succession suggests that the next stage after alder in these areas is likely to be dominated by shrubs, not trees. This final step, however, has not been demonstrated.

When conifers become established before or simultaneously with alder, mixed species stands may result. If alder density is not too great, these conifer seedlings eventually grow through the alder. If nitrogen limits conifer growth, the conifers can benefit from the nitrogen fixed by the alder, eventually dominating and then replacing the alder. Research has demonstrated a variety of possible outcomes to mixed-species stand development, but we have only a rudimentary understanding of how density, proportion, spatial pattern, site characteristics, and species choice determine the outcome of any particular stand.

## Plantation Management

A large body of knowledge is necessary to the successful establishment of vigorous plantations of any species. For red alder, the plantation establishment process has moved in the last 10 years from being unheard of to routinely done. Several thousand acres of alder plantations are established every year in western Oregon and Washington, an amazing technological accomplishment.

Alder seed is collected annually, the collections being based on both basic genetic considerations of seed movement and physiological understandings of seed maturation. Seed can be stored for at least 10 years. Several nursery techniques for producing high-quality seedlings have been developed; all involve the common factors of seed stratification, low bed density, and inoculation with *Frankia*. Standards to describe the "target" seedling have been developed and tested. Bed sowing technology, container technology, and predicting germination in beds all need improvement.

Identification of good plantation sites for red alder has been improved through studies that have helped clarify the mesic-site, pioneer ecological niche of alder as well to predict tree survival and growth after planting. Before-planting treatments that reduce the competitive ability of associated plants are generally beneficial and, in some cases, essential. Planting is done in the spring after frost danger is passed and before the summer drought. In southern Washington, this planting window falls between mid-March and mid-April. First-year height growth can be 1 to 2 meters.

A region-wide analysis of dynamics of natural and managed red alder stands has produced a density management guide that provides the key to stand management decisions. The strength of this new guide is the breadth of the data base used in its construction. With it, a land manager can make the critical decisions that regulate the processes of mortality and growth. The guide now needs validation through long-term research.

Fertilizing alder with phosphorus may be beneficial, especially on wet soils. Like research on alder physiology, most of the nutritional research has been with seedlings or small saplings and so needs to be expanded to older, larger trees. One intriguing hypothesis is that the growth slow-down and senescence in natural stands of alder over forty years old may be due, in part, to inadequate nutrition; fertilizing with phosphorus might improve growth.

There has been little research on pruning. Decay studies, however, indicate that decay is compartmentalized in red alder very quickly so there is little concern for decay problems following pruning. Although early work (Berntsen 1961) indicated that pruning stimulated epicormic branches in a 21-year-old stand, we believe that pruning in younger plantations warrants investigation.

## Growth and Yield

Several growth and yield tools have been used to predict the growth of alder stands. An analysis of these tools using a region-wide data base has shown that only the *Empirical Yield Tables* (Chambers 1973) can be recommended for general use and, then, only for unmanaged stands. This last point highlights the primary need in growth and yield research: studies in managed stands. We know that height growth, both of newly planted seedlings and following thinning, changes with management practices, but we cannot predict the magnitude of these changes or how they relate to site index or relative density. We also know that form changes with spacing control, but have no volume equations developed from managed stands. Because of these unknowns, we do not have good predictions of growth or yield from managed stands.

## Health and Protection

The rapid deterioration of alder logs is well known, but living alder trees are surprisingly decay-resistant to stem injury and natural branch pruning. Alder supports a variety of insects, some of which can reach epidemic proportions. The observed epidemics have lasted only a few years. Red alder suffers damage from the large herbivores in its range, but the impact is usually localized, patchy, and restricted to young trees. There has been no systematic evaluation of the extent of any of these kinds of losses, so we have no knowledge of the real importance of these agents to alder survival, growth, or wood quality.

## Interactions with Other Resources

Basic investigations have shown the habitat and food source associations between alder and some wildlife species; few critical dependencies have been noted. The frequent association between red alder and riparian zones makes it difficult to separate associations just with alder from other riparian effects. Studies in mixed conifer-alder stands have shown higher wildlife diversity than monocultures of either species. Studies to date have focused on birds and small mammals, so studies with other species groups are needed. Studies of the functional interactions between all wildlife species and habitat components are needed to separate the often confounded effects of factors like plant species, stand age, topography, and slope position.

Relatively little is known about water quality and fisheries. Taste and odor of domestic water have been affected by heavy annual alder litter fall. The contribution of alder to ground-water nitrates that enter into streams is being studied, but alder's role in general stream nutrient dynamics is unknown. Winter light is greater on streams with a deciduous canopy than on those under a coniferous canopy. The rapid decay of alder trunks is generally thought to make it a poor source of large woody debris for enhancing stream structure and fish habitat. The brief and qualitative nature of these comments makes it clear that there are many research needs in alder-stream system interactions.

## Wood Quality and Economics

Two lumber recovery studies have shown the dramatic rise in average value per board foot, and thus in log value, with increases in log diameter, a reflection of the rapid increase in percentage of high-value lumber with diameter. Results of these alder studies are particularly apparent when alder and Douglas-fir are compared. Douglas-fir has a gradually increasing response curve of average value per board foot as log diameter increases. Red alder has a steeper curve, indicating that management practices like pruning (to reduce knot density) and thinning (to increase log diameter) will probably be good investments. These studies have not examined the link between growing conditions and the quality of wood recovered from logs.

No lumber quality recovery studies or economic analyses of managed stands have been made because there are no mature managed stands. The need is clearly great, and some preliminary conclusions might be drawn from analysis of existing young plantations and reconstruction analysis in the few old spacing trials.

## Conclusions

Forty years of research with red alder has resulted in a substantial body of information about the biology and management of the species. Remarkably, this background knowledge was created in advance of widespread management—a unique occurrence in the history of forestry. Several thousand acres of red alder are now established annually in the Pacific Northwest. While this new practice is expected to be expanded, the finite wild alder resource is rapidly being harvested. Without a continued flow of information from research to speed plantation production, the alder resource and its dependent industry essentially will be lost.

Revolutionary new forest management practices are being demanded. Maintaining the health of soil-plant-water ecosystems has the highest priority. Sustaining biological and structural diversity is widely understood and expected. How best to manage riparian vegetation is yet to be decided, and forest products must be produced in the best manner under new approaches to forest management. We will continue to learn about and use, to the world's benefit, "the alder, whose fat shadow nourisheth."

## Acknowledgment

This is publication 2922 of the Forest Research Laboratory, Oregon State University.

## Literature Cited

Berntsen, C. M. 1961. Pruning and epicormic branching in red alder. J. For. 59(9):675-676.

Briggs, D. G., D. S. DeBell, and W. A. Atkinson, *comps.* 1978. Utilization and management of alder. USDA For. Serv., Gen. Tech. Rep. PNW-70.

Chambers, C. J. 1974. Empirical yield tables for predominantly alder stands in western Washington. Washington State Dept. of Natural Resources, DNR Rep. 31. Olympia, WA.

# List of Contributors

Alan A. Ager
USDA Forest Service
Umatilla National Forest
2517 Hailey Ave.
Pendleton, OR 97801

Glenn R. Ahrens
Forest Science Department
Oregon State University
Corvallis, OR 97331

Eric A. Allen
Pacific Forestry Centre
506 W. Burnside Road
Victoria, BC V8Z 1M5
Canada

Dwight D. Baker
Panlabs, Inc.
11804 North Creek Parkway, South
Bothell, WA 98011-8805

Dan Binkley
Department of Forest Sciences
Colorado State University
Ft. Collins, CO 80523

B. T. Bormann
USDA Forest Service
Pacific Northwest Research Station
Corvallis, OR 97331

S. S. N. Chan
USDA Forest Service
Pacific Northwest Research Station
Corvallis, OR 97330

Elizabeth C. Cole
Forest Science Department
Oregon State University
Corvallis, OR 97331

Paul J. Courtin
British Columbia Ministry of Forests
4595 Canada Way
Burnaby, BC V5G 4L9
Canada

Kermit Cromack, Jr.
Forest Science Department
Oregon State University
Corvallis, OR 97331

Dean S. DeBell
Forestry Sciences Laboratory
USDA Forest Service
Pacific Northwest Research Station
3625 93rd Ave. SW
Olympia, WA 98512

Alexander Dobkowski
Weyerhaeuser Company
Longview, WA 98632

Paul F. Figueroa
Weyerhaeuser Company
Centralia, WA 98531

Peter A. Giordano
Forest Science Department
Oregon State University
Corvallis, OR 97331

Constance A. Harrington
Forestry Sciences Laboratory
USDA Forest Service
Pacific Northwest Research Station
3625 93rd Ave. SW
Olympia, WA 98512

Tim B. Harrington
Forestry Department
University of Georgia
Athens, GA 30605

David E. Hibbs
Forest Science Department
Oregon State University
Corvallis, OR 97331

C. Y. Li
USDA Forest Service
Pacific Northwest Research Station
Forestry Sciences Laboratory
Corvallis, OR 97331

William C. McComb
Forest Science Department
Oregon State University
Corvallis, OR 97331

Jim McGrath
Wind River Nursery
USDA Forest Service
Carson, WA 98610

Randy Molina
USDA Forest Service
Pacific Northwest Research Stations
Forestry Sciences Laboratory
Corvallis, OR 97331

David Myrold
Crops and Soil Science Department
Oregon State University
Corvallis, OR 97331

Michael Newton
Forest Science Department
Oregon State University
Corvallis, OR 97331

Marlin E. Plank
Forestry Sciences Laboratory
Pacific Northwest Research Station
P. O. Box 3890
Portland, OR 97208

Klaus J. Puettmann
Department of Forest Resources
115 Green Hall
University of Minnesota
St. Paul, MN 55108

M. A. Radwan
Forestry Sciences Laboratory
USDA Forest Service
Pacific Northwest Research Station
3625 93rd Ave. SW
Olympia, WA 98512

W. O. Russell, III
Forest Science Department
Oregon State University
Corvallis, OR 97331

Lauri J. Shainsky
Forestry Sciences Laboratory
USDA Forest Service
Pacific Northwest Research Station,
P.O. Box 3890
Portland, OR 97208

Reinhard F. Stettler
College of Forest Resources
University of Washington, AR-10
Seattle, WA 98195

Yasuomi Tanaka
Weyerhaeuser Company
Centralia, WA 98531

Robert F. Tarrant
Forestry Sciences Laboratory
USDA Forest Service
Pacific Northwest Research Station
3625 93rd Ave. SW
Olympia, WA 98512

Susan Willits
Forestry Sciences Laboratory
Pacific Northwest Research Station
P. O. Box 3890
Portland, OR 97208

Barbara J. Yoder
Forest Science Department
Oregon State University
Corvallis, OR 97331

John C. Zasada
USDA Forest Service
North Central Forest Experiment Station
Box 898
Rhinelander, WI 54501

# Index